AF595348

Biomimetic Radical Chemistry and Applications 2021

Biomimetic Radical Chemistry and Applications 2021

Editor

Chryssostomos Chatgilialoglu

MDPI • Basel • Beijing • Wuhan • Barcelona • Belgrade • Manchester • Tokyo • Cluj • Tianjin

Editor
Chryssostomos
Chatgilialoglu
ISOF
Consiglio Nazionale delle
Ricerche
Bologna
Italy

Editorial Office
MDPI
St. Alban-Anlage 66
4052 Basel, Switzerland

This is a reprint of articles from the Special Issue published online in the open access journal *Molecules* (ISSN 1420-3049) (available at: www.mdpi.com/journal/molecules/special_issues/radical_2021).

For citation purposes, cite each article independently as indicated on the article page online and as indicated below:

LastName, A.A.; LastName, B.B.; LastName, C.C. Article Title. *Journal Name* **Year**, *Volume Number*, Page Range.

ISBN 978-3-0365-4350-5 (Hbk)
ISBN 978-3-0365-4349-9 (PDF)

Contents

About the Editor

Chryssostomos Chatgilialoglu

Dr. Chryssostomos Chatgilialoglu is the Research Director at the Italian National Research Council (CNR) in Bologna and Visiting Professor at the Center for Advanced Technologies, Adam Mickiewicz University, Poznan (Poland). He is also the Co-Founder and President of the company Lipinutragen. He achieved his doctorate degree in Industrial Chemistry at Bologna University in 1976 and completed his postdoctoral studies at York University (UK) and the National Research Council of Canada, Ottawa. From March 2014 to May 2016, he was appointed as the Director of the Institute of Nanoscience and Nanotechnology at the NCSR "Demokritos"in Athens (Greece). He was Chairman of the COST Actions Free Radicals in Chemical Biology (2007–2011) and Biomimetic Radical Chemistry (2013–2016).

He is the author or co-author of more than 310 publications in peer-reviewed journals and 36 book chapters, and the author and editor of several books, including {Organosilanes in Radical Chemistry}, Wiley 2004; {Encyclopedia of Radicals in Chemistry, Biology and Materials}, Wiley 2012; and {Membrane Lipidomics for Personalized Health}, Wiley 2015. He was invited to be the Guest Editor for Special Issues in several scientific journals, and invited to speak over 260 times at congresses and institutions.

His research group is active in the field of free radical chemistry, addressing applications in life sciences. In recent years, he has developed biomimetic chemistry of radical stress and associated biomarkers. The discovery of endogenous formation of trans-lipids, and research on 5′,8-cyclopurine DNA lesions and fatty-acid-based lipidomics has attracted worldwide attention. He is responsible for introducing tris(trimethylsilyl)silane as a radical-based reducing agent, and for this he was the winner of the Fluka Prize "Reagent of the Year 1990".

MDPI

Editorial

Biomimetic Radical Chemistry and Applications

Chryssostomos Chatgilialoglu [1,2]

[1] ISOF, Consiglio Nazionale delle Ricerche, 40129 Bologna, Italy; chrys@isof.cnr.it; Tel.: +39-051-639-8309
[2] Center of Advanced Technologies, Adam Mickiewicz University, 61-712 Poznań, Poland

Abstract: Some of the most interesting aspects of free radical chemistry that emerged in the last two decades are radical enzyme mechanisms, cell signaling cascades, antioxidant activities, and free radical-induced damage of biomolecules. In addition, identification of modified biomolecules opened the way for the evaluation of in vivo damage through biomarkers. When studying free radical-based chemical mechanisms, it is very important to establish biomimetic models, which allow the experiments to be performed in a simplified environment, but suitably designed to be in strict connection with cellular conditions. The 28 papers (11 reviews and 17 articles) published in the two Special Issues of *Molecules* on "Biomimetic Radical Chemistry and Applications (2019 and 2021)" show a remarkable range of research in this area. The biomimetic approach is presented with new insights and reviews of the current knowledge in the field of radical-based processes relevant to health, such as biomolecular damages and repair, signaling and biomarkers, biotechnological applications, and novel synthetic approaches.

Citation: Chatgilialoglu, C. Biomimetic Radical Chemistry and Applications. *Molecules* **2022**, *27*, 2042. https://doi.org/10.3390/molecules27072042

Received: 16 February 2022
Accepted: 16 March 2022
Published: 22 March 2022

1. Introduction

Free radicals have attracted considerable attention in various research areas, including organic synthesis, material science, atmospheric chemistry, radiation chemistry, pharmacology, biology, and medicine [1]. Free radicals are generated in the biological environment as a result of normal intracellular metabolism and function as physiological signaling species that participate in the modulation of apoptosis, stress responses, and proliferation [2]. The enormous importance of free radical chemistry for a variety of biological events, including ageing and inflammation, has motivated studies to understanding the related mechanistic steps at the molecular level. Therefore, the estimation of the type and extent of damages, as well as mechanisms and efficiency of protective and repair systems, are important subjects in life sciences.

Modelling free radical reactivity of biological systems is a crucial research area. Some of the most interesting aspects of free radical chemistry that have emerged in the last two decades are radical enzyme mechanisms, cell signaling cascades, antioxidant activities, and biomarkers of free radical damage to biomolecules [1,2]. In the latter case, identification of modified biomolecules has a diagnostic value for the evaluation of in vivo damages. When studying free radical-based chemical mechanisms, biomimetic chemistry and the design of related biomimetic models come into play to perform experiments in a controlled environment, strictly connected with cellular conditions. Figure 1 shows the connections of biomarkers identification through biomimetic radical chemistry and analytical protocols of biomolecule modifications, as well as their extension to clinical research such as ageing, inflammation, cancer, obesity, and other pathologies.

The papers published in the two Special Issues of *Molecules* on "Biomimetic Radical Chemistry and Applications (2019 and 2021)" show the strong interdisciplinary context with a remarkable range of research in this area. Several subjects are presented, with 17 articles and 11 reviews written by specialists in the fields.

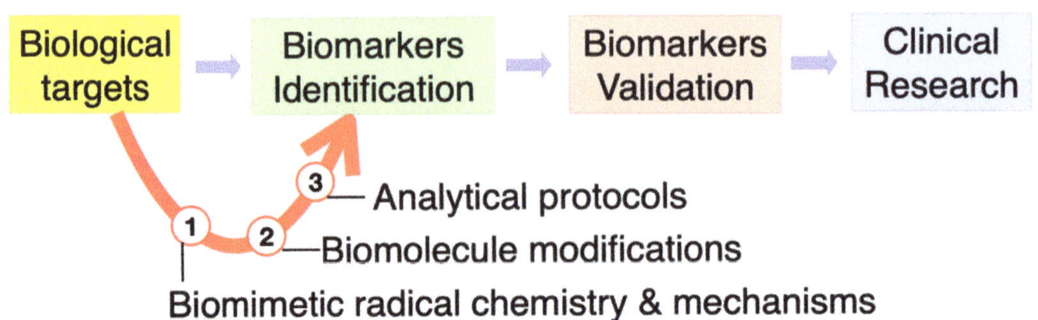

Figure 1. Omics technologies and the role of biomimetic radical chemistry in biomarkers discovery.

2. Reactive Oxygen/Nitrogen Species (ROS/RNS) Network

The free radical history in biology and medicine up to 1990 had three main entries: the 'free radical theory of aging' based on the damage and recycling involving reactive oxygen species (ROS) [3], the properties of the enzyme superoxide dismutase (SOD) [4], and the antioxidant network including the role of vitamins [2]. Now days, it is well documented that ROS, reactive nitrogen species (RNS), and reactive sulfur species (RSS) are produced in a wide range of physiological processes and are also responsible for a variety of pathological processes. Indeed, the overproduction of ROS/RNS/RSS has been linked with the etiology of various diseases, and antioxidant defense mechanisms are essential to protect against them [5,6].

Figure 2 summarizes the main feature of the ROS/RNS network, including molecules such as hydrogen peroxide (H_2O_2), hypochlorous acid (HOCl) and peroxynitrite ($ONOO^-$), as well as radicals such as superoxide radical anion ($O_2^{\bullet-}$), nitric oxide ($NO^\bullet$), hydroxyl radical ($HO^\bullet$), nitrogen dioxide ($NO_2^\bullet$), and the carbonate radical anion ($CO_3^{\bullet-}$). Aerobic life would not be possible without the enzymes SODs and catalase (CAT) that transform superoxide to water and oxygen. Nitric oxide synthase (NOS) is a class of enzymes that induce the formation of nitric oxide ($NO^\bullet$). Under physiological conditions, concentrations of ~0.1 nM $O_2^{\bullet-}$ and ~10 nM $NO^\bullet$ play a role in regulating the activation of transcription factors, cell proliferation and apoptosis. During the inflammatory response, their concentration can increase up to a 100-fold excess. These two radicals are the precursors of a variety of other reactive species.

Figure 2 shows the main pathways by which other biologically important free radicals can be produced, either via H_2O_2 or as a consequence of $ONOO^-$ formation from $O_2^{\bullet-}$ and $NO^\bullet$ [1,2,5,6]. H_2O_2 is at the crossroad of several pathways; the main ones are reported in Figure 2. Myeloperoxidase (MPO) uses H_2O_2 and anions like Cl^-, Br^-, SCN^- and NO_2^- to generate hypochlorous acid (HOCl) or HOBr, HOSCN and $NO_2^\bullet$, respectively. H_2O_2 transformation to highly reactive $HO^\bullet$ occurs by the Fenton reaction (Fe^{2+} and H_2O_2), the Haber–Weiss reaction ($O_2^{\bullet-}$ and H_2O_2), and reduction of previous formed HOCl by $O_2^{\bullet-}$. $ONOO^-$ exists in equilibrium with its protonated form ($pK_a = 6.6$), which spontaneously decomposes to $NO_2^\bullet$ and $HO^\bullet$. Other $ONOO^-$ also reacts with CO_2 and the resulting adduct rapidly decomposes to $NO_2^\bullet$ and $CO_3^{\bullet-}$. Although $O_2^{\bullet-}$ is very unreactive in typical free radical reactions, such as hydrogen atom abstraction or addition, its successors generate the most reactive $HO^\bullet$. The diffusion distance of $HO^\bullet$ is very small because of their high reactivity with all types of biomolecules and, consequently, there is a low probability to be intercepted by antioxidants [7].

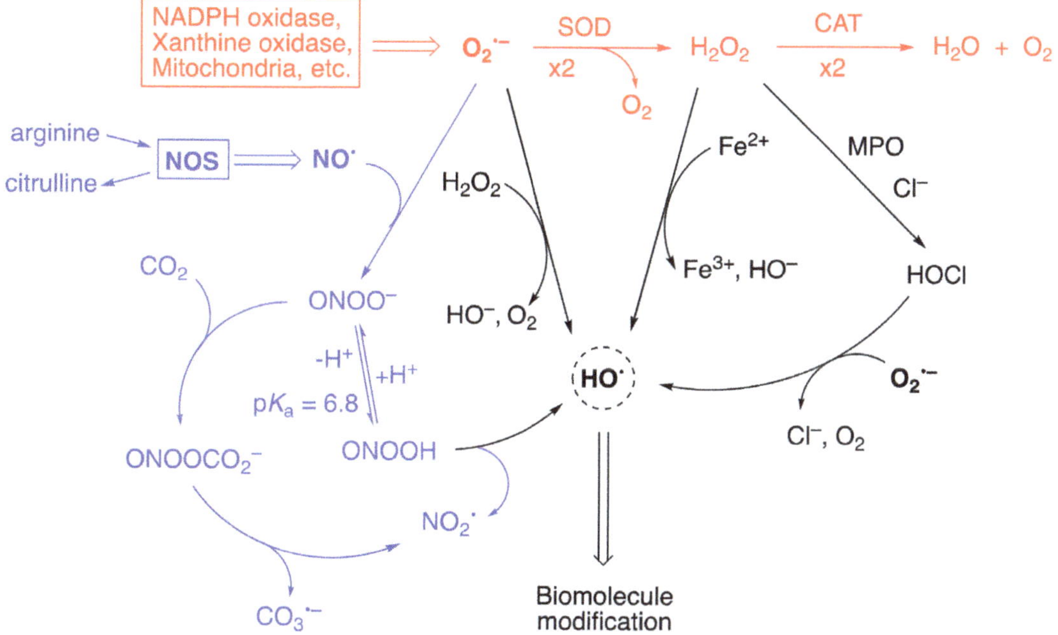

Figure 2. Pathways of interaction of superoxide and nitric oxide in biological systems. NOS: nitric oxide synthase; SOD: superoxide dismutase; CAT: catalase; MPO: myeloperoxidase.

3. Brief Overview of the Two Special Issues

3.1. Targeting DNA Damage and Repair

Three reviews dealing with oxidative DNA damage and repair appeared in the two Special Issues [8–10]. One electron oxidation gives rise to a radical cation whose charge (hole) can migrate through DNA covering several hundreds of Å, eventually leading to irreversible oxidative damage; an overview of work on the dynamics of hole transfer in DNA is reported [8]. Among the four common DNA bases (A, G, T, and C), G is the most readily oxidized to the G radical cation ($G^{\bullet+}$), which is also the putative initial intermediate in the oxidative DNA damage. Upon formation of $G^{\bullet+}$, fast deprotonation occurs by loss of a proton to give the guanyl radical $G(\text{-}H)^{\bullet}$, also named the guanine radical or neutral guanine radical. An overview of the one-electron oxidation of the GC pair and the complex mechanism of deprotonation vs. hydration steps of $GC^{\bullet+}$ pair is given. The role of the two $G(\text{-}H)^{\bullet}$ tautomers in single- and double-stranded oligonucleotides and the G-quadruplex, the supramolecular arrangement that attracts interest for its biological consequences, are discussed, including the importance of biomarkers of guanine DNA damage [9]. To maintain genomic stability and integrity, double-strand DNA has to be replicated in a strictly regulated manner, ensuring the accuracy of its copy number and its integrity. DNA damage-induced replication stress, the formation of DNA secondary structures, peculiar epigenetic modifications and cellular responses to the stress and their impact on the instability of the genome and epigenome, mainly in eukaryotic cells, have been summarized [10].

Six articles report new experimental data on targeting DNA damage [11–16]. Guanine radicals generated in single, double, and G-quadruplex oligonucleotides are studied by nanosecond transient absorption spectroscopy [11]. The time needed to establish electronic resonant conditions for charge transfer in oxidized DNA has been evaluated by molecular dynamics simulations followed by QM/MM computations, which include counterions and a realistic solvation shell [12]. Among the reactive oxygen species (ROS), the hydroxyl radical ($HO^{\bullet}$) is the most reactive toward any biomolecule including DNA (cf. Figure 2). Although the majority of the purine DNA lesions like 8-oxo-purine (8-oxo-Pu) are generated by various ROS (including $HO^{\bullet}$), the formation of 5′,8-cyclopurine (cPu) lesions in vitro and in vivo relies exclusively on the $HO^{\bullet}$ attack. Indeed, recent research focused on the purine DNA damage by $HO^{\bullet}$ with emphasis on mechanistic aspects for the various lesion formation and their interconnections [17]. New insights into the reaction paths of $HO^{\bullet}$ with genetic material and the formation of 8-oxo-Pu and cPu lesions vs. oxygen concentration have been reported in detail [13,18]. It is worth mentioning a recent article on inflammatory bowel disorders (IBD), showing the role of cPu as a biomarker [19]. Radiosensitizing properties of substituted uridines are of great importance for radiotherapy. The influence of the type of halogen atom in the radio-sensitizing properties of 5-halo-4-thio-2-deoxyuridine has been addressed; contrary to the 5-iodo derivative that is an efficient agent, the 5-bromo does not show radio-sensitizing properties at all [14]. Several classes of copper complexes prepared in the past were explored for their chemical nuclease activity both biologically and chemically [20]. Two studies include the mechanism of copper artificial metallonuclease to induce superoxide-mediated cleavage via the minor groove [15], and the DNA binding and quantitative cleavage activity of the $[Cu(TPMA)(N,N)]^{2+}$ class (where TPMA = tris-2-pyridylmethylamine) using a DNA electrochemical biosensor [16].

3.2. Protein Modifications and Enzymatic Activity

In the area of proteins, four reviews contribute to the two Special Issues [21–24]. The formation of covalently linked peptides and proteins plays a key role in many biological processes, both physiologically and pathologically. In one review, it is summarized: the spectrum of crosslinks currently known to be formed on proteins, including the mechanisms of their formation, experimental approaches to the detection, and identification and characterization of these species [21]. Reversible crosslinks, driven by the formation of disulfide bridges, appear to play a key role in cell signaling events, primarily as a result of reversible thiol–disulfide switches or related species. Irreversible protein crosslinks are mostly unwanted processes, occurring during metabolism, that can accumulate in aging or have been associated with the onset and development of pathological conditions and human diseases. Their recognition as reliable biomarkers of several pathologies, particularly neurodegenerative disorders, is an important field of molecular diagnostics in medicine [21,25]. Another review provides a thorough description of the role of phosphatidylethanolamine-derived protein adducts and effects on membrane properties [22]. The aim was to highlight this new area of research and to encourage a more nuanced investigation of the complex nature of the new lipid-mediated mechanism in the modification of membrane protein functions under oxidative stress.

When proteins are pharmaceutical compounds, such as insulin or human growth hormone, their degradation can occur by radical species that are generated from pharmaceutical excipients. Polysorbate is prone to generate peroxyl radicals that can trigger oxidative degradation [26]. In this review, the involvement of thiyl radicals in pharmaceutical protein degradation through hydrogen atom transfer, electron transfer, and addition reactions is reported [23]. Another review reports the work done in the chemical labelling of proteins using biomimetic radical chemistry [24]. This is inspired by the occurrence of radical reactions in an aqueous environment, such as for enzymatic catalysis or photoreactions. Such reactivity occurs selectively on specific amino acids (nucleophilic residues such as lysine or cysteine) by means of electrophilic compounds that allow site-selective protein labeling. This is an important field of chemical biology.

An overview of reductive dihydroxylation catalyzed by IspH, an enzyme involved in the biosynthesis of isoprenoids, has been reviewed [27]. IspH is an oxygen sensitive [4Fe-4S] metalloenzyme that catalyzes $2H^{+}/2e^{-}$ reductions and water elimination by involving nonconventional bioinorganic and bioorganometallic intermediates. This review focuses on the IspH mechanism, discussing the results that have been obtained in the last decades using an approach combining chemistry, enzymology, crystallography, spectroscopies, and docking calculations. A section about the inhibitors of IspH discovered up to now is also reported. The presented results constitute a useful and rational approach to inaugurate the design and development of new potential chemotherapeutics against pathogenic organisms.

Three original articles reported research related to biotechnological applications of proteins: (i) Therapeutic uses of natural peptide somatostatin is limited by its very short biological half-life of 1–2 min. The entrapment of this peptide in a lipid formulation allowed to retard release in aqueous medium and human plasma. Furthermore, a new radical reactivity was discovered arising from the interaction between this sulfur-containing peptide and its liposomal formulation [28]. (ii) Physico-chemical evidence showed that riboflavin together with hyaluronic acid play a role in the treatment of corneal cross-linking treatment of keratoconus by UVA light. Spin trapping experiments on collagen/hyaluronic acid/riboflavin solutions evidenced the formation of reactive oxygen species (ROS) by electron paramagnetic resonance measurements. Riboflavin under UVA irradiation generates ROS that can induce damages, whereas hyaluronic acid has a protecting role [29]. (iii) The formation and stabilization of gold nanoparticles in bovine serum albumin (BSA) solution has been discovered. Physico-chemical studies showed that the size of nanoparticles increases slowly with time, resulting in nanoparticles of different morphologies, and the stabilization is obtained through the interaction of sulfur-containing amino acid residues of albumin [30].

3.3. *Various Bioinspired Mechanistic Studies and Applications*

Three reviews deal with confocal microscopy-based detection [31], mechanistic studies of methionine (Met) oxidation [32], and recent applications of chemiluminescence (CL) [33]. One of them focuses on confocal microscopy-based detection of profound alterations in the plasma membrane, membranes of insulin granules and lipid droplets in single beta cells under various nutritional load conditions. The combination of whole cell lipidomics analysis and single cell confocal imaging of fluidity and micropolarity provides insight into stress-induced lipid turnover in subcellular organelles of pancreatic beta cells [31].

Oxidation of methionine (Met) is an important reaction that plays a key role in protein modifications during oxidative stress and aging. An overview of the transient species detection in one-electron oxidation of Met derivatives by various time-resolved techniques is presented [32]. Mechanistic aspects of Met oxidation in various structural environments (e.g., peptide) and at various pH by one-electron oxidants (including $HO^{\bullet}$ radical) are summarized and discussed. Neighboring group participation seems to be an essential parameter which controls one-electron oxidation of methionine. The observed transient species are precursors of final products [32].

The phenomenon of chemiluminescence (CL) can take place both in natural and artificial chemical systems and has been utilized in a variety of applications. A review reports on recent research in this area [33]. In this context, the CL role in the development of efficient therapeutic platforms is also discussed in relation to the reactive oxygen species (ROS) and singlet oxygen ($^1O_2$) produced as final products. The CL prospects in imaging, biomimetic organic and radical chemistry, and therapeutics are critically presented in respect to the persisting challenges and limitations of the existing strategies to date [33].

Heme iron and non-heme dimanganese catalases (CAT) protect biological systems against oxidative damage caused by hydrogen peroxide (cf. Figure 2). In order to gain more insight into the mechanism of these curious enzyme reactions, two original articles reported on the metal complexes as catalase mimics: a mononuclear non-heme oxoiron(IV) complex mediated H_2O_2 dismutation into O_2 and H_2O in aqueous solution [34], and a non-heme diiron-peroxo complex which shows a catalase-like reactivity [35].

The conversion of ribonucleosides to 2′-deoxyribonucleosides is catalyzed by ribonucleoside reductase enzymes in nature. One of the key steps in this complex radical mechanism is the reduction of the 3′-ketodeoxynucleotide by a pair of cysteine residues, providing the electrons via a disulfide radical anion ($RSSR^{\bullet-}$) in the active site of the enzyme [36]. Experimental conditions were found to obtain the bioinspired conversion of ketones to corresponding alcohols with high-yield by the intermediacy of disulfide radical anion of cysteine (CysSSCys)$^{\bullet -}$ in water [37].

Mechanistic studies of radical processes in pharmacological applications, which also inspire biological mechanisms, are represented by various original articles: the oxidation of 8-thioguanosine and 2-thiouracil by photolytic and radiolytic conditions [38,39]; bioinspired radical-based synthetic strategies toward anomeric spironucleosides as potential inhibitors of glycogen phosphorylase and for the preparation of azido-derivatives via a radical azidoalkylation of alkenes [40,41]; and the synthesis of two new iron-porphyrin-based catalysts inspired by naturally occurring proteins, such as horseradish peroxidase, hemoglobin, and cytochrome P450, tested for atom transfer radical polymerization (ATRP), obtaining polymers with specific properties [42].

4. Conclusions

The two Special Issues give the reader a wide overview of biomimetic radical chemistry, where molecular mechanisms have been defined and molecular libraries of products are also developed to be used for the discovery of some relevant biological processes. The biomimetic approach is a convenient tool, since achievements in free radical mechanisms can be easily transferred to a better comprehension of the radical-based biological pathways in living organisms, triggering advancements in health and diseases. In addition, identification of modified biomolecules paves the way for molecular libraries and the evaluation of in vivo damage through biomarkers.

The two Special Issues cover aspects of free radical chemistry in biological events, revealed using biomimetic chemical models. These include: catalytic pathways and mechanisms of radical enzymes, prebiotic chemistry, radical-induced DNA lesions or protein modifications, with further development concerning analytical protocols, repair processes, biological consequences, lipid peroxidation and isomerization, and defense systems based on antioxidants, as well as bio-inspired synthetic strategies.

Funding: This research received no external funding.

Institutional Review Board Statement: Not applicable.

Informed Consent Statement: Not applicable.

Data Availability Statement: Not applicable.

Acknowledgments: I would like to thank the European Cooperation in Science and Technology (COST) to have given me the opportunity to act as the Chair of the Action CM1201—Biomimetic Radical Chemistry, held between 2013–2016 (https://www.cost.eu/actions/CM1201/, accessed on 21 March 2022). During these years, a mix of scientific expertise participated actively to develop the main topics of the Action. The large majority of the papers in these two Special Issues are by the scientists participating to this COST Action, and I would like to acknowledge their support and collaboration.

Conflicts of Interest: The author declares no conflict of interest.

References

1. Chatgilialoglu, C.; Studer, A. (Eds.) *Encyclopedia of Radicals in Chemistry, Biology and Materials*; Wiley: Chichester, UK, 2012.
2. Halliwell, B.; Gutteridge, J.M.C. *Free Radicals in Biology and Medicine*, 5th ed.; Oxford University Press: Oxford, UK, 2015.
3. Harman, D. Aging: A theory based on free radical and radiation chemistry. *J. Gerontol.* **1956**, *11*, 298–300. [CrossRef]
4. Fridovich, I. Superoxide radical and superoxide dismutases. *Annu. Rev. Biochem.* **1995**, *64*, 97–112. [CrossRef]
5. Sies, H.; Berndt, C.; Jones, D.P. Oxidative stress. *Annu. Rev. Biochem.* **2017**, *86*, 715–748. [CrossRef] [PubMed]
6. Sies, H.; Jones, D.P. Reactive oxygen species (ROS) as pleiotropic physiological signalling agents. *Nat. Rev. Mol. Cell. Biol.* **2020**, *21*, 363–383. [CrossRef] [PubMed]
7. Winterbourn, C.C. Reconciling the chemistry and biology of reactive oxygen species. *Nat. Chem. Biol.* **2008**, *4*, 278–286. [CrossRef] [PubMed]
8. Peluso, A.; Caruso, T.; Landi, A.; Capobianco, A. The Dynamics of Hole Transfer in DNA. *Molecules* **2019**, *24*, 4044. [CrossRef]
9. Chatgilialoglu, C. The Two Faces of the Guanyl Radical: Molecular Context and Behavior. *Molecules* **2021**, *26*, 3511. [CrossRef]

10. Tsegay, P.S.; Lai, Y.; Liu, Y. Replication Stress and Consequential Instability of the Genome and Epigenome. *Molecules* **2019**, *24*, 3870. [CrossRef]
11. Balanikas, E.; Banyasz, A.; Baldacchino, G.; Markovitsi, D. Populations and Dynamics of Guanine Radicals in DNA strands—Direct versus Indirect Generation. *Molecules* **2019**, *24*, 2347. [CrossRef]
12. Landi, A.; Capobianco, A.; Peluso, A. The Time Scale of Electronic Resonance in Oxidized DNA as Modulated by Solvent Response: An MD/QM-MM Study. *Molecules* **2021**, *26*, 5497. [CrossRef]
13. Chatgilialoglu, C.; Krokidis, M.G.; Masi, A.; Barata-Vallejo, S.; Ferreri, C.; Terzidis, M.A.; Szreder, T.; Bobrowski, K. New Insights into the Reaction Paths of Hydroxyl Radicals with Purine Moieties in DNA and Double-Stranded Oligodeoxynucleotides. *Molecules* **2019**, *24*, 3860. [CrossRef]
14. Spisz, P.; Zdrowowicz, M.; Makurat, S.; Kozak, W.; Skotnicki, K.; Bobrowski, K.; Rak, J. Why Does the Type of Halogen Atom Matter for the Radiosensitizing Properties of 5-Halogen Substituted 4-Thio-2′-Deoxyuridines? *Molecules* **2019**, *24*, 2819. [CrossRef]
15. Molphy, Z.; McKee, V.; Kellett, A. Copper *bis*-Dipyridoquinoxaline Is a Potent DNA Intercalator that Induces Superoxide-Mediated Cleavage via the Minor Groove. *Molecules* **2019**, *24*, 4301. [CrossRef]
16. Banasiak, A.; Zuin Fantoni, N.; Kellett, A.; Colleran, J. Mapping the DNA Damaging Effects of Polypyridyl Copper Complexes with DNA Electrochemical Biosensors. *Molecules* **2022**, *27*, 645. [CrossRef] [PubMed]
17. Chatgilialoglu, C.; Ferreri, C.; Krokidis, M.G.; Masi, A.; Terzidis, M.A. On the relevance of hydroxyl radical to purine DNA damage. *Free Radic. Res.* **2021**, *55*, 384–404. [CrossRef]
18. Chatgilialoglu, C.; Eriksson, L.A.; Krokidis, M.G.; Masi, A.; Wang, S.-D.; Zhang, R. Oxygen dependent purine lesions in double-stranded oligodeoxynucleotides: Kinetic and computational studies highlight the mechanism for 5′,8-cyplopurine formation. *J. Am. Chem. Soc.* **2020**, *142*, 5825–5833. [CrossRef]
19. Masi, A.; Fortini, P.; Krokidis, M.G.; De Angelis, P.; Guglielmi, V.; Chatgilialoglu, C. Increased levels of 5′,8-cyclopurine DNA lesions in inflammatory bowel diseases. *Redox Biol.* **2020**, *34*, 101562. [CrossRef]
20. Fantoni, N.Z.; Brown, T.; Kellett, A. DNA-Targeted Metallodrugs: An Untapped Source of Artificial Gene Editing Technology. *ChemBioChem* **2021**, *22*, 2184–2205. [CrossRef] [PubMed]
21. Fuentes-Lemus, E.; Hägglund, P.; López-Alarcón, C.; Davies, M.J. Oxidative Crosslinking of Peptides and Proteins: Mechanisms of Formation, Detection, Characterization and Quantification. *Molecules* **2022**, *27*, 15. [CrossRef]
22. Pohl, E.E.; Jovanovic, O. The Role of Phosphatidylethanolamine Adducts in Modification of the Activity of Membrane Proteins under Oxidative Stress. *Molecules* **2019**, *24*, 4545. [CrossRef] [PubMed]
23. Schöneich, C. Thiyl Radical Reactions in the Chemical Degradation of Pharmaceutical Proteins. *Molecules* **2019**, *24*, 4357. [CrossRef] [PubMed]
24. Sato, S.; Nakamura, H. Protein Chemical Labeling Using Biomimetic Radical Chemistry. *Molecules* **2019**, *24*, 3980. [CrossRef] [PubMed]
25. Hawkins, C.L.; Davies, M.J. Detection, identification and quantification of oxidative protein modifications. *J. Biol. Chem.* **2019**, *294*, 19683–19708. [CrossRef] [PubMed]
26. Harmon, P.A.; Kosuda, K.; Nelson, E.; Mowery, M.; Reed, R.A. A Novel Peroxy Radical Based Oxidative. Stressing System for Ranking the Oxidizability of Drug Substances. *J. Pharm. Sci.* **2006**, *95*, 2014–2028. [CrossRef]
27. Jobelius, H.; Bianchino, G.I.; Borel, F.; Chaignon, P.; Seemann, M. The Reductive Dehydroxylation Catalyzed by IspH, a Source of Inspiration for the Development of Novel Anti-Infectives. *Molecules* **2022**, *27*, 708. [CrossRef]
28. Larocca, A.V.; Toniolo, G.; Tortorella, S.; Krokidis, M.G.; Menounou, G.; Di Bella, G.; Chatgilialoglu, C.; Ferreri, C. The Entrapment of Somatostatin in a Lipid Formulation: Retarded Release and Free Radical Reactivity. *Molecules* **2019**, *24*, 3085. [CrossRef] [PubMed]
29. Constantin, M.M.; Corbu, C.G.; Mocanu, S.; Popescu, E.I.; Micutz, M.; Staicu, T.; Şomoghi, R.; Trică, B.; Popa, V.T.; Precupas, A.; et al. Model Systems for Evidencing the Mediator Role of Riboflavin in the UVA Cross-Linking Treatment of Keratoconus. *Molecules* **2022**, *27*, 190. [CrossRef] [PubMed]
30. Matei, I.; Buta, C.M.; Turcu, I.M.; Culita, D.; Munteanu, C.; Ionita, G. Formation and Stabilization of Gold Nanoparticles in Bovine Serum Albumin Solution. *Molecules* **2019**, *24*, 3395. [CrossRef]
31. Maulucci, G.; Cohen, O.; Daniel, B.; Ferreri, C.; Sasson, S. The Combination of Whole Cell Lipidomics Analysis and Single Cell Confocal Imaging of Fluidity and Micropolarity Provides Insight into Stress-Induced Lipid Turnover in Subcellular Organelles of Pancreatic Beta Cells. *Molecules* **2019**, *24*, 3742. [CrossRef]
32. Marciniak, B.; Bobrowski, K. Photo- and Radiation-Induced One-Electron Oxidation of Methionine in Various Structural Environments Studied by Time-Resolved Techniques. *Molecules* **2022**, *27*, 1028. [CrossRef]
33. Tzani, M.A.; Gioftsidou, D.K.; Kallitsakis, M.G.; Pliatsios, N.V.; Kalogiouri, N.P.; Angaridis, P.A.; Lykakis, I.N.; Terzidis, M.A. Direct and Indirect Chemiluminescence: Reactions, Mechanisms and Challenges. *Molecules* **2021**, *26*, 7664. [CrossRef] [PubMed]
34. Kripli, B.; Sólyom, B.; Speier, G.; Kaizer, J. Stability and Catalase-Like Activity of a Mononuclear Non-Heme Oxoiron(IV) Complex in Aqueous Solution. *Molecules* **2019**, *24*, 3236. [CrossRef] [PubMed]
35. Lakk-Bogáth, D.; Török, P.; Csendes, F.V.; Keszei, S.; Gantner, B.; Kaizer, J. Disproportionation of H_2O_2 Mediated by Diiron-Peroxo Complexes as Catalase Mimics. *Molecules* **2021**, *26*, 4501. [CrossRef] [PubMed]
36. Minnihan, E.C.; Nocera, D.G.; Stubbe, J. Reversible, Long-Range Radical Transfer in *E. coli* Class Ia Ribonucleotide Reductase. *Acc. Chem. Res.* **2013**, *46*, 2524–2535. [CrossRef]
37. Barata-Vallejo, S.; Skotnicki, K.; Ferreri, C.; Marciniak, B.; Bobrowski, K.; Chatgilialoglu, C. Biomimetic Ketone Reduction by Disulfide Radical Anion. *Molecules* **2021**, *26*, 5429. [CrossRef]

38. Taras-Goslinska, K.; Vetica, F.; Barata-Vallejo, S.; Triantakostanti, V.; Marciniak, B.; Chatgilialoglu, C. Converging Fate of the Oxidation and Reduction of 8-Thioguanosine. *Molecules* **2019**, *24*, 3143. [CrossRef]
39. Skotnicki, K.; Taras-Goslinska, K.; Janik, I.; Bobrowski, K. Radiation Induced One-Electron Oxidation of 2-Thiouracil in Aqueous Solutions. *Molecules* **2019**, *24*, 4402. [CrossRef]
40. Stathi, A.; Mamais, M.; Chrysina, E.D.; Gimisis, T. Anomeric Spironucleosides of β-d-Glucopyranosyl Uracil as Potential Inhibitors of Glycogen Phosphorylase. *Molecules* **2019**, *24*, 2327. [CrossRef]
41. Millius, N.; Lapointe, G.; Renaud, P. Two-Step Azidoalkenylation of Terminal Alkenes Using Iodomethyl Sulfones. *Molecules* **2019**, *24*, 4184. [CrossRef]
42. Fu, L.; Simakova, A.; Park, S.; Wang, Y.; Fantin, M.; Matyjaszewski, K. Axially Ligated Mesohemins as Bio-Mimicking Catalysts for Atom Transfer Radical Polymerization. *Molecules* **2019**, *24*, 3969. [CrossRef]

MDPI

Review

The Two Faces of the Guanyl Radical: Molecular Context and Behavior

Chryssostomos Chatgilialoglu [1,2]

[1] ISOF, Consiglio Nazionale delle Ricerche, 40129 Bologna, Italy; chrys@isof.cnr.it; Tel.: +39-051-6398309
[2] Center of Advanced Technologies, Adam Mickiewicz University, 61-712 Poznań, Poland

Abstract: The guanyl radical or neutral guanine radical $G(-H)^{\bullet}$ results from the loss of a hydrogen atom ($H^{\bullet}$) or an electron/proton (e^-/H^+) couple from the guanine structures (G). The guanyl radical exists in two tautomeric forms. As the modes of formation of the two tautomers, their relationship and reactivity at the nucleoside level are subjects of intense research and are discussed in a holistic manner, including time-resolved spectroscopies, product studies, and relevant theoretical calculations. Particular attention is given to the one-electron oxidation of the GC pair and the complex mechanism of the deprotonation vs. hydration step of $GC^{\bullet+}$ pair. The role of the two $G(-H)^{\bullet}$ tautomers in single- and double-stranded oligonucleotides and the G-quadruplex, the supramolecular arrangement that attracts interest for its biological consequences, are considered. The importance of biomarkers of guanine DNA damage is also addressed.

Keywords: guanine; guanyl radical; tautomerism; guanine radical cation; oligonucleotides; DNA; G-quadruplex; time-resolved spectroscopies; reactive oxygen species (ROS); oxidation

Citation: Chatgilialoglu, C. The Two Faces of the Guanyl Radical: Molecular Context and Behavior. *Molecules* **2021**, *26*, 3511. https://doi.org/10.3390/molecules26123511

Academic Editor: Nicola Borbone

Received: 6 April 2021
Accepted: 1 June 2021
Published: 9 June 2021

1. The Guanine Sink

The free radical chemistry associated with guanine (Gua) and its derivatives, guanosine (Guo), 2′-deoxyguanosine (dGuo), guanosine-5′-monophosphate (GMP), and 2′-deoxyguanosine-5′-monophosphate (dGMP), is of particular interest due to its biological relevance. Indeed, this reactivity is by far the most involved in oxidative DNA damage, initiated by reactive oxygen species (ROS) generated by normal cellular metabolism, or by exogenous sources such as ionizing or ultraviolet irradiation [1–3]. In the thirty years since the discovery of long-range charge transport in DNA, as reviewed by Barton and coworkers [4–6], a large body of experimental data have accumulated showing that the one-electron oxidation of DNA produces a hole that can migrate through the double helix with the final destination at the G sites [4–10]. Among the four common DNA bases (A, G, T, and C), G is the most readily oxidized to the G radical cation ($G^{\bullet+}$), which is also the putative initial intermediate in the oxidative DNA damage. Thus, there have been numerous studies on the formation and behavior of $G^{\bullet+}$ as a nucleoside and in oligonucleotides (ODNs) using spectroscopic and product studies. Upon formation of $G^{\bullet+}$, fast deprotonation occurs by the loss of a proton to give the guanyl radical $G(-H)^{\bullet}$, also named the guanine radical or neutral guanine radical. The main objective of this review is to summarize the behavior of $G(-H)^{\bullet}$ in nucleosides and nucleotides and give an overview of its formation in a macromolecular environment, such as single-stranded (ss), double-stranded (ds) ODNs, and G-quadruplex arrangements.

2. The Two Tautomers of the Guanyl Radical

The guanyl radical $G(-H)^{\bullet}$ corresponds to any intermediate that has lost an H-atom or an electron/proton (e^-/H^+) couple from the G moiety. $G(-H)^{\bullet}$ is one of the most recurrent and intensively studied transient species. The reason is obvious, as dGuo (**1**, R = 2′-deoxyribose) is one of the building blocks of DNA with the lowest oxidation potential (Figure 1). The chemistry of $G(-H)^{\bullet}$ depends on the molecular context where it belongs (free base, nucleoside in ribo or deoxyribo forms, ss- or ds-ODNs including multi-G sequences, RNA, DNA, and G-quadruplex). Although numerous mechanistic studies have been performed during recent years, the emerging picture is still incomplete, with several unanswered questions and divergent opinions. The extrapolation of its chemistry from simple nucleosides to more complex environments (e.g., DNA) can be misleading. This situation often generates confusion to the newcomers in the field. In the present article, the existing dichotomies among mechanistic insights are examined and critically discussed to give a reasoned overview of the guanyl radical structure and behavior.

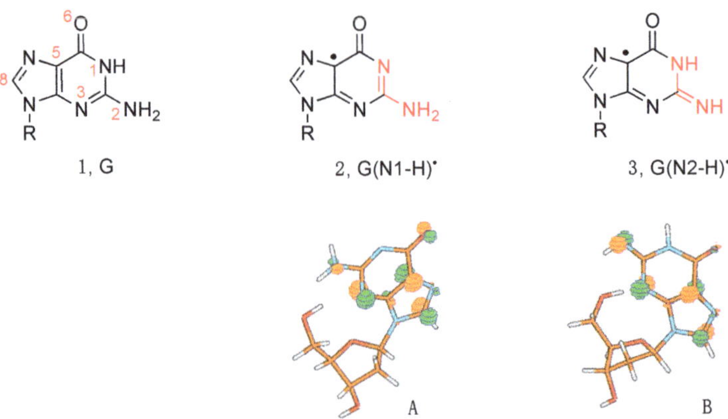

Figure 1. Structure **1** represents the guanine moiety in guanosine (Guo), 2′-deoxyguanosine (dGuo), guanosine-5′- monophosphate (GMP), 2′-deoxyguanosine-5′-monophosphate (dGMP), or 3′,5′-cyclic guanosine monophosphate (cGMP); the atom numbering of the guanine moiety is given in red. The corresponding guanyl radicals G(N1-H)• (**2**) or G(N2-H)• (**3**); the part of the molecule involved in tautomerization is highlighted in red. SOMO of the two forms of guanyl radical G(N1-H)• (**A**) and G(N2-H)• (**B**) computed at the B3LYP/6-311G** level (taken from ref [11]).

Figure 1 shows two tautomeric structures of G(-H)•: form **2** or G(N1-H)• that has lost H from the N1 position, and form **3** or G(N2-H)• that has lost H from the N2 position. The acronyms G(N1-H)• and G(N2-H)• have been commonly used in the recent literature and are used throughout this review; alternative acronyms have been used: G•, N1G•, $(G\text{-}H_1)^\bullet$ or the aminic form for G(N1-H)•, and *N*(2)G•, N2G•, $(G\text{-}H_2)^\bullet$, $G(N^2\text{-}H)^\bullet$ or the iminic form for G(N2-H)•. Figure 1 also shows the distribution of the unpaired electron in the SOMO of G(N1-H)• (**A**) and G(N2-H)• (**B**) computed at the B3LYP/6-311G** level [11]. In **A**, the unpaired electron is mainly localized at the N3, O6, C5, and C8 atoms (spin densities of 0.30, 0.34, 0.21, and 0.27, respectively), whereas, in **B**, the unpaired electron is mainly localized at the N3, C5, C8, and N2 atoms (spin densities of 0.34, 0.29, 0.22, and 0.39, respectively) [11]. As each tautomer has various resonance forms (discussed in some detail in reference [11]), the reported drawings for radicals **2** and **3**, where the unpaired electron is placed in the C5 position, are chosen for a facile distinction of the two tautomers (Figure 1).

3. One-Electron Oxidation of Guanine Derivatives

In the guanine derivatives **1** (G), the two pK_a values of 2.5 and 9.3 are related to the protonation and deprotonation steps, respectively [12,13]. Early work by pulse radiolysis studies with both conductometric and optical detection showed that the one-electron oxidations of Guo and dGuo by the powerful $SO_4^{\bullet-}$ or the milder $Br_2^{\bullet-}$ oxidants (see Appendix A) afford the corresponding radical cation $G^{\bullet+}$, which is in agreement with the reduction potentials of the reactants $E_7(\mathbf{2/1})$ = 1.29 V, $E°(SO_4^{\bullet-}/SO_4^{2-}) \approx 2.43$ V, and $E°(Br_2^{\bullet-}/2Br^-) = 1.66$ V [12,14,15]. The determined pK_a value of 3.9 in $G^{\bullet+}$ is associated with the loss of a proton from N1 to give the guanyl radical G(N1-H)$^•$ (Figure 2); at high pH, G(N1-H)$^•$ further deprotonates to produce G(-2H)$^{•-}$ with pK_a 10.7 [12,13]. Therefore, the one-electron oxidation of **1** in the pH range 5–8 affords the guanyl radical G(N1-H)$^•$. The deprotonation rate constant was measured at pH 7 by pulse radiolysis to be 1.8×10^7 s^{-1} at r.t., showing a $k(H_2O)/k(D_2O)$ of 1.7 [16] and by laser flash photolysis to be 1.5×10^7 s^{-1} with the activation energy of 15.1 ± 1.5 kJ mol^{-1} [17]. DFT calculations on solvation models were used to calculate the deprotonation potential energy profile [17]. The one-electron oxidation of Guo by $Br_2^{•-}$ was investigated by pulse radiolysis coupled with transient electrochemistry; the intermediate G(N1-H)$^•$ (at 28 ± 4 μM concentration) decays by a second-order rate of 3.4×10^7 M^{-1} s^{-1}, and there is competition between the oxidation and reduction of G(N1-H)$^•$, i.e., the ability of the radical to be either oxidized or reduced, presumably via disproportionation rather than the dimerization process [18].

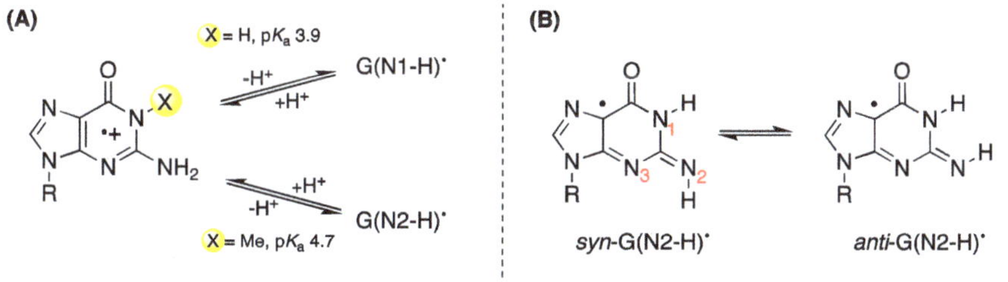

Figure 2. (**A**) The deprotonation of G$^{•+}$ (X = H) and 1-Me-G$^{•+}$ (X = Me) affording G(N1-H)$^•$ and G(N2-H)$^•$, respectively. (**B**) Conformational isomers of G(N2-H)$^•$ in *syn* and *anti* with respect to N3 atom; the earlier *syn*- and *anti*-conformations were defined with respect to N1-H [11] and, as *syn* and *anti* with respect to N3 have been commonly used in the recent literature, these notations are adopted in the text.

On the other hand, when the N1-H moiety is replaced by N1-Me, the deprotonation occurs from the exocyclic NH_2 group and associates with a pK_a value of 4.7 (Figure 2A) [14]. Indeed, theoretical calculations have suggested that deprotonation from the N2-H moiety is competitive with deprotonation from N1-H. Both EPR and UV-visible spectra of the one-electron-oxidized dGuo by $Cl_2^{•-}$ are reported for each of the species: G$^{•+}$, G(-H)$^•$, and G(-2H)$^{•-}$ in aqueous medium at 77 K [19]. The experimental data corroborated by DFT calculations indicated that G(N1-H)$^•$ is favored over G(N2-H)$^•$ in environments with high dielectric constants, such as water. Furthermore, the two conformational isomers of G(N2-H)$^•$ in which the remaining N2–H is either *syn* or *anti* with respect to the guanine N3 atom were calculated, finding that the *syn*-conformer is lower in energy that the *anti*-conformer is (Figure 2B) [19]; see also the three-dimensional structure **B** in Figure 1 that refers to the *syn*-conformer [11].

The oxidation of GMP and dGMP has also been reported in some detail. It has been reported that the presence of the phosphate group in dGMP$^{•+}$ does not significantly affect the prototropic equilibria of the guanine moiety. ENDOR studies carried out using X-ray-irradiated single crystals of dGMP showed that the deprotonation of G$^{•+}$ occurs from N2 rather than the N1 position with the formation of G(N2-H)$^•$ [20]. More recently, the structure and reactivity of G$^{•+}$ and G(-H)$^•$ have been investigated in aqueous solutions by using pulse radiolysis (Raman and UV−visible detection) from the 5′-dGMP oxidation by $SO_4^{•-}$ at pH 7.4 [21], or by the photo-oxidation of 5′-GMP by the 3,3′,4,4′-benzophenone tetracarboxylic acid (TCBP) triplet within the pH range of 2−12 using laser flash photolysis and time-resolved chemically induced dynamic nuclear polarization (CIDNP) [22–24]. There is strong spectroscopic evidence that G(N1-H)$^•$ is further transformed to a new species in the range of pH 7.4–8.0 [21,23,24]. The formation of this species shows a pH dependence, suggesting that it is the G radical cation, named (G$^{•+}$)′, and is tentatively assigned to the protonation at the N7 position of G(N1-H)$^•$ with a rate constant of 8.1×10^6 s^{-1}. Although this suggestion is strongly criticized based on theoretical calculations [25], further support is provided by time-resolved CIDNP [23,24]. From the full analysis of the pH-dependent CIDNP kinetics, the protonation and deprotonation behavior was quantitatively characterized, giving $pK_a = 8.0 \pm 0.2$ of the radical $(G^{\bullet+})'$ [24]. The same spectroscopical method suggested the formation of the radical dication $G^{\bullet++}$ with $pK_a = 2.1$ [22]. The above-described experiments are based on the oxidation of GMP or dGMP and the corresponding guanyl radical. It is well-known that GMP and dGMP with respect to Guo and dGuo have two extra pK_a values of 0.3 and 6.3 that correspond to the deprotonation of the phosphate moiety, viz. $P(O)(OH)_2$ and $P(O)(OH)O^-$ [26]. Although the observed pK_a values around 2 and 8 could be associated with the deprotonation of the phosphate moiety in the one-electron oxidation of GMP or dGMP, in the earlier report, the authors mentioned that the results are qualitatively the same by replacing GMP with Guo [23]. More detailed analysis on Guo is necessary to clarify this aspect.

4. One-Electron Reduction of 8-Haloguanine Derivatives

The first directly observed differences of the two tautomeric forms of $G(\text{-H})^\bullet$ came from a pulse radiolysis study. Using the reaction of hydrated electrons (e_{aq}^-) with 8-haloguanosine (halo = Cl, Br, and I), the two observed short-lived intermediates, which show a substantial difference in their absorption spectra around 620 nm, were assigned to the two $G(\text{-H})^\bullet$ tautomers [11]. The identification of various transient species and transition states has also been addressed computationally by means of time-dependent DFT (TD-B3LYP/6-311G**//B1B95/6-31+G**) calculations [11]. Figure 3 shows the transient UV-visible spectra of the reaction of 8-Br-Guo (**4**) with e_{aq}^- to produce the less stable $G(N2\text{-H})^\bullet$ (in red color), followed by a tautomerization to give $G(N1\text{-H})^\bullet$ (in blue color), for which an identical spectrum was obtained by the one-electron oxidation of Guo (**1**). Theory suggests that the electron adduct of 8-Br-Guo protonated at C8 forms a loose π-complex with the Br atom situated above the molecular plane, and is prompt to eject Br^-. Moreover, the transition state **5** for the elimination of *anti*-H from the NH_2 group was found to be much more stable than the transition states for eliminations of *syn*-H by 18.0 kJ mol^{-1} and of N1-H by 14.1 kJ mol^{-1}. Presumably, the observed spectrum is the *syn*-$G(N2\text{-H})^\bullet$ radical [11]. The band around 620 nm was also predicted very well by the DFT-TD calculations [27], i.e., $\lambda = 616$ nm, $f = 0.045$, and transition {spin}: pπ(out-of-phase N2′, N1, O6′) $\rightarrow$ SOMO (out-of-phase N2′, N3, C5, C8) {b}.

The Arrhenius parameters for this tautomerization, the solvent kinetic isotope effect ($k(H_2O)/k(D_2O)$), and the calculations at the PCM/B1B95/6-31+G**//B1B95/6-31+G** level suggested a complex transition state **6** for the water-assisted tautomerization [11,28,29]; the reaction mechanism of e_{aq}^- with **4** reported in 2000 [29] was revised in 2005 [28], and the revision did not affect the experimental results (the kinetic data). For the ribose series, the rate constants for the transformation $G(N2\text{-H})^\bullet \rightarrow G(N1\text{-H})^\bullet$ were measured in H_2O and D_2O, $k(H_2O) = 5.0 \times 10^4$ s^{-1} and $k(D_2O) = 6.2 \times 10^3$ s^{-1}, respectively, with a solvent kinetic isotope effect of 8.0 [29]. The Arrhenius parameters, $\log(A/s^{-1}) = 8.7 \pm 0.4$ and $E_a = 23.0 \pm 2.5$ kJ mol^{-1}, were measured in the temperature range of 5.8–50.3 °C [28]. The tautomerization is accelerated by phosphate and H^+, with the rate constants $k_{phosp} = 1.4 \times 10^8$ M^{-1} s^{-1} at pH~7 and $k_{H+} \cong 1 \times 10^{10}$ M^{-1} s^{-1} at pH 4, respectively [29]. The $G(N2\text{-H})^\bullet$ is unreactive toward methyl viologen (MV2+) and molecular oxygen, whereas it is able to effect the oxidation of *N,N,N′,N′*-tetramethyl-p-phenylenediamine (TMPD) with a rate constant of $k = 1.7 \times 10^9$ M^{-1} s^{-1} [29]. Figure 3 also shows the calculated transition state **6** for the water-assisted tautomerization computed at the PCM/B1B95/6-31+G**//B1B95/6-31+G** level. The computed value of 21.3 kJ mol^{-1} is in very good agreement with the experimental value of 23.0 kJ mol^{-1} [11]. Other authors [19] calculated an analogous reaction using the sugar moiety instead of 9-Me at the B3LYP/6-31G(d) level and found an activation barrier of 78.2 kJ mol^{-1}, confirming that the choice of theory and the location of the water molecule in the reagents have a profound effect on the outcome. Pulse radiolysis also revealed the "instantaneous" formation of $G^{\bullet+}$ or $G(\text{-2H})^{\bullet-}$ in the reaction of e_{aq}^- with 8-Br-Guo in acid or basic solutions, and the observed pK_a values of 3.8 and 10.4 matched well with the reported values of 3.9 and 10.7 for the oxidized Guo, respectively [29].

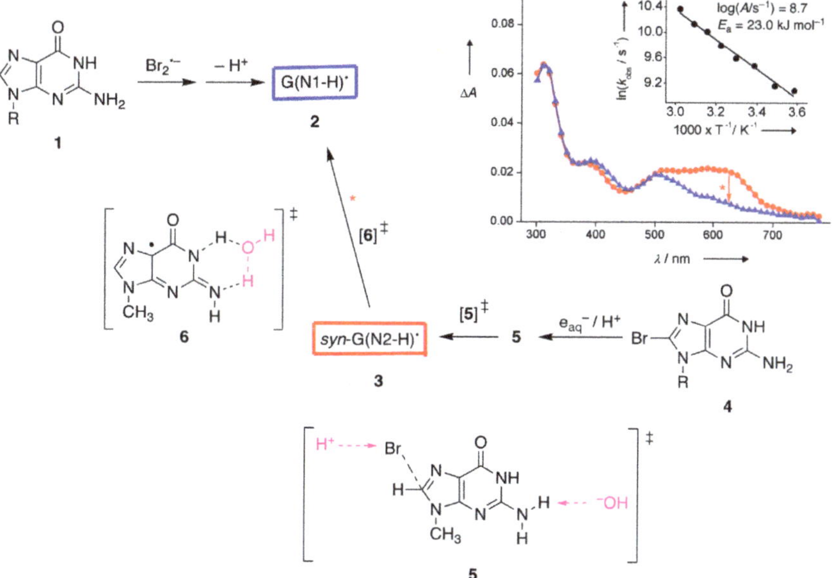

Figure 3. Oxidation of guanosine **1** by $Br_2^{•-}$ and the formation of transient guanyl radical **2** (in blue color). The reduction of 8-bromoguanosine **4** by e_{aq}^- and the transient guanyl radical **3** taken in 2 μs after the pulse (in red color); by monitoring the reaction at 620 nm (see red arrow), radical **3** is transformed to **2** with a first-order rate constant of 5×10^4 s^{-1} at room temperature with Arrhenius parameters reported in the insert (adapted from [11,28]). The transition state **5**, for the deprotonation from NH_2 by means of HO^- during the elimination of Br^- from the π-complex as HBr, and the transition state **6**, for the H_2O-assisted tautomerization, were calculated at the PCM/B1B95/6-31+G**//B1B95/6-31+G** level.

We discussed above the electron-coupled proton transfer (ECPT) reactions of 8-Br-Guo (**4**) at pH~7. It is also worth mentioning the pulse radiolysis experiments under acidic conditions [29]. At pH 3, the typical spectrum of $G^{•+}$ was generated in two steps: the first one was complete in less than 1 μs and the second one in ~8 μs, corresponding to the reactions of e_{aq}^- and $H^•$, respectively. In the γ-radiolysis study of 8-Br-Guo in aqueous solutions followed by product studies, the formation of Guo as a single product at various pH was shown. In D_2O solutions, the quantitative incorporation of deuterium at the 8-position was also observed [29].

A variety of substituted 8-bromoguanine derivatives was also examined. For the tautomerization *syn*-G(N2-H)$^•$ → G(N1-H)$^•$ (R = 2′-deoxyribose), the same behavior is observed with an identical rate constant ($k = 5.0 \times 10^4$ s^{-1}) when starting from 8-bromo-2′-deoxyguanosine [11]. It is worth mentioning the redox reactions of guanine derivatives having methylation at the N1 position: the oxidation of 1-methyl guanosine (1-Me-Guo, **7**) by $Br_2^{•-}$ and the reduction of 8-Br-1-Me-Guo (**9**) by e_{aq}^- showed identical absorption spectra having a band at 610–620 nm, which decays via second-order kinetics (Figure 4). The transient is consistent with the 1-Me-G(N2-H)$^•$ radical (**8**) and the absence of tautomerization due to the N1-Me group [11,28]. The ESR and UV-visible spectra of 1-Me-$G^{•+}$ and its corresponding deprotonated 1-Me-G(N2–H)$^•$ are also recorded from the one-electron-oxidized 1-Me-dGuo by $Cl_2^{•-}$ at 77 K at pDs ~5 and ~8 in 7.5 M of LiCl glass/D_2O, respectively [30]. Identical spectra were observed in the deoxyribose derivative, and the distinction of *syn*- and the *anti*-conformation is not possible experimentally [30].

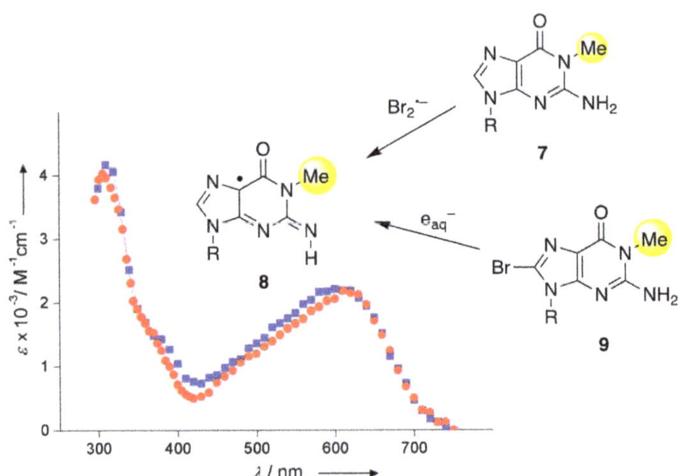

Figure 4. Absorption spectrum (red circles) refers to the oxidation of **7** by $Br_2^{\bullet-}$ and is assigned to the G(N2-H)$^\bullet$ radical (**8**); for the ε values, $G(Br_2^{\bullet-})$ = 0.62 μmol J^{-1} was assumed (taken from [14]). The absorption spectrum (blue square) refers to the reduction of **9** by e_{aq}^- and is assigned to the same G(N2-H)$^\bullet$, having presumably the *syn*-conformation; for the ε values, $G(e_{aq}^-)$ = 0.27 μmol J^{-1} was assumed (taken from [28]). See Appendix A for radiation chemical yields.

A thorough comparative computational analysis on the absorption spectra of three guanine-derived radicals (i.e., $G^{\bullet+}$, G(N1-H)$^\bullet$, and *syn*-G(N2-H)$^\bullet$) in the methyl guanine model was recently reported, indicating that the use of five water molecules of the first solvation shell, the Polarizable Continuum Model (PCM) by TD-DFT based approaches, and the incorporation of vibronic effects provide a good agreement between experiments and theory [31]. The substitution in position 9 of hydrogen in guanosine with a methyl group produces small but noticeable changes in the spectra.

5. Reaction of Hydroxyl Radical with Guanine Derivatives

The reduction potential of $HO^\bullet$ is very high (+2.73 V in acid media and +1.90 V at pH 7), but the direct electron-transfer is still rarely observed in $HO^\bullet$ reactions [13]. This is also the case for the reaction of $HO^\bullet$ with guanine derivatives (**1**), where the redox properties are favorable for electron transfer but H-atom abstraction and addition are observed.

The reactions of $HO^\bullet$ with Guo or dGuo occur with rate constants of ca. 5×10^9 M^{-1} s^{-1}. An ambident reactivity has been observed for the reaction of $HO^\bullet$ with the G moiety in aqueous medium. In 2000, it was reported that the addition to the base occurs at the C4 position (~65%) to give an adduct, which is the precursor of G(N1-H)$^\bullet$ via a dehydration path and at the C8 position (~17%) [32]. The main path (~65%) was based on the oxidizing properties of the transient species (toward TMPD with k = ~2×10^9 M^{-1} s^{-1}). A few years later, the reaction was revisited by us finding that the main path (~65%) is represented by hydrogen abstraction from the exocyclic NH_2 group affording G(N2-H)$^\bullet$, the oxidizing species that is the precursor of G(N1-H)$^\bullet$ via the tautomerization path [27,33].

The revision was based on the results of a detailed pulse radiolysis with Guo (**1**) and 1-Me-Guo (**7**) and their bromo derivatives 8-Br-Guo (**4**) and 8-Br-1-Me-Guo (**9**) [27,33]. The absorption spectra for the reaction of $HO^\bullet$ with **1** and **7** recorded ~1 μs after the pulse are reported in Figure 5a,b (black), respectively. The same figures report the absorption spectra obtained from the reactions of e_{aq}^- with **4** and **9** (red) described in the previous section. The ε values of the black spectra arising from the reaction with $HO^\bullet$ radicals were calculated using $G = 0.61$ μmol J^{-1}, as both $HO^{\bullet}$ and $H^{\bullet}$ species are scavenged by the guanosine derivatives (cf. Appendix A). On the other hand, both red spectra were calculated using $G = 0.27$ μmol J^{-1} of e_{aq}^{-}, choosing a percentage of intensity of the corresponding species in order to match the maximum of the spectra around 600 nm. The overlap in the range of 500–700 nm was excellent, choosing 65% and 55% for G(N2-H)$^{\bullet}$ and 1-Me-G(N2-H)$^{\bullet}$, respectively, suggesting that $HO^{\bullet}$ radicals react 65% with Guo (Figure 6) and 55% with 1-Me-Guo by H-atom abstraction from the exocyclic NH_2 group. The fate of the former is the tautomerization by pseudo-first-order kinetics ($k_{taut} = 2.3 \times 10^4$ s^{-1}) to give the tautomer G(N1-H)$^{\bullet}$, whereas the latter follows a second-order decay because the N1 position is blocked. The discussed spectra in the range of 500–700 nm do not contain the contribution from the H-atom reactions, because the H-atom adducts do not absorb at wavelengths above 500 nm [34]. The validation of this analysis is further confirmed when the exocyclic NH_2 moiety is replaced by NHEt or NEt_2; the tautomerization G(N2-H)$^{\bullet}$ $\rightarrow$ G(N1-H)$^{\bullet}$ occurs only in the first case with a rate constant of 3.6×10^4 s^{-1}, whereas, in the second case, a guanyl radical is not formed at all [27,33]. Furthermore, DFT-TD calculations strongly support the structural assignment, as no computed optical transitions were found for the C4-OH adduct of dGuo above 500 nm [27,33]. The insets in Figure 5a,b show the corresponding difference between the two spectra. Both of them show a strong band around 300 nm and various smaller broad bands up to 500 nm. These spectra are expected to be the sum of contributions from the 8-HO and 8-H adducts, as also predicted by the DFT-TD calculations [27,33].

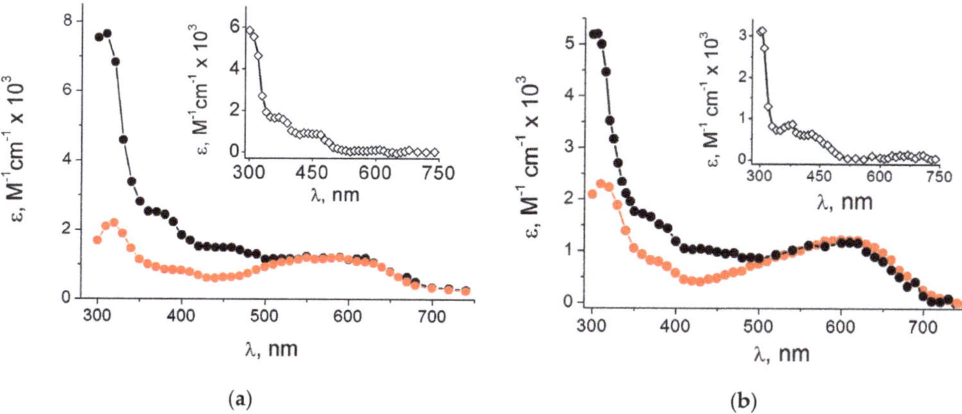

Figure 5. (**a**) Absorption spectra obtained from the reaction of $HO^{\bullet}$ with Guo (**1**) recorded 1 μs (black) after the pulse and 65% of the intensity of the absorption spectra obtained from the reaction of e_{aq}^{-} with 8-Br-Guo (**4**) recorded 1 μs (red) after the pulse (inset: the spectrum resulting from the subtraction of red from black) (taken from [27]); (**b**) absorption spectra obtained from the reaction of $HO^{\bullet}$ with 1-Me-Guo (**7**) recorded 1.2 μs (black) after the pulse and 55% of the intensity of the absorption spectra obtained from the reaction of e_{aq}^{-} with 8-Br-1-Me-Guo (**9**) recorded 10 μs (red) after the pulse (inset: the spectrum resulting from the subtraction of red from black) (taken from [33]).

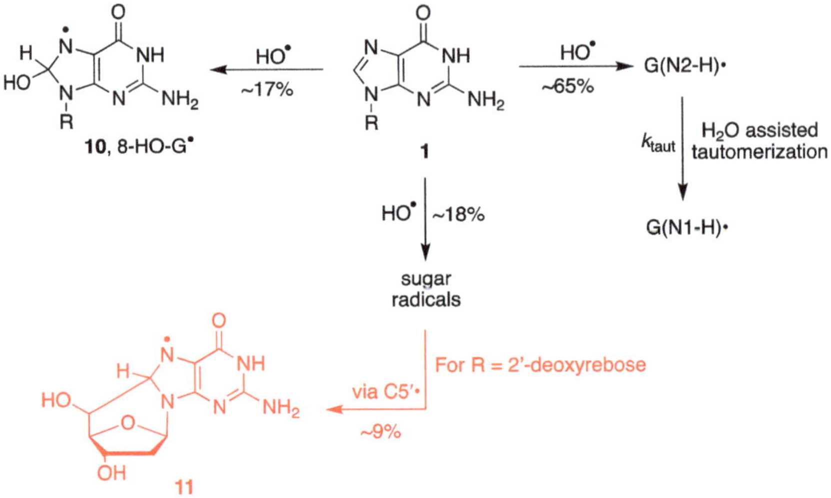

Figure 6. Reaction of $HO^•$ with guanine moiety of **1** in aqueous medium. The main path (~65%) is the H-atom abstraction from the exocyclic NH_2 group to give G(N2-H)$^•$ followed by water-assisted tautomerization G(N1-H)$^•$. The two minor paths are the addition at the C8 position giving the adduct **10** (~17%) and the hydrogen abstraction from the sugar moiety (~18%). When R = 2′-deoxyrebose, about half of the H-atom abstraction from the sugar occurs at the H5′ positions, followed by cyclization with the formation of radical **11** (adapted from [33]).

It is also worth mentioning the substitution effect at the C8 position [33]. By replacing the hydrogen of C8–H by an alkyl group, the rate constant of tautomerization is similar ($3.0 \times 10^4\ s^{-1}$), whereas a 15-fold increase is observed by replacing H with Br ($3.6 \times 10^5\ s^{-1}$). It is worth underlining that the rate constant of tautomerization is found to be 2 times slower than that reported for the reaction of bromo-derivative **4** in Figure 3. The difference in experimental conditions (N_2O-saturated aqueous solutions vs. Ar-purged aqueous solutions and 0.25 M *t*-BuOH at pH~7) and, more importantly, the overlap of various transient spectra in the reaction of $HO^•$ may be the origin of such a difference.

The overall mechanism for the reaction of $HO^•$ with Guo and dGuo is illustrated in Figure 6, with the main path (~65%) being the H-atom abstraction from the NH_2 group and the two minor paths, the formation of the adduct radical **10** (17%) and the H-atom abstraction from the sugar moiety (~18%) [27,33]. A half percentage of this latter path occurs at the H5′ positions and, in the dGuo case, is followed by radical cyclization with the formation of cyclo-derivative **11** [33,35].

It is worth mentioning that time-resolved optical transient absorption spectra for the reaction of $HO^•$ with guanine (Gua, free base) was investigated using the pulse radiolysis technique [36]. The reaction was performed at pH 4.6 in order to have mainly the neutral molecule (Gua has three pK_a values: 3.3, 9.4, and 12.4). The transient absorption spectra were different from those reported for the reaction of $HO^•$ with dGuo and Guo, with the 620 nm band particularly being absent. On the basis of the spectral characteristics, reactivity, and quantum chemical calculations, the absorbing species at 330 and 300 nm have been assigned to the C4-OH and C8-OH adducts, respectively [36]. Indeed, a significant difference (more than 200 mV) exists between the oxidation potentials of Gua and Guo or dGuo, where the π electrons play a significant role in the oxidation process, together with the lone electron pair on N9 in the case of the unsubstituted imidazole ring [37]. The reaction of $HO^•$ with Gua in an aqueous environment is also addressed by density functional theory (B3LYP) [38]. The calculations suggested the formation of an ion-pair intermediate ($G^{•+}$—OH^-) that deprotonates to form H_2O and a neutral G radical, favoring G(N1–H)$^•$ over G(N2–H)$^•$. The CAM-B3LYP calculation of their UV-visible spectra predicted an absorption around 620 nm [38], which is in contrast with the mentioned experimental results [36]. The theoretical work on the reaction of HO• with Gua [38], often used as a reference for the reaction of the HO• radical with Guo or dGuo or even DNA, is not appropriate. Dizdaroglu in two reviews of 2012 collected and discussed the reactivity of HO• with guanine derivatives without taking into consideration the guanine moiety in different molecular contexts, such as Gua vs. dGuo, and they erroneously questioned the mechanism reported in Figure 6 [39,40].

6. Photochemical Precursors

An alternative approach for guanyl radical generation is based on the photolabile modifier of the guanine moiety. Synthetic precursors **12** and **13** (R = deoxyribose) were designed to generate G(N1-H)• through the homolysis of the N–O bond, and compound **14** was proposed for the selective photo-generation of G(N2-H)• (Figure 7). The G(N1-H)• radical was confirmed for **12** by continuous photolysis product analysis and trapping studies in water, as well as laser flash photolysis experiments [41]. dGuo was formed quantitatively when the precursor **13** is photolyzed under anaerobic conditions in phosphate buffer (pH = 7.2) and in the presence of an excess of β-mercaptoethanol [42].

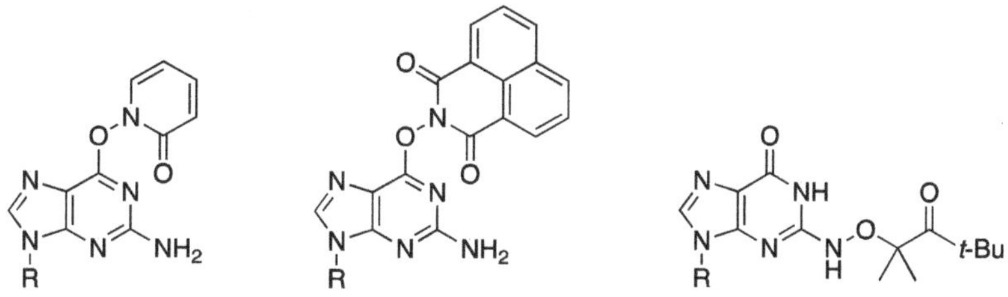

Figure 7. Photochemical precursors **12** and **13** for the formation of G(N1-H)•, and **14** for the formation of G(N2-H)•.

Upon 355 nm laser flash photolysis of **14** (1 mM) and acetophenone (30 mM) in aqueous buffer (pH 7.0)/acetonitrile (1:1, *v*/*v*), the formation of G(N2-H)• is proposed [43]. A broad band between 550 and 700 nm with a maximum at 650 nm was observed, and the authors hypothesized that the broad band contains peaks at 610 and 650 nm belonging to the different conformational isomers of G(N2-H)•, in which the remaining N–H bond is either *syn* or *anti* with respect to the guanine N3 atom (cf. Figure 2B). Under their experimental conditions, a variety of possible transients is possible and the observed spectra are an overlap of various species with an over-interpretation regarding the G(N2-H)• radical, which is probably present in a relative low percentage. Indeed, their recorded absorption spectra are in contrast with those reported previously, and no tautomerization was observed within hundreds of microseconds in these experiments. Based on these questionable results, it was suggested that the hydrogen atom abstraction from dG is unlikely to be a major pathway when HO• reacts with dG (see Figures 5 and 6) [43]. Additionally, these authors did not consider that H_2O-assisted tautomerization depends on the solvent-composition. Water or phosphate buffer (pH 7.0) is not equivalent to the mixture aqueous buffer (pH 7.0)/acetonitrile (1:1, *v*/*v*). The microheterogeneities of water–acetonitrile mixtures are well documented (see e.g., refs [44–46]). The independent generation of the G(N2-H)• radical by photochemical precursors for time-resolved spectroscopic studies needs to be addressed by further research.

7. The Fate of Guanyl Radical in Nucleosides

As shown above, guanine can be oxidized to $G^{\bullet+}$ by a variety of oxidants such as $SO_4^{\bullet-}$, $Br_2^{\bullet-}$, $Cl_2^{\bullet-}$, and $CO_3^{\bullet-}$, including various metal complexes. The carbonate radical anion ($CO_3^{\bullet-}$) is an important reactive species under physiological conditions. The pair CO_2/HCO_3^- is an active buffer maintaining physiological pH, and HCO_3^- exists in millimolar levels in vivo. The reaction of CO_2 with peroxynitrite ($ONOO^-$) generates $CO_3^{\bullet-}$ through the unstable nitrosoperoxycarbonate [47]. The reaction of $HO^\bullet$ with HCO_3^- produces $CO_3^{\bullet-}$. The Fenton reaction in the presence of HCO_3^- affords $CO_3^{\bullet-}$ rather than the $HO^\bullet$ radical [48]. The reduction potential of $CO_3^{\bullet-}/CO_3$ is 1.59 V; therefore, $CO_3^{\bullet-}$ is a milder single-electron oxidant that abstracts an electron from guanine moieties [49]. The oxidation mechanism and kinetics of dGuo (**1**) by the carbonate radical anion ($CO_3^{\bullet-}$) have also been investigated theoretically [50].

The iron-induced Fenton oxidation of dGuo (**1**) in the presence of bicarbonate buffer and in its absence (phosphate buffer) affords different distribution products consistent with $CO_3^{\bullet-}$ and $HO^\bullet$ reactive intermediates, respectively [51]. Figure 8 shows the overall mechanism of oxidation of dGuo by $HO^\bullet$ and $CO_3^{\bullet-}$. The two oxidants have some common intermediates generated by different routes, e.g., (N1-H)$^\bullet$ and 8-HO-G$^\bullet$ radicals. The 8-HO-G$^\bullet$ is the common precursor of 8-oxo-G and Fapy-G. The Fapy-G formation requires ring-opening followed by one-electron reduction or vice versa (i.e., one-electron reduction followed by ring-opening). The formation of Fapy-G depends on the oxygen concentration and the redox environment [52,53]. 8-oxo-G, with a reduction potential 0.55 V lower than that of G [54], undergoes further oxidation in the presence of one-electron oxidants. Two main products derive from the oxidation of 8-oxo-G, namely spiroiminodihydantoin (Sp) and 5-guanidinohydantoin-2′-deoxyribose (Gh) [55]. From the oxidation of dGuo under aerobic conditions or in the presence of $CO_3^{\bullet-}$, two other products are formed via C5 paths (Figure 8), viz. 5-carboxamido-5-formamido-2- iminohydantoin-2′-deoxyribonucleoside (2Ih) and 2-amino-5-[2-deoxyribose]-4H-imidazol-4-one (Iz). The latter is further hydrolyzed to 2,2-diamino-4-[(2-deoxyribose)amino]-5(2H)-oxazolone (Z) [56,57].

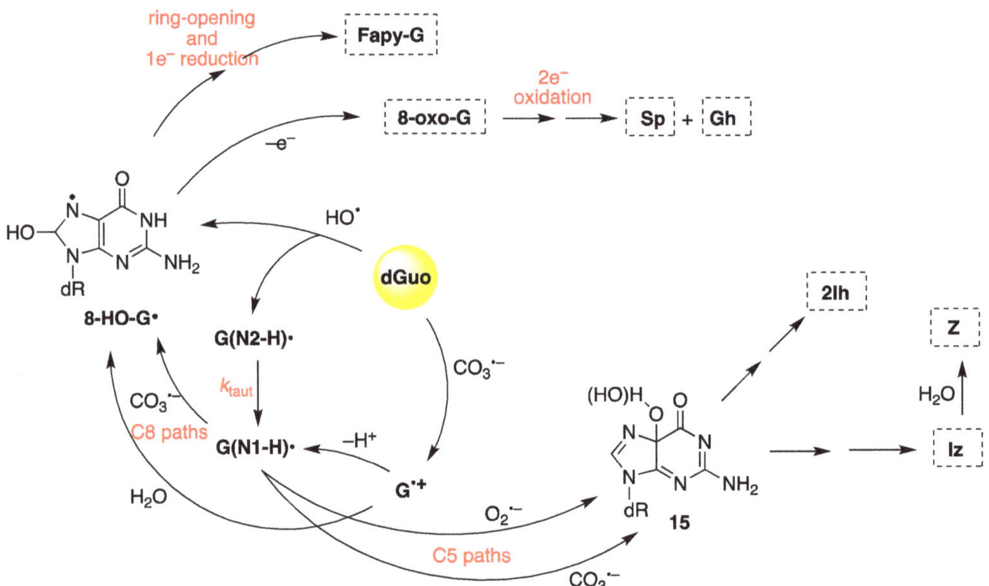

Figure 8. The mechanism of dGuo (**1**) oxidation by hydroxyl radical ($HO^\bullet$) or carbonate radical anion ($CO_3^{\bullet-}$); for the chemical structures of 8-oxo-G, Fapy-G, Sp, Gh, 2Ih, Iz, and Z, see Figure 15.

8. Guanyl Radicals in ss-ODNs and ds-ODNs

The mode of formation and the kinetics of decay of $G^{\bullet+}$ depend strongly on ODN secondary structures. In this section, we consider ss- and ds-ODNs, whereas the ODNs arranged in G-quadruplexes are reported in the next section.

The one-electron oxidation of ss-ODN (TCGCT) by $CO_3^{\bullet-}$ and $SO_4^{\bullet-}$ was studied by laser flash photolysis in some detail [58]. Both reactions lead to the formation of the guanine radical cation $G^{\bullet+}$ and/or its deprotonation product G(N1-H)$^{\bullet}$ depending on pH (pK_a = 3.9) (Figure 9). At pH 2.5, the $G^{\bullet+}$ radical is hydrated within 3 ms, forming 8-oxo-dG in the ODN. At pH 7.0, G(N1-H)$^{\bullet}$ reacts only slowly with H_2O and lives for ~70 ms, ultimately giving intrastrand cross-link adducts. Alternatively, it can be further oxidized by reaction with $CO_3^{\bullet-}$, generating the two-electron oxidation products 8-oxo-dG (via C8 addition **16**) and 2Ih (via C5 addition **17**). The same group has studied the one-electron oxidation of a 30mer ss-ODN containing eleven Gs in comparison with the same strand in ds-ODN, and these results are discussed below.

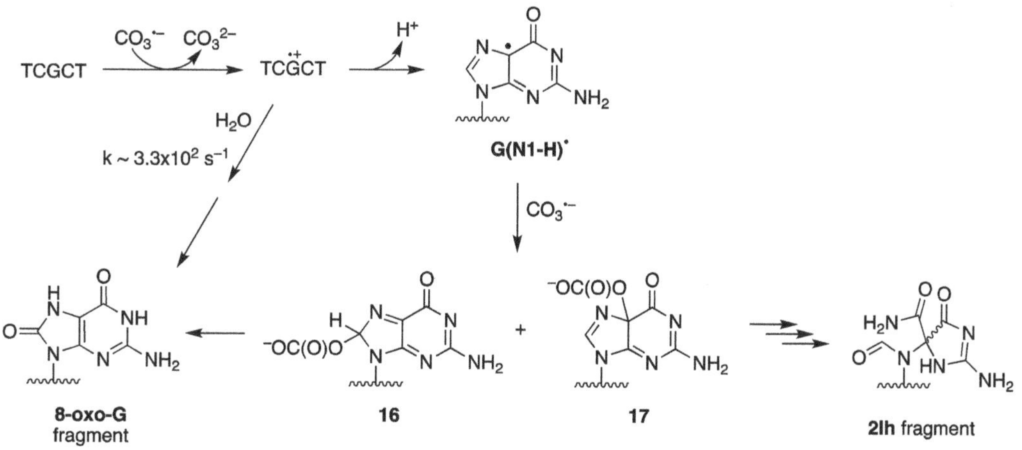

Figure 9. Mechanistic scheme of one-electron oxidation of ss-ODN (TCGCT) by carbonate radical anion $CO_3^{\bullet-}$ (adapted from ref. [54].

Since early works, it has been postulated that in DNA, the proton is not directly lost from $G^{\bullet+}$ to the aqueous phase but remains within the hydrogen-bonded $G^{\bullet+}$:::C pair located toward the cytosine N3 atom [12]. ESR studies of the one-electron oxidation of a variety of ds-ODNs indicated that the proton is entirely transferred [G(N1-H)$^{\bullet}$:::CH$^+$] at 77 K [59], whereas a prototropic equilibrium [$G^{\bullet+}$:::C $\leftrightharpoons$ G(N1-H)$^{\bullet}$:::CH$^+$] should be established at ambient temperature. Most of the experimental studies described below on ds-ODNs used $SO_4^{\bullet-}$ as the oxidizing agent. It should be recalled that $SO_4^{\bullet-}$ reacts with all four nucleobases (G, C, A, and T) with rate constants that are close to diffusion-controlled rates [13], although the primary damage is localized at Gs, having the lowest reduction potential (cf. Section 1) [4–10].

The one-electron oxidation by the powerful $SO_4^{\bullet-}$ of a variety of ds-ODNs containing G:::C pairs has been reported in pulse radiolysis [16,60] or laser flash photolysis [61] studies. Pulse radiolysis with optical detection showed that different ds-ODNs containing G, GG, and GGG sequences afford the corresponding radical cation $G^{\bullet+}$:::C. In the earlier work [16], a biphasic decay of $G^{+\bullet}$:::C was assigned to the shift in the N1 proton in $G^{+\bullet}$ to its partner C, followed by the release of the proton into the solution. The same group revisited these reactions by analyzing the optical spectra and the kinetics of eleven ds-ODNs with different sequences (11mer to 13mer ds-ODNs) [60]. The revised results concern the monophasic decay of [$G^{\bullet+}$:::C $\leftrightharpoons$ G(N1-H)$^{\bullet}$:::CH$^+$] associated with the release of the proton into solution. Based on pK_a values, the population of positive charges is estimated to be ~1:3 in favor of cytosine [12]. The rate constant for deprotonation was dependent on the ds-ODN sequence, especially on the sequence of bases adjacent to the guanine base, varying in the range of 0.3–2 $\times$ 10^7 s^{-1} by monitoring the transient. In the ds-ODN of the 13mer $A_5G_3A_5$, the first transient decays with k = 4.5 $\times$ 10^6 s^{-1} in H_2O, showing a $k(H_2O)/k(D_2O)$ of

3.8-fold. Interestingly, by replacing C with 5-Me-C or 5-Br-C in ds-ODN, the rate constant of deprotonation increased or decreased, respectively. Overall, the proposed mechanism consists of path A in Figure 10 [60].

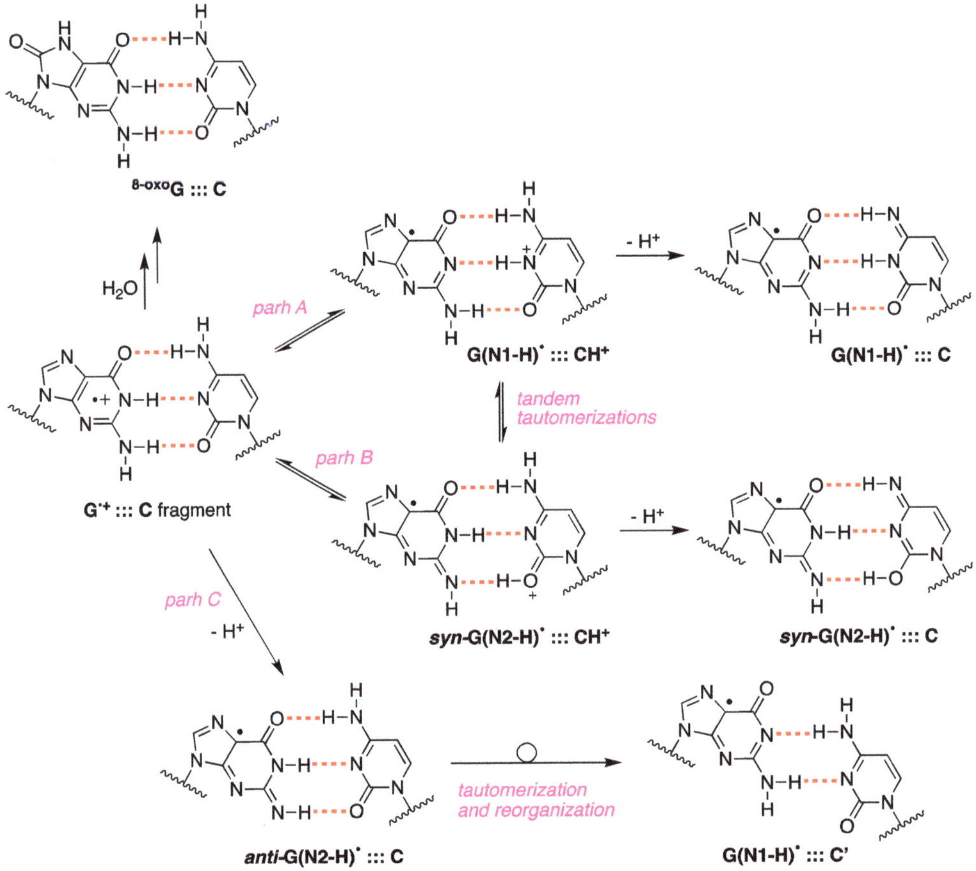

Figure 10. Mechanistic possibilities of $G^{\bullet+}$:::C deprotonation in ds-ODNs and its reaction with water to give $^{8\text{-oxo}}$G:::C.

Upon our discovery of the first directly observed differences of the two tautomeric forms of one-electron-oxidized Guo or dGuo (see Figure 3), we proposed that an alternative scenario of $G^{\bullet+}$:::C deprotonation involving the tautomer G(N2-H)$^\bullet$ should also be considered in future studies; in particular, using time-resolved studies, the monitoring of the absorbance changes around 620 nm is important, as G(N2-H)$^\bullet$ has a distinct absorption [11]. Figure 10 shows the proposed paths B and C that involve *syn*-G(N2-H)$^\bullet$ and *anti*-G(N2-H)$^\bullet$, respectively (cf. Figure 2B). In path C, the protons can escape directly into the bulk solution forming *anti*-G(N2-H)$^\bullet$, whereas, in path B, a proton-transfer equilibrium involves the *syn*-G(N2-H)$^\bullet$ conformer that is calculated to be more stable. We were occasionally criticized for such a proposal, but, today, it is gratifying to see that such a "second face" of the guanyl radical, i.e., G(N2-H)$^\bullet$, is indeed recalled as an intermediate in several oxidation reactions of secondary structures of DNA (vide infra).

For the first time, the presence of *anti*-G(N2–H)$^\bullet$:::C in a ds-ODN is shown to be easily distinguished from the other prototropic forms, owing to its readily observable nitrogen hyperfine coupling by electron spin resonance (ESR) [30]. Indeed, the one-oxidation of d[TGCGCGCA]$_2$ has been studied in detail by ESR and UV-visible spectral analysis;

at pH $\geq$ 7, the initial site of deprotonation is found to be at N1, forming G(N1–H)$^{\bullet}$:::C at 155 K, and upon annealing to 175 K, the site of deprotonation to the solvent shifts to an equilibrium mixture of G(N1–H)$^{\bullet}$:::C and *anti*-G(N2–H)$^{\bullet}$:::C. This ESR identification is supported by the visible absorption at 630 nm, which is characteristic for G(N2–H)$^{\bullet}$ radicals.

The one-electron oxidation of 30mer ODNs by $SO_4^{\bullet-}$, shown in Figure 11, was studied in ss-ODNs or ds-ODN by time-resolved absorption spectroscopy and quantification of the 8-oxo-G lesion as the final product [61]. Figure 11 shows the dependence of the yields of 8-oxo-dG in ds-ODN vs. ss-ODNs on the number of successive 308 nm laser pulses. The yield of 8-oxo-G in ds-ODN is ~7 times higher than that in ss-ODNs, and the initial 8-oxo-G yields are ~1 μM per laser pulse that corresponds to ~20% per G(-H)$^{\bullet}$ radical generated. Therefore, the secondary structure of ds-ODN is a crucial factor that enhances the formation of 8-oxo-G lesions from G(N1-H)$^{\bullet}$ radicals. The transient species observed after the complete decay of $SO_4^{\bullet-}$ (5 μs) for both ds-ODN and ss-ODN were very similar and assigned to G(N1-H)$^{\bullet}$. However, the kinetics of G(N1-H)$^{\bullet}$ decay were quite different; in ds-ODN, the decay is biphasic with one component decaying with a lifetime of ~2.2 ms and the other one with a lifetime of ~0.18 s, whereas, in ss-ODN, the decay is monophasic with a ~0.28 s lifetime. The millisecond decay component in ds-ODN is correlated with the enhancement of 8-oxo-G yields, which are ~7 times greater than those in ss-ODN. In ds-ODN, the authors proposed that the equilibrium [G$^{\bullet+}$:::C $\leftrightarrows$ G(N1-H)$^{\bullet}$:::CH^{+}] allows for the hydration followed by the formation of $^{8\text{-oxo}}$G:::C (see Figure 10). By contrast, in ss-ODN, the deprotonation of G$^{\bullet+}$ and the irreversible escape of the proton into the aqueous phase compete more effectively with the hydration mechanism, thus diminishing the yield of 8-oxo-G. In order to accommodate the fast deprotonation (see above) [60], they invoked path C in Figure 10, followed by a tautomerization of G(N2-H)$^{\bullet}$ to G(N1-H)$^{\bullet}$ coupled with base displacement and reorganization of the hydrogen-bonding network [61].

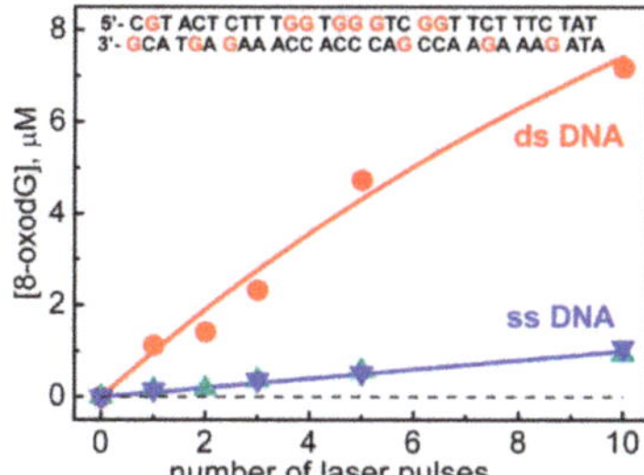

Figure 11. Dependence of the yields of 8-oxodGuo lesions in ds-ODN vs. ss-ODN on the number of successive 308 nm laser pulses. Taken from [58].

It was demonstrated that absorption of a single low-energy UV photon by ds-ODNs at 265 nm generates guanyl radicals, with an appreciable quantum yield (>10^{-3}). The transient species detected after 3 μs are identified as guanyl radicals G(N1-H)$^{\bullet}$, which decay with a half-life of 2.5 ms. The experimental results corroborated by theoretical studies suggest nonvertical processes, associated with the relaxation of electronic excited states [62–64]. Similar results were obtained with ct-DNA [64]. The same ss-ODNs and ds-ODN reported in Figure 11 were studied by photoionization using ultra-purified chemicals [63]. Although the spectra on the microsecond timescale are the same as those reported [61], the radical lifetimes were found to be much shorter. It was also shown that base pairing slows down the radical lifetime, and this was attributed to the fact that larger conformational motions in single strands allow the systems to adopt more rapidly reactive conformations [63].

It is also worth citing the pulse radiolysis studies using the selenite radical, $SeO_3^{\bullet-}$ of E^0 1.68 V, oxidizing DNA with a rate constant of 3.5 × 10^7 M^{-1} s^{-1} [65]. The one-electron oxidation of ct-DNA by $SeO_3^{\bullet-}$ involves the G:::C pair in a different way, as it was proposed via a concerted formation of a cytosyl radical followed by proton loss from the amine substituent of cytosine into the major groove and the H-atom transfer to the guanyl radical G(N1-H)$^{\bullet}$:::C [65]. The neutral radical, *anti*-G(N2-H)$^{\bullet}$:::C, in DNA was formed directly through H-atom abstraction from its amine moiety by the benzotriazinyl radical, and a DFT calculation corroborated such a pathway; the E_7 value of 1.22 V for *anti*-G(N2-H)$^{\bullet}$:::C has been determined by analysis of both absorption and kinetic data [66]. Moreover, an investigation of the aniline radical cation, $(ArNH_2)^{\bullet+}$, with ds-ODN or DNA suggested the formation of the guanyl radical in an energetically more stable "slipped" structure, similar to the one reported in Figure 10 (bottom-right) [67].

Much attention has been focused on computing the properties of the $G^{\bullet+}$:::C pair fragment of DNA (e.g., the proton transfer equilibria, hydrated cation, tautomerism or redox behavior) by DFT calculations including solvent effects. The theoretical calculations have been abundant during the last two decades in this area, and it is not the scope of this review to examine the whole scenario of the reported calculations. Only a few with more attention on recent publications are reported; in particular, the influence of hydration on the proton transfer $G^{\bullet+}$:::C pair [68–70]; the calculations for a single GC base pair, suggesting a very slight favor of *anti*-G(N2–H)$^{\bullet}$:::C by 1.7 kJ mol^{-1} over the G(N1–H)$^{\bullet}$:::C structure [30]; and the calculations providing reference absorption spectra for guanine radicals in duplexes and the computed spectra that predict changes in transient absorption spectra expected for hole localization, as well as for deprotonation (to cytosine and bulk water) and hydration of the radical cation [62,71].

It is worth mentioning that oxyl radicals such as $O_2^{\bullet-}$, $NO_2^{\bullet}$, and $CO_3^{\bullet-}$ react with G(N1-H)$^{\bullet}$ of ss-ODNs or ds-ODNs in cross-radical termination mode and, in all cases, the combination occurs via the analogous C8 and C5 paths reported in Figure 8 with the formation of expected end-products [72–74]. The ratio of the C5/C8 products can depend on the macromolecular secondary structure, e.g., in the case of $NO_2^{\bullet}$, the ratio of C5/C8 addition decreases from 2.1−2.6 in ss-ODNs to 0.8−1.1 in ds-ODNs [74]. The influence of the local base sequence and secondary structure on G oxidation by riboflavin-mediated photooxidation, and nitrosoperoxycarbonate with the formation of 8-oxo-G, Z, and 8-NO_2-G lesions are also reported [75].

The radical chemistry reported for Guo or dGuo is often transferred to complex macromolecular arrangements such as the secondary DNA structure. From this section, it is clear that the one-electron oxidation of ds-ODNs or DNA is a complex matter with many variants, including chemical methods of generation, time-resolved spectroscopic approaches, base sequences, and thermochemical cascade. Mechanistic proposals are based on multidisciplinary knowledge, where DFT calculations can often help for a better understanding of the favored pathways but must be combined with the experimental evidence.

9. Guanyl Radicals in G-Quadruplex

The structure of the G-quadruplex consists of stacked G-quartets where each G-quartet is a planar array of four Hoogsteen-bonded guanines stabilized by metal cations [76]. The existence of $G^{\bullet+}$ in the G-quadruplex and their deprotonation from the NH_2 moiety of guanines was first reported in a study performed via photosensitization (Figure 12A) [77]. The four G-quadruplexes buffered using Na^+ or K^+, i.e., Tel22/Na^+, TG4T/K^+, G4T4G4/Na^+, and the TBA/K^+, are used, that adopt three of them antiparallel and TG4T/K^+ parallel orientations (see Figure 12B). The one-electron oxidation was carried out by $SO_4^{\bullet-}$ using laser flash photolysis for the transient detection. The deprotonation rate of $G^{\bullet+}$ within G-quadruplexes Tel22/Na^+, TG4T/K^+, and G4T4G4/Na^+ (k = ~2×10^5 s^{-1}) was found to be 1−2 orders of magnitude slower compared to $G^{\bullet+}$ from dGuo or ds-ODNs with the formation of the G(N2-H)$^{\bullet}$ radical in *anti*-conformation (Figure 12A), whereas in TBA/K^+, the deprotonation occurred from the G in the loop affording G(N1−H)$^{\bullet}$ with a rate constant of 1.4×10^6 s^{-1} [77].

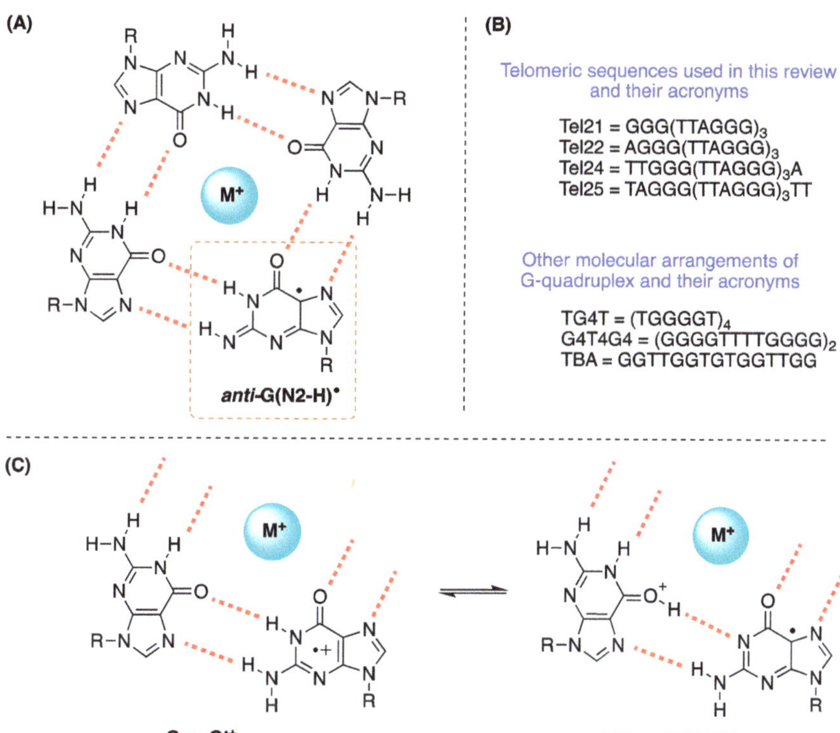

Figure 12. (**A**) Drawing of G-quartet containing the guanyl radical *anti*-G(N2-H)• (cf. Figure 2B); (**B**) the ss-ODNs sequences used in this review (Sections 9 and 10) in G-quadruplex structures; (**C**) fragments of G-quartet; the proposed equilibrium in one-electron-oxidized G quartet [78].

In antithesis with the above-mentioned work [77], the one-electron oxidation of Tel25, buffered using Na^+ or K^+, by $SO_4^{\bullet-}$ was also investigated by time-resolved spectroscopy and determined the end-product 8-oxo-G. In neutral aqueous solutions (pH 7.0), the observed transient is G(N1-H)• after the complete decay of $SO_4^{\bullet-}$ (~10 μs after the actinic laser flash). In both systems, the G(-H)• decay is biphasic with one component decaying with a lifetime of ~0.1 ms, and the other one with a lifetime of 20–30 ms. The fast decay component (~0.1 ms) in G-quadruplexes is correlated with the formation of 8-oxo-G lesions. The authors proposed that in G-quadruplexes, G(-H)• radicals retain their radical cation character by sharing the N1-proton with the O6-atom of G in the [$G^{\bullet+}$:::G] Hoogsteen base pair (Figure 12C); this equilibrium [G:::$G^{\bullet+}$ ⇆ GH^+:::G(N1-H)•] leads to the hydration of $G^{\bullet+}$ within the millisecond time domain, and is followed by the formation of the 8-oxo-G lesions [78].

The first study on the UV-induced ionization of G-quadruplexes was reported by Markovitsi and co-workers [79]. Using nanosecond time-resolved spectroscopy corroborated by TD-DFT calculations and the telomeric sequence Tel21 (cf. Figure 12B) in buffered solutions containing Na^+ cations, they showed that irradiation at 266 nm, corresponding to an energy significantly lower than the guanine ionization potential, provokes one-photon ionization in aqueous solution, affording the guanyl radical. The fate of guanine radicals is followed over five orders of magnitude of time: (i) up to 30 ns 100% $G^{\bullet+}$, (ii) in 3 μs 50% $G^{\bullet+}$ and 50% *anti*-G(N2-H)• that become 35% $G^{\bullet+}$ and 50% G(N2-H)• in 20 μs; (iii) in 5 ms 50% G(N1-H)• that reaches 6% in 180 ms. The two main features are a long-lived $G^{\bullet+}$ with respect to ds-ODNs due to the secondary structure and the detection of the two guanyl radicals, including the tautomerization *anti*-G(N2-H)• → G(N1-H)•, which takes place on the millisecond timescale. The calculations indicated that *anti*-G(N2-H)$^{\bullet}$ is 12.1 kJ mol^{-1} more stable than the G(N1-H)$^{\bullet}$ in the simulated environment of Tel21/Na^{+} [79]. Similar behaviors were also reported for Tel25/Na^{+} [63] and TG4T/Na^{+} [80], although their guanines adopt, respectively, antiparallel and parallel orientations in respect of the glycosidic bonds. By replacing Na^{+} with K^{+} in Tel21 [81] and TG4T [82], some important changes were observed: the quantum yield of one-photon ionization at 266 nm was twice as high in both systems and the *anti*-G(N2-H)$^{\bullet}$ radical decayed faster. Moreover, in the case of TG4T/K^{+}, the tautomerization step G(N2-H)$^{\bullet}$ $\rightarrow$ G(N1-H)$^{\bullet}$ is suppressed. Such a behavior shows that the nature of the metal ion plays an important role in the reaction mechanism.

It is also worth mentioning the formation of 8-oxo-G in the comparative study on Tel21/Na^{+} and TG4T/Na^{+}, where for the former, the yield of 8-oxo-G is much higher than that of the latter [79,80]. As the fate of G$^{\bullet+}$ is split between hydration vs. deprotonation, the authors suggested that the G$^{\bullet+}$ population survives on the millisecond timescale in Tel21/Na^{+}, whereas the deprotonation is much faster in TG4T/Na^{+}, thus rendering the reaction path leading to 8-oxodG less probable. Markovitsi and co-workers recently summarized their work on the guanyl radicals in ds-ODNs and G-quadruplexes [64,83].

Two main questions remain open on the guanyl radical in G-quadruplexes: (i) what is the mechanism of tautomerization *anti*-G(N2-H)$^{\bullet}$ $\rightarrow$ G(N1-H)$^{\bullet}$, and (ii) based on the results described above, the major oxidative damage is expected to arise from deprotonated radicals; what are the end-products of the two guanyl radicals and their relation with the 8-oxo-G lesion. Theoretical calculations combined with time-resolved methodologies and product studies will certainly help understand such pending radical mechanisms.

10. Reaction of Hydroxyl Radical with ODNs and DNA

The reaction of HO$^{\bullet}$ with G in the macromolecular environment is worth a special comment, because the behavior is quite different from the reaction of HO$^{\bullet}$ with simple Guo or dGuo (Section 5). The four common nucleosides of DNA (dGuo, dAdo, dCyt, and Thy) react with HO$^{\bullet}$ at close to diffusion-controlled rates (4–7 $\times$ 10^{9} M^{-1} s^{-1}), with the pyrimidine nucleosides being slightly higher [13]. The sites of DNA attack by diffusible HO$^{\bullet}$ radicals are known to be either the hydrogen atom abstraction from the 2′-deoxyribose units or the addition to the base moieties, with the latter being the predominant one (accounting for 85–90% of attacked sites) [13].

The reaction of HO$^{\bullet}$ with the 21-mer ds-ODNs reported in Figure 13A shows that increasing concentrations of oxygen lead to elevated levels of 8-oxo-G. The 8-oxo-G levels vary from 1.35 8-oxo-dG/10^{6} dG/Gy in the absence of O_2 up to 4.60 for 2.01 $\times$ 10^{-4} M of O_2 (15%) [84]. The mechanism of the formation of 8-oxo-G by the reaction of HO$^{\bullet}$ with ct-DNA and ds-ODNs in the presence of oxygen has been investigated in detail [85–87]. It was demonstrated that three pathways contribute to the formation of 8-oxo-dG, as shown in Figure 14. The direct addition of HO$^{\bullet}$ to the C8 position of the guanine moiety to form 8-HO-G$^{\bullet}$ accounts for the minor path (~5%), whereas the remaining ~95% of 8-oxo-dG are produced by two types of reactions in approximately equal amounts involving DNA radicals, that is, one-electron oxidation (probably by a variety of reactions not very well understood) followed by hydration reaction (~45%) or an intramolecular addition of a transiently generated pyrimidine peroxyl radical onto the C8 of a vicinal guanine base (~50%). As mentioned earlier, the Fapy-G lesions are chemically related to 8-oxo-G as they are generated from the same intermediate 8-HO-dG$^{\bullet}$ (Figure 14). The formation of Fapy-G is dependent on the oxygen concentration and the redox environment. [52,53].

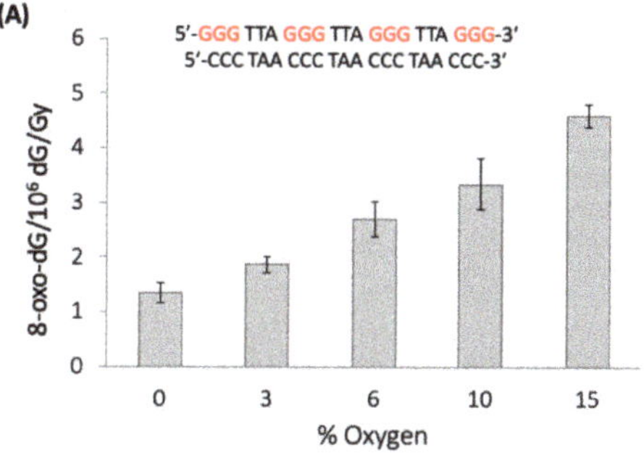

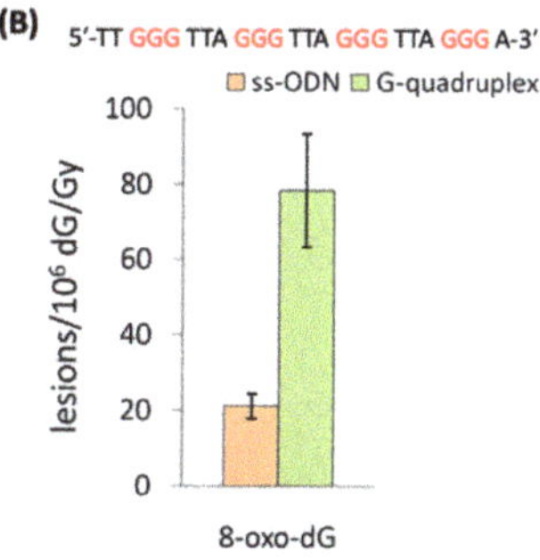

Figure 13. (**A**) Radiation-induced formation of 8-oxo-dG in ds-ODNs at increasing concentrations of oxygen; data from [84]. (**B**) Radiation-induced formation of 8-oxo-dG lesions in ss-ODN (nonfolded state) and G-quadruplex (folded state) in deareated conditions; the G-quadruplex was prepared by dissolving the oligonucleotides in 10 mM of KH_2PO_4 buffer containing 70 mM of $KClO_4$ (pH 7.0). Values represent the mean ± SD per 10^6 dG per Gy of γ-irradiation in the range from 0 to 60 Gy of irradiation.

A pulse radiolysis study reported the reaction of four 12mer ONDs with four or two Gs subjected, either as ss-ODNs or ds-ODNs, to the reaction with $HO^\bullet$ radicals using an intensified charge-coupled device (ICCD) as an alternative to the photomultiplier tube (PMT) for transient spectra measurements [88]. The characteristic band >600 nm of the guanyl radical G(N2-H)$^\bullet$ was present in ss-ODNs and disappeared on the microsecond timescale, but it was absent in ds-ODNs.

Figure 14. Reaction mechanism for the formation of 8-oxo-G in ds-ODN. For the hydration reaction of $G^{\bullet+}$ and identification of 8-HO-G$^\bullet$ by electron spin resonance (ESR) in DNA, see reference [89].

The impact of G-quadruplex folding with respect to unfolded sequences in ss-ODNs on the formation of 8-oxo-G lesions was also investigated [90]. In particular, the unfolded sequences of TG4T and mutated Tel24, as well as the G-quadruplexes of TG4T/K^+ and Tel24/K^+ (see Figure 12B), were exposed to $HO^\bullet$ radicals generated by γ-radiolysis, and the 8-oxo-G was then quantified via stable isotope dilution LC-MS/MS analysis. Figure 13B shows the results of Tel24, where 8-oxo-G lesions formation in the folded G-quadruplex is ~4 folds higher than the unfolded sequences (78.4 vs. 21.2 of 8-oxo-G/10^6 dG/Gy, respectively). In the TG4T system, the 8-oxo-G lesions formation in tetramolecular parallel-stranded G-quadruplex is ~2.8 folds higher than the monomolecular sequences (39.9 vs. 14.4 of 8-oxo-G/10^6 dG/Gy, respectively). Therefore, the self-organization of guanines in the G-quadruplex tetrads produced ~3- and ~4-fold increases in 8-oxo-G yields, indicating that the G-quadruplex is significantly more susceptible to $HO^\bullet$ oxidation. The mechanism of the formation of 8-oxo-G by $HO^\bullet$ in G-quadruplexes is not clear at present, with the addition at the C8 position being the most reasonable key step.

11. Biomarkers of Guanine DNA Damage

Reactive oxygen and nitrogen species (ROS/RNS) are produced in a wide range of physiological processes and are also responsible for a variety of pathological processes [1–3]. Persistent oxidative stress developed at sites of chronic inflammation is characterized by overproduction of ROS/RNS, which targets cellular DNA, and, as a consequence, may lead to mutations and cancer. For example, the 8-oxo-G lesion is genotoxic [91], and failure to remove this lesion before replication induces the G:::C → T:::A transversion mutation.

The ROS/RNS network includes molecules such as hydrogen peroxide (H_2O_2), hypochlorous acid (HOCl), and peroxynitrite ($ONOO^-$), as well as radicals such as the superoxide radical anion ($O_2^{\bullet-}$), nitric oxide ($NO^\bullet$), hydroxyl radical ($HO^\bullet$), nitrogen dioxide ($NO_2^\bullet$), and carbonate radical anion ($CO_3^{\bullet-}$). Most of these species can react with DNA, and the nucleobase guanine with the lowest reduction potential is the dominant site for oxidation within DNA through the hole transfer [4–10]. Analytical protocols and, in particular, liquid chromatography–tandem mass spectrometry (LC-MS/MS) allow the identification of the DNA lesions with high-accuracy in cellular DNA after enzymatic digestion, as reviewed recently [92,93]. The main guanine-derived lesions observed in vitro and/or in vivo from oxidatively generated DNA damage are collected in Figure 15 and divided in four groups, depending on the precursor. In model DNA studies, the relative yields of these lesions depend on the reaction context, and several comparative studies with various types of oxidants are reported [94].

8-Oxo-G and Fapy-G are chemically connected because they derive from the same radical precursor (8-hydroxyl radical adduct, 8-HO-$G^\bullet$) and their comparisons have been extensively treated [52,92]. Fapy-G due to its instability is mainly detected as the free base modification (i.e., Fapy-Gua), and both GC-MS/MS and LC-MS/MS techniques with the isotope dilution approach have been used for the measurements in vitro and in vivo [95]. A large number of publications deal with these lesions not only in vitro but also in vivo [96,97]. There are a few studies at the macromolecular level in vitro for the formation of Ih, Iz, Z, Sp, and Gh lesions. The Sp and Gh lesions derive from further oxidation products of 8-oxo-G, whereas the d2Ih, dIz, and dZ lesions are the final products of the further reaction of $G^{\bullet+}$ or G(N1-H)$^\bullet$ with oxyl radicals generated under aerobic conditions in vitro. Comparative analyses of the four oxidized G lesions (8-oxo-G, Z, Sp, and Gh) from the reaction of ct-DNA with $HO^\bullet$, $ONOO^-$, and $^1O_2$ [98], as well as of four oxidized G lesions (8-oxo-G, Z, and 8-NO_2-G) from the reaction of ds-ODNs with $HO^\bullet$ and $NO_2^\bullet/CO_3^{\bullet-}$ [75], were examined using LC–MS/MS and isotopomeric internal standards. There are a few studies in murine models reporting the detection of dSp, dGh, and dZ [99].

5′,8-Cyclo-2′-deoxyguanosine (cdG) in its 5′R and 5′S diastereoisomeric forms results from the C5′ radical cyclization on the C8 position of G (see Figure 5) [92,93]. Although the guanine DNA lesions mentioned above are generated by various ROS (including $HO^\bullet$), the formation of 5′R-cdG and 5′S-cdG lesions in vitro and in vivo relies exclusively on the $HO^\bullet$ attack. Recent results on the quantification of 5′R-cdG, 5′S-cdG, and 8-oxo-G lesions in various types of biological specimens associated with the cellular repair efficiency, as well as with distinct pathologies, have been reported, providing some insights on their biological significance [92].

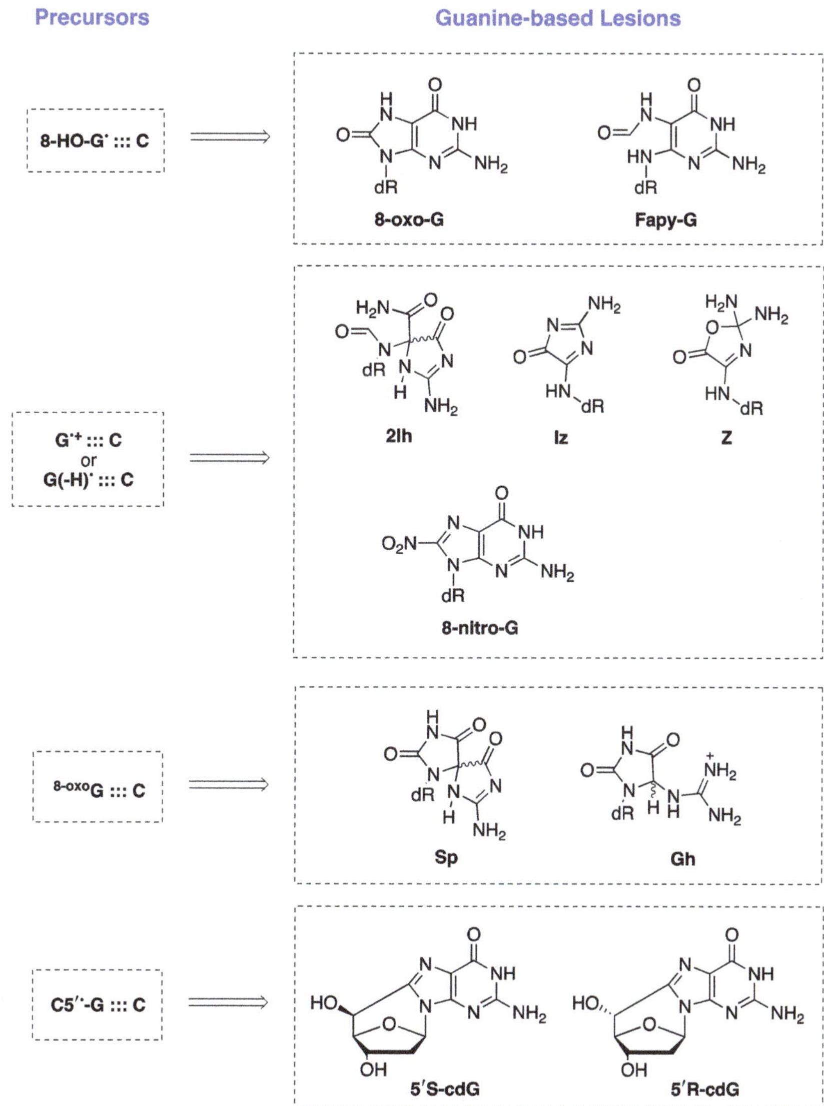

Figure 15. (**Left**) The precursor of the lesions in ds-ODNs or DNA. (**Right**) Guanine-derived lesions detected in in vitro experiments by the reaction of reactive oxygen species (ROS) with ds-ODNs or naked DNA. Most of these lesions have also been found after oxidatively induced DNA damage in vivo.

Base excision repair (BER) is the major enzymatic pathway involved in the processing of base damages such as 8-oxo-G and Fapy-G by creating an abasic site [100]. Nucleotide excision repair (NER) pathways are the major cellular pathway for the repair of bulky adducts and other helix-distorting lesions such as 5′R-cdG and 5′S-cdG lesions [93]. Evidence has also been provided that the removal of some oxidatively generated purine DNA lesions involves the overlap of both pathways [101,102].

12. Conclusions

The one-electron oxidation of the guanine moiety ($G^{\bullet+}$) is the most relevant process generating transient precursors involved in the oxidative damage of DNA with a variety of end-products, including the 8-oxo-dG lesion. There is a plethora of experimental and theoretical papers dealing with the well-understood long-range charge transport in DNA with the final destination of the G sites. Upon formation of $G^{\bullet+}$, two main paths are recognized: (i) a nucleophilic attack, such as the hydration reaction with the formation of 8-HOG$^{\bullet}$, and (ii) the prototropic equilibrium [$G^{\bullet+}$:::C ⇆ G(-H)$^{\bullet}$:::CH^{+}]. Both are at the crossroads of various pathways to the end-products. Furthermore, the one-electron oxidation of the G-quadruplex, with their planar array stabilized by metal cations, represent another context of $G^{\bullet+}$ fate.

For such a complex chemistry, it is obvious that the single nucleosides such as Guo or dGuo and the free base Gua have been the prototypes for experiments and theoretical calculations. However, neither Gua represents dGuo nor dGuo represents ODNs in their various supramolecular arrangements, and the extrapolation of its free radical chemistry can generate confusion. For example, for the introduction of the HO$^{\bullet}$ radical reactivity toward DNA, the main path reported is the addition at the C4 position followed by dehydration. There is no evidence for the existence of such a reaction. Overall, the reaction of HO$^{\bullet}$ with the guanine moiety is strictly dependent on the molecular context. The chemistry changes completely on going from the free base (Gua) to the nucleoside (dGuo) and from dGuo to DNA or the G-quadruplex, as we described herein. Several authors have often reported the calculations of HO$^{\bullet}$ with Gua for introducing the DNA damage, and this caused confusion in such a multidisciplinary field.

The two faces of the G$^{\bullet}$ radical have been reproduced by various methods and at different levels of complexity. The paradigm of one-electron-oxidized guanosines formation under the one-electron reduction of 8-bromo-guanosines has been the driving force to obtain the two faces of G$^{\bullet}$, i.e., G(N2-H)$^{\bullet}$ and G(N1-H)$^{\bullet}$, and their connection via tautomerization. This allowed the revision of the reaction of the HO$^{\bullet}$ radical with guanosine derivatives, where the main path is the H-atom abstraction from the exocyclic NH_2 group. Both faces of the G$^{\bullet}$ radical have been invoked in order to interpret the time-resolved spectroscopies at ds-ODNs, DNA, or G-quadruplexes ODNs and to associate them to 8-oxo-G yields. This review presents a reasonable overview of the G$^{\bullet}$ structure depending on molecular context and behavior.

The end-products (or lesions) of guanine DNA have been identified and successfully connected to the various mechanistic steps. Now, the comparative quantification of various purine lesions appears periodically and are related to health conditions. For example, 8-oxo-G and, to minor extent, Fapy-G have been measured in various human specimens and rationalized with various pathologies. Analogous data for the other lesions in Figure 15 are limited or absent. In the last two years, our group has been very active in the detection of 5′,8-cyclopurine lesions (including the 5′S-cdG and 5′R-cdG reported in Figure 15), providing results from cells, animals, and humans. Further perspectives of transdisciplinary research combining chemistry, mechanisms, and lesions formation can be foreseen in order to reach a satisfactory comprehension of the guanine radical involvement in DNA damage and human health.

Funding: This research received no external funding.

Institutional Review Board Statement: Not applicable.

Informed Consent Statement: Not applicable.

Data Availability Statement: Not applicable.

Acknowledgments: I would like to thank Krzysztof Bobrowski, Carla Ferreri, Maurizio Guerra and Dimitra Markovitsi for helpful discussions.

Conflicts of Interest: The author declares no conflict of interest.

Appendix A

In this section, some fundamental information regarding the methods of generation of the reactive species reported in this review by ionizing radiation are given in order to facilitate the nonexpert readers. The reader is referred to the literature for further information [103,104].

Radiolysis of neutral water leads to the species e_{aq}^-, $HO^•$, and $H^•$, as shown in Reaction 1, where the values in parentheses represent the radiation chemical yields (*G*) in units of $\mu mol\ J^{-1}$.

$$H_2O + \gamma\text{-irr} \rightarrow e_{aq}^- \ (0.27),\ HO^• \ (0.28),\ H^• \ (0.062) \quad \text{(A1)}$$

The reactions of $HO^•$ with the substrates were studied by irradiating N_2O-saturated solution (~0.02 M of N_2O); under these conditions, e_{aq}^- are converted into the $HO^•$ radical via Reaction 2 ($k_2 = 9.1 \times 10^9\ M^{-1}\ s^{-1}$), with $G(HO^•) = 0.55\ \mu mol\ J^{-1}$; i.e., $HO^•$ radicals and $H^•$ atoms account for 90 and 10%, respectively, of the reactive species.

$$e_{aq}^- + N_2O + H_2O \rightarrow HO^- + N_2 + HO^• \quad \text{(A2)}$$

The reactions of e_{aq}^- with the substrates were studied by irradiating deoxygenated solutions containing 0.25 M of *t*-BuOH. In the presence of *t*-BuOH, $HO^•$ radicals are scavenged (Reaction 3, $k_3 = 6.0 \times 10^8\ M^{-1}\ s^{-1}$), whereas $H^•$ may be trapped only partially (Reaction 4, $k_4 = 1.1 \times 10^6\ M^{-1}\ s^{-1}$).

$$HO^• + t\text{-BuOH} \rightarrow (CH_3)_2C(OH)CH_2^• + H_2O \quad \text{(A3)}$$

$$H^• + t\text{-BuOH} \rightarrow (CH_3)_2C(OH)CH_2^• + H_2 \quad \text{(A4)}$$

The sulfate radical anion, $SO_4^{•-}$, was generated by irradiating Ar-purged solutions containing 10 mM of peroxydisulfate $S_2O_8^{2-}$ and ~0.1 M of *t*-BuOH as a $HO^•$ radical scavenger (Reaction 5, $k_5 = 1.2 \times 10^{10}\ M^{-1}\ s^{-1}$).

$$e_{aq}^- + S_2O_8^{2-} \rightarrow SO_4^{2-} + SO_4^{•-} \quad \text{(A5)}$$

The dibromine radical anion, $Br_2^{•-}$, was generated by irradiating N_2O-saturated solutions containing 0.1 M of KBr. Under these conditions, at pH 7, the $HO^•$ radical is converted into $Br_2^{•-}$, through the process shown in Reaction 6 ($k_6 = 1.1 \times 10^{10}\ M^{-1}\ s^{-1}$).

$$HO^• + 2Br^- \rightarrow Br_2^{•-} + HO^- \quad \text{(A6)}$$

References

1. Dizdaroglu, M.; Lloyd, R.S. *DNA Damage, DNA Repair and Disease*; Royal Society of Chemistry: Croydon, UK, 2021.
2. Markkanen, E. Not breathing is not an option: How to deal with oxidative DNA damage. *DNA Repair* **2017**, *59*, 82–105. [CrossRef] [PubMed]
3. Tubbs, A.; Nussenzweig, A. Endogenous DNA damage as a source of genomic instability in cancer. *Cell* **2017**, *168*, 644–656. [CrossRef] [PubMed]
4. Genereux, J.C.; Barton, J.K. Mechanisms for DNA Charge Transport. *Chem. Rev.* **2010**, *110*, 1642–1662. [CrossRef] [PubMed]
5. Sontz, P.A.; Muren, N.B.; Barton, J.K. DNA Charge Transport for Sensing and Signaling. *Acc. Chem. Res.* **2012**, *45*, 1792–1800. [CrossRef]
6. Zwang, T.J.; Tse, E.C.M.; Barton, J.K. Sensing DNA through DNA Charge Transport. *ACS Chem. Biol.* **2018**, *13*, 1799–1809. [CrossRef]
7. Kawai, K.; Majima, T. Hole Transfer Kinetics of DNA. *Acc. Chem. Res.* **2013**, *46*, 2616–2625. [CrossRef]
8. Fujitsuka, M.; Majima, T. Charge Transfer Dynamics in DNA Revealed by Time-Resolved Spectroscopy. *Chem. Sci.* **2017**, *8*, 1752–1762. [CrossRef]
9. Lewis, F.D.; Young, R.M.; Wasielewski, M.R. Tracking Photoinduced Charge Separation in DNA: From Start to Finish. *Acc. Chem. Res.* **2018**, *51*, 1746–1754. [CrossRef]
10. Peluso, A.; Caruso, T.; Landi, A.; Capobianco, A. The Dynamics of Hole Transfer in DNA. *Molecules* **2019**, *24*, 4044. [CrossRef]

11. Chatgilialoglu, C.; Caminal, C.; Altieri, A.; Vougioukalakis, G.C.; Mulazzani, Q.G.; Gimisis, T.; Guerra, M. Tautomerism in the guanyl radicals. *J. Am. Chem. Soc.* **2006**, *128*, 13796–13805. [CrossRef]
12. Steenken, S. Purine bases, nucleosides, and nucleotides: Aqueous solution redox chemistry and transformation reactions of their radical cations and e^- and OH adducts. *Chem. Rev.* **1989**, *89*, 503–520. [CrossRef]
13. Von Sonntag, C. *Free-Radical-Induced DNA Damage and Its Repair, A Chemical Perspective*; Springer: Berlin, Germany, 2006.
14. Candeias, L.P.; Steenken, S. Structure and acid-base properties of one-electron-oxidized deoxyguanosine, guanosine, and 1-methylguanosine. *J. Am. Chem. Soc.* **1989**, *111*, 1094–1099. [CrossRef]
15. Steenken, S.; Jovanovic, S.V. How easily oxidizable Is DNA? One-electron reduction potentials of adenosine and guanosine radicals in aqueous solution. *J. Am. Chem. Soc.* **1997**, *119*, 617–618. [CrossRef]
16. Kobayashi, K.; Tagawa, S. Direct Observation of Guanine Radical Cation Deprotonation in Duplex DNA Using Pulse Radiolysis. *J. Am. Chem. Soc.* **2003**, *125*, 10213–10218. [CrossRef]
17. Zhang, X.; Jie, J.; Song, D.; Su, H. Deprotonation of Guanine Radical Cation $G^{\bullet+}$ Mediated by the Protonated Water Cluster. *J. Phys. Chem. A* **2020**, *124*, 6076–6083. [CrossRef]
18. Latus, A.; Alam, M.S.; Mostafavi, M.; Marignier, J.-L.; Maisonhaute, E. Guanosine radical reactivity explored by pulse radiolysis coupled with transient electrochemistry. *Chem. Commun.* **2015**, *51*, 9089–9092. [CrossRef]
19. Adhikary, A.; Kumar, A.; Becker, D.; Sevilla, M.D. The Guanine Cation Radical: Investigation of Deprotonation States by ESR and DFT. *J. Phys. Chem. B* **2006**, *110*, 24171–24180. [CrossRef]
20. Hole, E.O.; Sagstuen, E.; Nelson, W.H.; Close, D.M. The Structure of the guanine cation: ESR/ENDOR of cyclic guanosine monophosphate single crystals after X irradiation at 10 K. *Radiat. Res.* **1992**, *129*, 1–10. [CrossRef]
21. Choi, J.; Yang, C.; Fujitsuka, M.; Tojo, S.; Ihee, H.; Majima, T. Proton Transfer of Guanine Radical Cations Studied by Time-Resolved Resonance Raman Spectroscopy Combined with Pulse Radiolysis. *J. Phys. Chem. Lett.* **2015**, *6*, 5045–5050. [CrossRef]
22. Saprygina, N.N.; Morozova, O.B.; Abramova, T.V.; Grampp, G.; Yurkovskaya, A.V. Oxidation of Purine Nucleotides by Triplet 3,3′,4,4′-Benzophenone Tetracarboxylic Acid in Aqueous Solution: pH-Dependence. *J. Phys. Chem. A* **2014**, *118*, 4966–4974. [CrossRef]
23. Morozova, O.B.; Fishman, N.N.; Yurkovskaya, A.V. Indirect NMR detection of transient guanosyl radical protonation in neutral aqueous solution. *Phys. Chem. Chem. Phys.* **2017**, *19*, 21262–21266. [CrossRef]
24. Morozova, O.B.; Fishman, N.N.; Yurkovskaya, A.V. Kinetics of Reversible Protonation of Transient Neutral Guanine Radical in Neutral Aqueous Solution. *ChemPhysChem* **2018**, *19*, 2696–2702. [CrossRef]
25. Sevilla, M.D.; Kumar, A.; Adhikary, A. Comment on Proton Transfer of Guanine Radical Cations Studied by Time-Resolved Resonance Raman Spectroscopy Combined with Pulse Radiolysis. *J. Phys. Chem. B* **2016**, *120*, 2984–2986. [CrossRef]
26. Knobloch, B.; Sigel, H.; Okruszek, A.; Sigel, R.K.O. Acid–base properties of the nucleic-acid model 2′-deoxyguanylyl(5′→3′)-2′-deoxy-5′-guanylate, $d(pGpG)^{3-}$, and of related guanine derivatives. *Org. Biomolec. Chem.* **2006**, *4*, 1085–1090. [CrossRef]
27. Chatgilialoglu, C.; D'Angelantonio, M.; Guerra, M.; Kaloudis, P.; Mulazzani, Q.G. A reevaluation of the ambident reactivity of guanine moiety towards hydroxyl radicals. *Angew. Chem. Int. Ed.* **2009**, *48*, 2214–2217. [CrossRef]
28. Chatgilialoglu, C.; Caminal, C.; Guerra, M.; Mulazzani, Q.G. Tautomers of one-electron oxidized guanosine. *Angew. Chem. Int. Ed.* **2005**, *44*, 6030–6032. [CrossRef]
29. Ioele, M.; Bazzanini, R.; Chatgilialoglu, C.; Mulazzani, Q.G. Chemical radiation studies of 8-bromoguanosine in aqueous solutions. *J. Am. Chem. Soc.* **2000**, *122*, 1900–1907. [CrossRef]
30. Adhikary, A.; Kumar, A.; Munafo, S.A.; Khanduri, D.; Sevilla, M.D. Prototropic equilibria in DNA containing one-electron oxidized GC:intra-duplex vs. duplex to solvent deprotonation. *Phys. Chem. Chem. Phys.* **2010**, *12*, 5353–5368. [CrossRef]
31. Martínez-Fernández, L.; Cerezo, J.; Asha, H.; Santoro, F.; Coriani, S.; Improta, R. The absorption apectrum of guanine based radicals: A comparative computational analysis. *ChemPhotoChem* **2019**, *3*, 846–855. [CrossRef]
32. Candeias, L.P.; Steenken, S. Reaction of $HO^{\bullet}$ with guanine derivatives in aqueous solution: Formation of two different redox-active OH-adduct radicals and their unimolecular transformation reactions. Properties of G(-H). *Chem. Eur. J.* **2000**, *6*, 475–484. [CrossRef]
33. Chatgilialoglu, C.; D'Angelantonio, M.; Kciuk, G.; Bobrowski, K. New insights into the reaction paths of hydroxyl radicals with 2′-deoxyguanosine. *Chem. Res. Toxicol.* **2011**, *24*, 2200–2206. [CrossRef] [PubMed]
34. D'Angelantonio, M.; Russo, M.; Kaloudis, P.; Mulazzani, Q.G.; Wardman, P.; Guerra, M.; Chatgilialoglu, C. Reaction of hydrated electrons with guanine derivatives: Tautomerism of intermediate species. *J. Phys. Chem. B* **2009**, *113*, 2170–2176. [CrossRef] [PubMed]
35. Chatgilialoglu, C.; Bazzanini, R.; Jimenez, L.B.; Miranda, M.A. (5′S)- and (5′R)-5′,8-cyclo-2′-deoxyguanosine: Mechanistic insights on the 2′-deoxyguanosin-5′-yl radical cyclization. *Chem. Res. Toxicol.* **2007**, *20*, 1820–1824. [CrossRef] [PubMed]
36. Phadatare, S.D.; Sharma, K.K.K.; Rao, B.S.M.; Naumov, S.; Sharma, G.K. Spectra Characterization of Guanine C4-OH Adduct: A Radiation and Quantum Chemical Study. *J. Phys. Chem. B* **2011**, *115*, 13650–13658. [CrossRef] [PubMed]
37. Liska, A.; Triskova, I.; Ludvik, J.; Trnkova, L. Oxidation potentials of guanine, guanosine and guanosine-5′-monophosphate: Theory and experiment. *Electrochim. Acta* **2019**, *318*, 108–119. [CrossRef]
38. Kumar, A.; Pottiboyina, V.; Sevilla, M.D. Hydroxyl Radical (OH·) Reaction with Guanine in an Aqueous Environment: A DFT Study. *J. Phys. Chem. B* **2011**, *115*, 15129–15137. [CrossRef] [PubMed]
39. Dizdaroglu, M.; Jaruga, P. Mechanism of free radical-induced damage to DNA. *Free Radic. Res.* **2012**, *46*, 382–419. [CrossRef]

40. Dizdaroglu, M. Oxidatively induced DNA damage: Mechanisms, repair and disease. *Cancer Lett.* **2012**, *327*, 26–47. [CrossRef]
41. Kaloudis, P.; Paris, C.; Vrantza, D.; Encinas, S.; Perez-Ruiz, R.; Miranda, M.A.; Gimisis, T. Photolabile N-hydroxypyrid-2(1H)-one derivatives of guanine nucleosides: A new method for independent guanine radical generation. *Org. Biomol. Chem.* **2009**, *7*, 4965–4972. [CrossRef]
42. Zheng, L.; Greenberg, M.M. Independent generation and reactivity of 2′-deoxyguanosin-N1-yl radical. *J. Org. Chem.* **2020**, *85*, 8665–8672. [CrossRef]
43. Zheng, l.; Xiaojuan Dai, X.; Su, H.; Greenberg, M.M. Independent generation and time-resolved detection of 2′-deoxyguanosin-N2-yl Radicals. *Angew. Chem. Int. Ed.* **2020**, *59*, 13406–13413. [CrossRef]
44. Marcus, Y.; Migron, Y. Polarity, Hydrogen Bonding, and Structure of Mixtures of Water and Cyanomethane. *J. Phys. Chem.* **1991**, *95*, 400–406. [CrossRef]
45. Mountain, R.D. Microstructure and Hydrogen Bonding in Water—Acetonitrile Mixtures. *J. Phys. Chem. B* **2010**, *114*, 16460–16464. [CrossRef]
46. Lewandowska-Andralojc, A.; Hug, G.L.; Marciniak, B.; Hörner, G.; Swiatla-Wojcik, D. Water-triggered photoinduced electron transfer in acetonitrile—Water binary solvent. Solvent microstructure-tuned reactivity of hydrophobic solutes. *J. Phys. Chem. B* **2020**, *124*, 5654–5664.
47. Pacher, P.; Beckman, J.S.; Liaudet, L. Nitric oxide and peroxynitrite in health and disease. *Physiol. Rev.* **2007**, *87*, 315–424. [CrossRef]
48. Illes, E.; Patra, S.G.; Marks, V.; Mizrahi, A.; Meyerstein, D. The Fe^{II}(citrate) Fenton reaction under physiological conditions. *J. Inorg. Biochem.* **2020**, *206*, 111018. [CrossRef]
49. Shafirovich, V.; Dourandin, A.; Huang, W.; Geacintov, N.E. The carbonate radical is a site-selective oxidizing agent of guanine in double-stranded oligonucleotides. *J. Biol. Chem.* **2001**, *276*, 24621–24626. [CrossRef]
50. Wang, Y.; An, P.; Li, S.; Zhou, L. The oxidation mechanism and kinetics of 2′-deoxyguanosine by carbonate radical anion. *Chem. Phys. Lett.* **2020**, *739*, 136982. [CrossRef]
51. Fleming, A.M.; Burrows, C.J. Iron Fenton oxidation of 2′-deoxyguanosine in physiological bicarbonate buffer yields products consistent with the reactive oxygen species carbonate radical anion not hydroxyl radical. *Chem. Commun.* **2020**, *56*, 9779–9782. [CrossRef]
52. Dizdaroglu, M.; Kirkali, G.; Jaruga, P. Formamidopyrimidines in DNA: Mechanisms of formation, repair, and biological effects. *Free Radic. Biol Med.* **2008**, *45*, 1610–1621. [CrossRef]
53. Greenberg, M.M. The formamidopyrimidines: Purine lesions formed in competition with 8-oxopurines from oxidative stress. *Acc. Chem. Res.* **2012**, *45*, 588–597. [CrossRef] [PubMed]
54. Steenken, S.; Jovanovic, S.V.; Bietti, M.; Bernhard, K. The trap depth (in DNA) of 8-oxo-7,8-dihydro-2′deoxyguanosine as derived from electron-transfer equilibria in aqueous solution. *J. Am. Chem. Soc.* **2000**, *122*, 2373–2374. [CrossRef]
55. Fleming, A.M.; Muller, J.G.; Dlouhy, A.C.; Burrows, C.J. Structural context effects in the oxidation of 8-oxo-7,8-dihydro-2′-deoxyguanosine to hydantoin products: Electrostatics, base stacking, and base pairing. *J. Am. Chem. Soc.* **2012**, *134*, 15091–15102. [CrossRef] [PubMed]
56. Fleming, A.M.; Muller, J.G.; Ji, I.; Burrows, C.J. Characterization of 2′-deoxyguanosine oxidation products observed in the Fenton-like system Cu(II)/H2O2/reductant in nucleoside and oligodeoxynucleotide contexts. *Org. Biomol. Chem.* **2011**, *9*, 3338–3348. [CrossRef]
57. Alshykhly, O.R.; Fleming, A.M.; Burrows, C.J. 5-Carboxamido-5-formamido-2-iminohydantoin, in Addition to 8-oxo-7,8-Dihydroguanine, Is the Major Product of the Iron-Fenton or X-ray Radiation-Induced Oxidation of Guanine under Aerobic Reducing Conditions in Nucleoside and DNA Contexts. *J. Org. Chem.* **2015**, *80*, 6996–7007. [CrossRef]
58. Rokhlenko, Y.; Geacintov, N.E.; Shafirovich, V. Lifetimes and reaction pathways of guanine radical cations and neutral guanine radicals in an oligonucleotide in aqueous solutions. *J. Am. Chem. Soc.* **2012**, *134*, 4955–4962. [CrossRef]
59. Adhikary, A.; Khanduri, D.; Sevilla, M.D. Direct observation of the hole protonation state and hole localization site in DNA-oligomers. *J. Am. Chem. Soc.* **2009**, *131*, 8614–8619. [CrossRef]
60. Kobayashi, K.; Yamagami, R.; Tagawa, S. Effect of base sequence and deprotonation of guanine cation radical in DNA. *J. Phys. Chem. B* **2008**, *112*, 10752–10757. [CrossRef]
61. Rokhlenko, Y.; Cadet, J.; Geacintov, N.E.; Shafirovich, V. Mechanistic aspects of hydration of guanine radical cations in DNA. *J. Am. Chem. Soc.* **2014**, *136*, 5956–5962. [CrossRef]
62. Banyasz, A.; Martínez-Fernández, L.; Improta, R.; Ketola, T.-M.; Balty, C.; Markovitsi, D. Radicals generated in alternating guanine cytosine duplexes by direct absorption of low-energy UV radiation. *Phys. Chem. Chem. Phys.* **2018**, *20*, 21381–21389. [CrossRef]
63. Balanikas, E.; Banyasz, A.; Baldacchino, G.; Markovitsi, D. Populations and Dynamics of Guanine Radicals in DNA strands—Direct versus Indirect Generation. *Molecules* **2019**, *24*, 2347. [CrossRef]
64. Balanikas, E.; Banyasz, A.; Douki, T.; Baldacchino, G.; Markovitsi, D. Guanine Radicals Induced in DNA by Low-Energy Photoionization. *Acc. Chem. Res.* **2020**, *53*, 1511–1519. [CrossRef]
65. Anderson, R.F.; Shinde, S.S.; Maroz, A. Cytosine-Gated Hole Creation and Transfer in DNA in Aqueous Solution. *J. Am. Chem. Soc.* **2006**, *128*, 15966–15967. [CrossRef]

66. Shinde, S.S.; Maroz, A.; Hay, M.P.; Anderson, R.F. One-Electron Reduction Potential of the Neutral Guanyl Radical in the GC Base Pair of Duplex DNA. *J. Am. Chem. Soc.* **2009**, *131*, 5203–5207. [CrossRef]
67. Anderson, R.F.; Shinde, S.S.; Maroz, A.; Reynisson, J. The reduction potential of the slipped GC base pair in one-electron oxidized duplex DNA. *Phys. Chem. Chem. Phys.* **2020**, *22*, 642–646. [CrossRef]
68. Jaeger, W.M.; Schaefer, H.F., III. Characterizing Radiation-Induced Oxidation of DNA by Way of the Monohydrated Ganine—Cytosine Radical Cation. *J. Phys. Chem. B* **2009**, *113*, 8142–8148. [CrossRef]
69. Kumar, A.; Sevilla, M.D. Influence of Hydration on Proton Transfer in the Guanine—Cytosine Radical Cation ($G^{\bullet+}$—C) Base Pair: A Density Functional Theory Study. *J. Phys. Chem. B* **2009**, *113*, 11359–11361. [CrossRef]
70. Steenken, S.; Reynisson, J. DFT calculations on the deprotonation site of the one-electron oxidised guanine–cytosine base pair. *Phys. Chem. Chem. Phys.* **2010**, *12*, 9088–9093. [CrossRef]
71. Kumar, A.; Sevilla, M.D. Excited States of One-Electron Oxidized Guanine-Cytosine Base Pair Radicals: A Time Dependent Density Functional Theory Study. *J. Phys. Chem. A* **2019**, *123*, 3098–3108. [CrossRef]
72. Misiaszek, R.; Crean, C.; Joffe, A.; Geacintov, N.E.; Shafirovich, V. Oxidative DNA damage associated with combination of guanine and superoxide radicals and repair mechanisms via radical trapping. *J. Biol. Chem.* **2004**, *279*, 32106–32115. [CrossRef]
73. Misiaszek, R.; Crean, C.; Geacintov, N.E.; Shafirovich, V. Combination of Nitrogen Dioxide Radicals with 8-Oxo-7,8-dihydroguanine and Guanine Radicals in DNA: Oxidation and Nitration End-Products. *J. Am. Chem. Soc.* **2005**, *127*, 2191–2200. [CrossRef]
74. Joffe, A.; Mock, S.; Yun, B.H.; Kolbanovskiy, A.; Geacintov, N.E.; Shafirovich, V. Oxidative Generation of Guanine Radicals by Carbonate Radicals and Their Reactions with Nitrogen Dioxide to Form Site Specific 5-Guanidino-4-nitroimidazole Lesions in Oligodeoxynucleotides. *Chem. Res. Toxicol.* **2003**, *16*, 966–973. [CrossRef]
75. Matter, B.; Seiler, C.L.; Murphy, K.; Ming, X.; Zhao, J.; Lindgren, B.; Jones, R.; Tretyakova, N. Mapping three guanine oxidation products along DNA following exposure to three types of reactive oxygen species. *Free Radic. Biol. Med.* **2018**, *121*, 180–189. [CrossRef]
76. Davis, J.T. G-Quartets 40 Years Later: From 5′-GMP to Molecular Biology and Supramolecular Chemistry. *Angew. Chem. Int. Ed.* **2004**, *43*, 668–698. [CrossRef]
77. Wu, L.D.; Liu, K.H.; Jie, J.L.; Song, D.; Su, H.M. Direct Observation of Guanine Radical Cation Deprotonation in G-Quadruplex DNA. *J. Am. Chem. Soc.* **2015**, *137*, 259–266. [CrossRef] [PubMed]
78. Merta, T.J.; Geacintov, N.E.; Shafirovich, V. Generation of 8-oxo-7,8-dihydroguanine in G-quadruplexes models of human telomere sequences by one-electron oxidation. *Photochem. Photobiol.* **2019**, *95*, 244–251. [CrossRef]
79. Banyasz, A.; Martinez-Fernandez, L.; Balty, C.; Perron, M.; Douki, T.; Improta, R.; Markovitsi, D. Absorption of Low-Energy UV Radiation by Human Telomere G-Quadruplexes Generates Long-Lived Guanine Radical Cations. *J. Am. Chem. Soc.* **2017**, *139*, 10561–10568. [CrossRef]
80. Banyasz, A.; Balanikas, E.; Martinez-Fernandez, L.; Baldacchino, G.; Douki, T.; Improta, R.; Markovitsi, D. Radicals generated in tetramolecular guanine quadruplexes by photo-ionization: Spectral and dynamical features. *J. Phys. Chem. B* **2019**, *123*, 4950–4957. [CrossRef]
81. Balanikas, E.; Banyasz, A.; Baldacchino, G.; Markovitsi, D. Guanine Radicals Generated in Telomeric G-quadruplexes by Direct Absorption of Low-Energy UV Photons: Effect of Potassium Ions. *Molecules* **2020**, *25*, 2094. [CrossRef]
82. Behmand, B.; Balanikas, E.; Martinez-Fernandez, L.; Improta, R.; Banyasz, A.; Baldacchino, G.; Markovitsi, D. Potassium Ions Enhance Guanine Radical Generation upon Absorption of Low-Energy Photons by G-quadruplexes and Modify Their Reactivity. *J. Phys. Chem. Lett.* **2020**, *11*, 1305–1309. [CrossRef]
83. Gustavsson, T.; Markovitsi, D. Fundamentals of the Intrinsic DNA Fluorescence. *Acc. Chem. Res.* **2021**, *54*, 1226–1235. [CrossRef] [PubMed]
84. Chatgilialoglu, C.; Eriksson, L.A.; Krokidis, M.G.; Masi, A.; Wang, S.-D.; Zhang, R. Oxygen dependent purine lesions in double-stranded oligodeoxynucleotides: Kinetic and computational studies highlight the mechanism for 5′,8-cyplopurine formation. *J. Am. Chem. Soc.* **2020**, *142*, 5825–5833. [CrossRef] [PubMed]
85. Bergeron, F.; Auvre, F.; Radicella, J.P.; Ravanat, J.-L. $HO^{\bullet}$ radicals induce an unexpected high proportion of tandem base lesions refractory to repair by DNA glycosylases. *Proc. Natl. Acad. Sci. USA* **2010**, *107*, 5528–5533. [CrossRef] [PubMed]
86. Ravanat, J.-L.; Breton, J.; Douki, T.; Gasparutto, D.; Grand, A.; Rachidi, W.; Sauvaigo, S. Radiation-mediated formation of complex damage to DNA: A chemical aspect overview. *Br. J. Radiol.* **2014**, *87*, 20130715. [CrossRef]
87. Robert, G.; Wagner, J.R. Tandem lesions arising from 5-(uracilyl)methyl peroxyl radical addition to guanine: Product analysis and mechanistic studies. *Chem. Res. Toxicol.* **2020**, *33*, 565–575. [CrossRef]
88. Chatgilialoglu, C.; Krokidis, M.G.; Masi, A.; Barata-Vallejo, S.; Ferreri, C.; Terzidis, M.A.; Szreder, T.; Bobrowski, K. New insights into the reaction paths of hydroxyl radicals with purine moieties in DNA and double-stranded oligonucleotides. *Molecules* **2019**, *24*, 3860. [CrossRef]
89. Shukla, L.I.; Adhikary, A.; Pazdro, R.; Becker, D.; Sevilla, M.D. Formation of 8-oxo-7,8-dihydroguanine-radicals in γ-irradiated DNA by multiple one-electron oxidations. *Nucleic Acids Res.* **2004**, *32*, 6565–6574. [CrossRef]
90. Terzidis, M.A.; Prisecaru, A.; Molphy, Z.; Barron, N.; Randazzo, A.; Dumont, E.; Krokidis, M.G.; Kellett, A.; Chatgilialoglu, C. Radical-induced purine lesion formation is dependent on DNA helical topology. *Free Radic. Res.* **2016**, *50*, S91–S101. [CrossRef]

91. Fleming, A.M.; Burrows, C.J. Interplay of Guanine Oxidation and G-Quadruplex Folding in Gene Promoters. *J. Am. Chem. Soc.* **2020**, *142*, 1115–1136. [CrossRef]
92. Chatgilialoglu, C.; Ferreri, C.; Krokidis, M.G.; Masi, A.; Terzidis, M.A. On the relevance of hydroxyl radical to purine DNA damage. *Free Radic. Res.* **2021**. [CrossRef]
93. Chatgilialoglu, C.; Ferreri, C.; Geacintov, N.E.; Krokidis, M.G.; Liu, Y.; Masi, A.; Shafirovich, V.; Terzidis, M.A.; Tsegay, P.S. 5′,8-Cyclopurine lesions in DNA damage: Chemical, analytical, biological and diagnostic significance. *Cells* **2019**, *8*, 513. [CrossRef]
94. Yu, Y.; Cui, Y.; Niedernhofer, L.J.; Wang, Y. Occurrence, biological consequences, and human health relevance of oxidative stress-induced DNA damage. *Chem. Res. Toxicol.* **2016**, *29*, 2008–2039. [CrossRef]
95. Jaruga, P.; Kirkali, G.; Dizdaroglu, M. Measurement of formamidopyrimidines in DNA. *Free Radic. Biol. Med.* **2008**, *45*, 1601–1609. [CrossRef]
96. Dizdaroglu, M. Oxidatively induced DNA damage and its repair in cancer. *Mutat. Res. Rev. Mutat. Res.* **2015**, *763*, 212–245. [CrossRef]
97. Cadet, J.; Davies, K.J.A.; Medeiros, M.H.G.; Di Mascio, P.; Wagner, J.R. Formation and repair of oxidatively generated damage in cellular DNA. *Free Radic. Biol. Med.* **2017**, *107*, 13–34. [CrossRef]
98. Cui, L.; Ye, W.; Prestwich, E.G.; Wishnok, J.S.; Taghizadeh, K.; Dedon, P.C.; Tannenbaum, S.R. Comparative analysis of four oxidized guanine lesions from reactions of DNA with peroxynitrite, single oxygen, and γ-radiation. *Chem. Res. Toxicol.* **2013**, *26*, 195–202. [CrossRef]
99. Fleming, A.M.; Burrows, C.J. Formation and processing of DNA damage substrates for the hNEIL enzymes. *Free Radic. Biol. Med.* **2017**, *107*, 35–52. [CrossRef]
100. Whitaker, A.M.; Schaich, M.A.; Smith, M.R.; Flynn, T.S.; Freudenthal, B.D. Base excision repair of oxidative DNA damage: From mechanism to disease. *Front. Biosci.* **2017**, *22*, 1493–1522.
101. Shafirovich, V.; Geacintov, N.E. Removal of oxidatively generated DNA damage by overlapping repair pathways. *Free Radic. Biol. Med.* **2017**, *107*, 53–61. [CrossRef]
102. Kumar, N.; Raja, S.; Van Houten, B. The involvement of nucleotide excision repair proteins in the removal of oxidative DNA damage. *Nucleic Acid Res.* **2020**, *48*, 11227–11243. [CrossRef]
103. Buxton, G.V.; Greenstock, C.L.; Helman, W.P.; Ross, A.B. Critical review of rate constants for reactions of hydrated electrons, hydrogen atoms and hydroxyl radicals ($^{\bullet}OH)/^{\bullet}O^{-}$) in aqueous solution. *J. Phys. Chem. Ref. Data* **1988**, *17*, 513–886. [CrossRef]
104. Bobrowski, K. Radiation induced radical reactions. In *Encyclopedia of Radicals in Chemistry, Biology and Materials*; Chatgilialoglu, C., Studer, A., Eds.; John Wiley: Hoboken, NJ, USA, 2012; Volume 1, pp. 395–432.

 molecules

MDPI

Article

The Time Scale of Electronic Resonance in Oxidized DNA as Modulated by Solvent Response: An MD/QM-MM Study

Alessandro Landi , Amedeo Capobianco * and Andrea Peluso

Dipartimento di Chimica e Biologia "A. Zambelli", Università di Salerno, Via Giovanni Paolo II, 132, I-84084 Fisciano, SA, Italy; alelandi1@unisa.it (A.L.); apeluso@unisa.it (A.P.)
* Correspondence: acapobianco@unisa.it

Abstract: The time needed to establish electronic resonant conditions for charge transfer in oxidized DNA has been evaluated by molecular dynamics simulations followed by QM/MM computations which include counterions and a realistic solvation shell. The solvent response is predicted to take *ca.* 800–1000 ps to bring two guanine sites into resonance, a range of values in reasonable agreement with the estimate previously obtained by a kinetic model able to correctly reproduce the observed yield ratios of oxidative damage for several sequences of oxidized DNA.

Keywords: DNA oxidation; DNA hole transfer; DNA; molecular dynamics; quantum dynamics; electron transfer; charge transfer; quantum coherence

Citation: Landi, A.; Capobianco, A.; Peluso, A. The Time Scale of Electronic Resonance in Oxidized DNA as Modulated by Solvent Response: An MD/QM-MM Study. *Molecules* **2021**, *26*, 5497. https://doi.org/10.3390/molecules26185497

Academic Editor: Kevin Critchley

Received: 4 August 2021
Accepted: 7 September 2021
Published: 10 September 2021

1. Introduction

Hole transport along one-electron oxidized DNA has been deeply investigated in the last decades [1–5], primarily for its biological role in mutagenesis, carcinogenesis, and ageing [6,7], and also because of its attractiveness for molecular electronics applications [8–11]. Indeed, owing to its unique redox and self assembling properties, DNA could be potentially used as an active material or a templating agent in the fabrication of biocompatible and biodegradable devices [12,13].

Hole transfer (HT) usually ends up on a guanine (G) site where the oxidative damage is preferentially observed, because guanine is the most prone to oxidation by removal of a single electron among natural occurring nucleobases. Furthermore, consecutive stacked guanines give rise to deeper hole-traps due to their lower hole site energy with respect to a single G [14,15]. DNA tracts made of adjacent adenines (A) are known to act as shuttle sites greatly favoring the hole transport, because adenine possesses a relatively low ionization energy and promotes intense stacking interactions which are able to stabilize the positive charge through the formation of delocalized polarons [16–21]. On the opposite, pyrimidine nucleobases (thymine, T, cytosine, C) act as barrier sites for hole transport due to their higher oxidation free energy [22].

Experimental evidence gathered in last years has shown that the dynamics of hole transfer in DNA is largely dependent on the specific sequence of DNA bases, being modulated by the different hole site energies and electronic coupling of nucleobases [15,19,23–25]. The kinetics of HT is characterized by two distinct regimes [26,27]: A short-range one with an exponential decay of hole-transfer rates as a function of the donor-acceptor (DA) distance, and a long-range regime, where HT rates exhibit a very weak distance dependence. Those results have usually been rationalized in terms of two different mechanisms: single step coherent hole tunneling (super-exchange or flickering resonance) for short DA distances and incoherent multistep hopping for long-range hole transport [28–33].

Recently, we proposed a unifying mechanism of hole transport in DNA, accounting for both short- and long-range regimes [34,35]. Our approach takes into account the manifold of fast coherent elementary electron-transfer processes which occur when the donor, i.e., the site where the hole has been injected and the acceptor, i.e., the site where the charge is eventually localized are brought into vibronic degeneracy by the response of solvent molecules and counterions. In detail, Figure 1, the proposed mechanism consists of four steps: (i) an activation step which brings a donor and the acceptor sites into resonance; (ii) an elementary electron transfer between resonant donor and acceptor groups; (iii) relaxation of non-equilibrium species (including the environment) to their minimum energy configurations; (iv) formation of oxidative damage products. In Figure 1, D^+(Bridge)A and D(Bridge)A^+ indicate the minimum energy structures with the charge localized on the donor (D) and the acceptor (A) groups; [D^+(Bridge)A]* and [D(Bridge)A^+]* denote the ensembles of structures in which D and A are in vibronic resonance, and P_D and P_A are the products of oxidative damage occurring at the D and A sites.

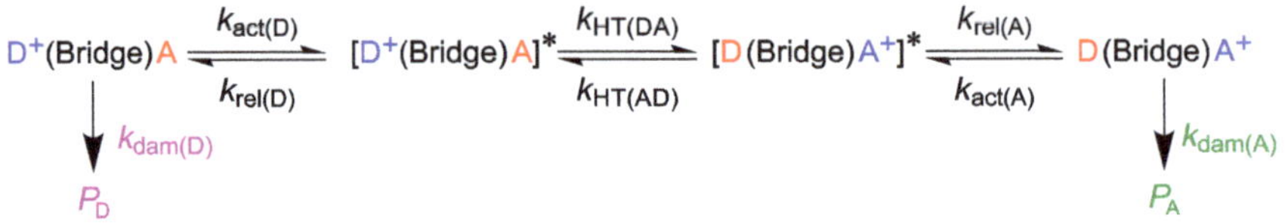

$$\begin{array}{ccccccc} D^+(\text{Bridge})A & \underset{k_{rel(D)}}{\overset{k_{act(D)}}{\rightleftharpoons}} & [D^+(\text{Bridge})A]^* & \underset{k_{HT(AD)}}{\overset{k_{HT(DA)}}{\rightleftharpoons}} & [D(\text{Bridge})A^+]^* & \underset{k_{act(A)}}{\overset{k_{rel(A)}}{\rightleftharpoons}} & D(\text{Bridge})A^+ \\ \downarrow k_{dam(D)} & & & & & & \downarrow k_{dam(A)} \\ P_D & & & & & & P_A \end{array}$$

Figure 1. The kinetic scheme for HT in DNA. D^+(Bridge)A and D(Bridge)A^+ indicate the initial and the final state, giving rise to damaged products P_D and P_A. [D^+(Bridge)A]* and [D(Bridge)A^+]* denote the ensembles of structures in which the hole donor and acceptor are in vibronic resonance with each other, so that $k_{HT}(DA) = k_{HT}(AD)$; k_{HT} has been computed by resolving the time dependent Schrödinger equation. k_{dam}'s and k_{rel}'s have been inferred from experimental data; no direct information is available for k_{act}'s.

k_{act}'s (step 1 and the reverse of step 3) are the rates for the activation stage that accounts for the time needed to bring [D^+(Bridge)A]* and [D(Bridge)A^+]* into degeneracy, condition achieved by collisions and environmental motions. An Arrhenius-like dependence with temperature is assumed to hold:

$$k_{act}(DA) = k^0_{act}(T) \times \exp\left(-\frac{\Delta E_{DA}}{k_B T}\right),$$

where ΔE_{DA} is the in situ electronic hole site energy difference between donor and acceptor and $k^0_{act}(T)$ is the rate constant for bringing into electronic resonance two sites possessing the same hole site energy.

Step 3 and the reverse of step 1 take into account the solvent response to a non-equilibrium charge distribution of the solute; pump-probe experiments in water solutions showed that solvent relaxation occurs in a few tens of femtoseconds, so that we set $k_{rel}(D) = k_{rel}(A) = 10^{13}\ s^{-1}$ [36,37]. Because, in actual experiments, the hole is injected at G and the damage is observed at $(G)_n$ tracts, k_{dam} have been set as the rates of deprotonation of $G\cdot^+$ and $GG\cdot^+$ radicals, i.e., 1×10^7 and $3 \times 10^6\ s^{-1}$, respectively [7,38]. Deprotonation of oxidized purine nucleobases inside DNA has been frequently observed [39,40] and it is commonly accepted as the first step of formation of the products of oxidative damage occurring at G sites [41–43].

The k_{HT}'s (step 2) were computed by quantum dynamics simulations of hole transfer considering the DNA sequence in its minimum-energy configuration, inasmuch as solvent configuration does not alter tunneling rates in resonance conditions. Because the hole donor and acceptor sites are in vibronic resonance with each other, $k_{HT}(DA) = k_{HT}(AD)$.

With the above parameters and a proper setting for k_{act} (see below), the distribution of oxidative damages predicted by the multistep mechanism of Figure 1 was found to be in excellent agreement with experimental observations for both the 5′-GGG$(T)_n$G-3′ sequences studied by Giese [26] and most of the oligonucleotides investigated by Schuster and coworkers in [44]. Notably, our mechanism allows for reconciling apparently discordant experimental observations. The $(A)_4$ tract in double stranded (ds) DNA is a paradigmatic example. Indeed, $(A)_4$ was found to act as a shuttle site in 5′-GGG$(T)_4$G-3′ double stranded (ds) DNA where the charge is injected at G-3′ and the damage is mostly observed at the GGG site. On the opposite, $(A)_4$ proved to be not effective for the hole transport in 5′-^{8}GTGTG$(T)_4$GTGTGG-3′ (^{8}G denotes 8-oxo-7,8-dihydroguanine) where the charge initially generated at the G-3′ of the GG unit yields a damage mainly localized at the 5′-G of the same GG unit, instead of migrating to ^{8}G which is a deeper trap than GG and even GGG. Our mechanism is able to rationalize both experimental facts. The $(A)_4$ tract is *per se* an efficient bridge, however the latter sequence contains several G sites possessing close hole energies, so that interference among probability amplitudes pertaining to indistinguishable hole paths has to be taken into account. In detail, for 5′-^{8}GTGTG$(T)_4$GTGTGG-3′ several nucleobases can be brought in resonance with the donor site, so that different indistinguishable paths all contribute to hole localization on the observed trap site, mainly the 5′-G of the GG tract [35].

A further advantage of the present model is that of accounting for solvent effects, thus providing absolute rates and yields of oxidative damage for hole transfer in solution.

In previous work, k_{act} was taken as an adjustable parameter in the kinetic model. In order to compare theoretical predictions with the experimental results, the set of ordinary differential equations (ODEs) of the kinetic scheme of Figure 1 was solved to compute the yield ratios of damaged products. Numerical resolutions of the set of ODEs demonstrated that the experimental yield ratios are compatible with the experimental outcomes only with k_{act} of the order of 10^{10} s^{-1}. Unfortunately, that parameter has no experimental counterpart, because no direct information about the time needed to bring two sites of DNA into resonance is available. Hence, a further investigation about the reliability of the inferred value for k_{act} appears to be a needed task. Interestingly, the resolution of the ODE equations has shown that the optimum value for k_{act} is almost independent of the nature and the length of the specific DNA sequence, indeed a single value, $k_{act} = 2 \times 10^{10}$ s^{-1} was found to ensure an excellent agreement between predicted and observed damaged product ratios for all the DNA sequences investigated in [35]. That observation suggests that a useful analysis of the timescale of the activation step of HT could be initially limited to a single sequence. Therefore, herein we have carried out a preliminary study based on classical molecular dynamics (MD) simulations and QM/MM computations aimed at estimating the order of magnitude of the time needed to bring the G and GGG sites of double stranded 5′-GGGTG-3′, the simplest among previously studied DNA sequences, into vibronic resonance.

2. Computational Details

The starting geometry of the ds 5′-GGGTG-3′ DNA nucleotide was generated by using the 3DNA package [45] with the standard parameters of calf thymus B-DNA. The geometry of the 3′ terminal G was replaced with the equilibrium geometry of G^+, i.e., G in its cationic form. The geometry of G^+ was optimized at the B3LYP/6-311++G(d,p) level of theory with the unrestricted formalism. Solvent (water) effects were included via the polarizable continuum model (PCM) [46]. MD simulations for the $GGGTG^+$ sequence were carried out by using the AMBER 18 suite of programs adopting the OL15 force field [47,48]. Explicit water molecules were treated by using the TIP3P force field [49]. An equilibrated box (truncated octahedron, radius = 22 Å) consisting of ≈15,000 water molecules was used, corresponding to a *ca.* 4 mM DNA. Na^+ counterions were added to balance the negative charges of the phosphate groups; the positive charge of the DNA double strand was neutralized by adding a Cl^- ion [50]. Each Na^+ ion was initially placed in the OPO plane of each phosphate unit, equidistant from formally charged oxygen atoms. The O–Na^+ distance was set at 14 Å, in such a way that no solute atom is closer than 10 Å to any box side, see [51]. That distance was chosen after carrying out a few preliminary tests, showing that for lower O–Na^+ distances, only the diffusion of Na^+ ions in the bulk was observed, without achieving properly converged simulations. That is a subtle point, because although it has now been well ascertained that starting position and mobility of ions are relevant parameters, strongly influencing the function and the structure of nucleic acids, modeling those effects in MD simulations is still a challenging task [52] and has indeed been termed as "one of Achilles heels of atomistic MD simulations of RNA" [53]. Difficulties of simulating ions stem from force-field approximations, use of small sizes of the simulation boxes, and issues with the periodic boundary conditions [54–56].

Pending further studies, the 'addions' facility of the Amber suite, based on the minimization of the Coulombic potential evaluated on a grid, was adopted to place the Cl^- ion.

The charges used for G^+ in MD simulations were obtained by summing the atomic charges internally stored in Amber for guanine [57] to the difference of the atomic RESP charge of G^+ with the corresponding ones of G [58]. The particle mesh Ewald (PME) method for the treatment of the long-range electrostatics and the SHAKE algorithm (with an integration time step of 2 fs) for restraining the length of chemical bonds involving hydrogen atoms were used in all simulations [59,60]. A cut-off distance of 24 Å was used for non-bonding interactions. Harmonic restraints ($k = 50$ kcal $mol^{-1}/Å^2$) were imposed to G^+-3′ residue in order to keep its geometry as close as possible to the starting geometry of G^+ during all simulations (*vide infra*).

An initial minimization of the water box and ions was performed by keeping the DNA strand fixed, followed by a further geometry optimization of the whole system. Geometry optimization was followed by an equilibration at constant volume from 0 to 300 K in 20 ps (steps of 2 fs). Longer equilibration times (up to 1 ns) were also considered, but they resulted in no appreciable differences in the subsequent MD simulations. Molecular dynamics were carried out on a 2 ns time scale adopting an NPT ensemble, at 300 K. Because we are not concerned here with the problem of sampling all the configurations, 2 ns appears to be a well suited time, certainly longer with respect to the expected HT time scale. Furthermore, internal base-pair opening events leading to DNA denaturation are estimated to occur in *ca.* 1–10 ms, far longer than time scale of hole transport in oxidized DNA [21,61].

For each production simulation (time step 2 fs), snapshots were retrieved every 0.5 ps by using the PTRAJ module of the Amber package. For each snapshot, two single-point two-layer ONIOM QM/MM calculations [62] have been carried out, in which the globally neutral 5′-GGGTG-3′ double strand has been considered. In the former ONIOM computation, only the G-3′ nucleobase was included in the QM region; in the latter, only the 5′-GGG nucleobase step was included in the QM region. Hole site energies of the G-3′ and 5′-GGG units were estimated in the framework of the vertical approximation, by adopting Koopman's theorem and evaluating the energy of the Kohn–Sham HOMO of the G and GGG moieties [63].

Dangling bonds due to the breaking of the covalent bonds between the C9 atom of G and the C1′ atom of deoxyribose were saturated by hydrogen atoms [64]. The QM layer was treated at the density functional level of theory by using the APFD functional in conjunction with the 6-31G* basis set [65]. The MM layer composed of the remaining nucleobases, the sugar-phosphate backbone, the surrounding ions (Na^+ and Cl^-), and the water molecules was treated at the AMBER/TIP3P MM level. The electron embedding scheme, in which the electron density of the QM layer is polarized by the MM part was adopted throughout. ONIOM computations were carried out by using the Gaussian 09 package [66].

A pictorial representation of the optimized system is given in Figure 2, Cartesian coordinates of the optimized structure are given in the Supplementary Materials.

The conformations of ds-5′-GGGTG-3′ were analyzed in terms of standard rigid body coordinates (see Figure 3) by using the 3DNA software [45,67,68].

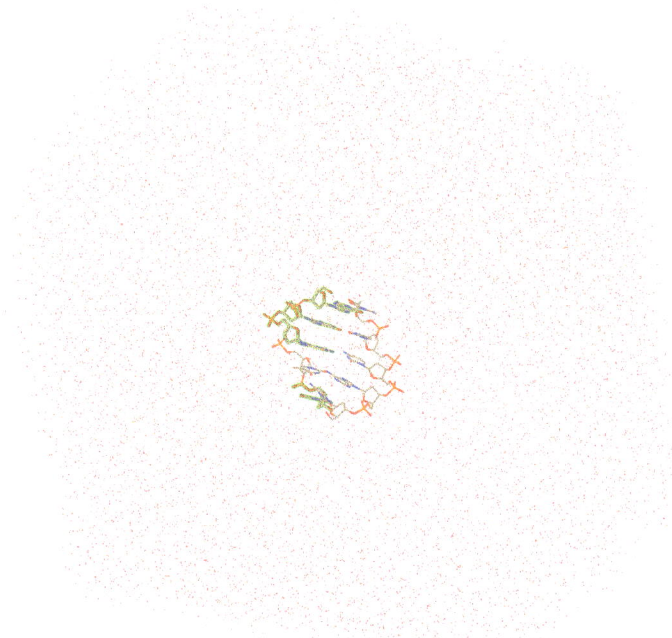

Figure 2. Optimized structure of 5′-GGGTG-3′ double strand immersed in a box consisting of *ca.* 15,000 water molecules. Hydrogen atoms have been omitted for clarity. Water molecules are depicted as red dots. Guanine nucleobases included in the QM layers of ONIOM computations are highlighted with a green border.

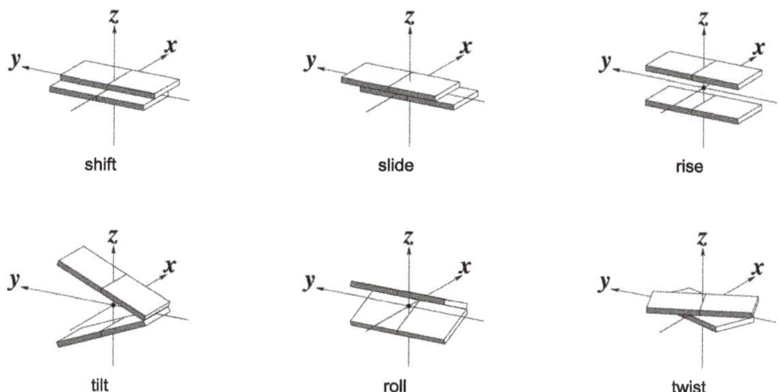

Figure 3. Graphical definitions of rigid-body coordinates used to describe the geometry of sequential base-pair steps.

3. Preliminary Considerations

Several MD and QM/MM studies dealing with HT in DNA were carried out in the last two decades. However, most simulations were performed on neutral DNA strands and were aimed at inferring the parameters that regulate the charge transfer in oxidized DNA, such as the hole site energies, the hopping integrals and their fluctuations originated by the dielectric environment. Therefore, quite long times were adopted in order to sample the whole conformational space [69–72]. The finalities of the present work are very different:

herein, we are interested at estimating the time needed to disrupt the local solvation pattern around the G^+-3′ nucleotide, corresponding to the loss of solvation energy of that site, eventually favoring the stabilization of the hole for the 5′-GGG^+ terminal. To that end, we had to resort to a globally charged DNA strand [51].

Although NMR studies show that proximal waters near sites on the surface DNA are much less mobile than bulk water, with residence times near the phosphate backbone of a few hundred picoseconds [73,74], the dynamical response of water to a change in solute charge distribution is a much faster process, occurring in a few tenths of fs [36,37]. Therefore, starting the simulation with the charge localized on G^+ appears to be a realistic choice. That condition, also met by adopting the equilibrium geometry of G^+, ensures an optimum hydration pattern for the ionized 5′-GGGTG-3′ oligonucleotide holding the hole at the G-3′ site. It must be further stressed that, herein, we are interested in the activation step (Figure 1), namely the step that brings 5′-GGG into resonance with G-3′, but with the charge still localized on G. Therefore, the hole must remain localized on the G-3′ moiety during the whole simulation. Only when solvent response to such a non-equilibrium charge distribution leads to resonance conditions, HT can take place. As long as the simulation proceeds, the GGG tract, i.e., the thermodynamic site for the hole, can acquire favorable solvation energy. That can been rationalized in the framework of the extended RRKM theory [75] inasmuch as the GGG stack possesses several more vibrational degrees of freedom than a single G. Of course, even G-3′ on which the charge is located, could change its nuclear configuration over time, thus contributing to achieve the equalization of the G and GGG hole site levels. However, in order to properly model that effect, a realistic force field for G^+ would be needed, ideally a quantum force field, due to the well-known difficulties of molecular mechanics in treating charged species. We have deliberately excluded that contribution because we are not sufficiently confident in the strength of the current classic force field in dealing with molecular ions. Therefore, rather than introducing further ambiguities, we have preferred to keep the geometry of G^+-3′ restrained during the simulation to further ensure that the charge remain localized at that site.

4. Results and Discussion

The rigid body coordinates of 5′-GGGTG-3′ are reported in Figures 4 and 5. Although the geometry of the 3′ ending G has been kept restrained to the one of G^+, the 5′-GGGTG-3′ double helix retains an almost regular structure, pretty close to B-DNA during the whole simulation. Indeed, slide and roll values (Figure 4) fall into the domain of standard B-DNA in *ca.* 50% of the simulation time, occasionally approaching the A-DNA zone of Calladine's diagram [76] for all the steps, but the 5′-TG-3′:3′-AC-5′ terminal residue of the oligomer, which exhibits the largest deviation from standard B-DNA, achieving roll values close to 20° for a large part of the simulation. However, this is not definitively new, indeed ending steps of DNA are known to exhibit a larger conformational heterogeneity than the internal positions [77,78]. Furthermore (Figure 5), both rise and twist remain close to their standard values (3.3 Å and 36°, respectively), a relevant point inasmuch as it is now well ascertained that electronic couplings for HT are mainly influenced by the fluctuations of the rise and twist coordinates [79,80].

The electrostatic potential evaluated at the mean point of the positively charged G-3′ site and the neutral 5′-GGG unit is reported as a function of time, up to 1 ps, in Figure 6.

A positive charge is well stabilized on both sites, especially the GGG stack, indeed the averaged potentials amount to -8.78 ± 0.27 V and -14.78 ± 0.57 V for G and GGG, respectively. Notably, the potential at GGG is seen to decrease during the simulation. The large variations of the potential observed at GGG roughly indicate how thermal fluctuations of the environment can comparatively stabilize a hole at that site. The first oscillation of the potential occurs in *ca.* 45 fs for both signals, suggesting the action of DNA vibrations. Indeed, the Fourier analysis of the signals of Figure 6 returns a spectrum of operating frequencies in the range 100 to 3000 cm^{-1}, thus demonstrating that both the DNA vibrational modes and backbone and solvent reorganization influence the electrostatic potential [81].

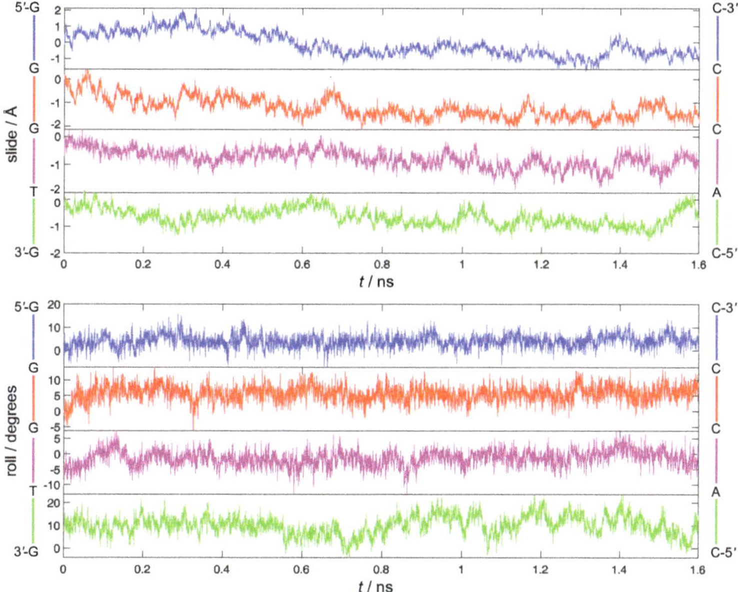

Figure 4. Time evolution of the four slide (Å, **top**) and roll (degrees, **bottom**) coordinates of ds-5′-GGGTG-3′ during 1.6 ns dynamics simulations in water.

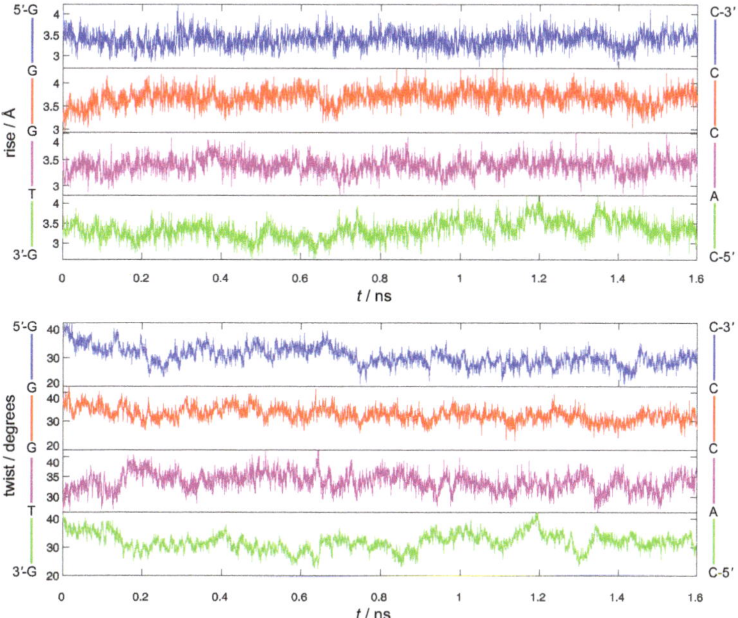

Figure 5. Time evolution of the four rise (Å, **top**) and twist (degrees, **bottom**) coordinates of ds-5′-GGGTG-3′ during 1.6 ns dynamics simulations in water.

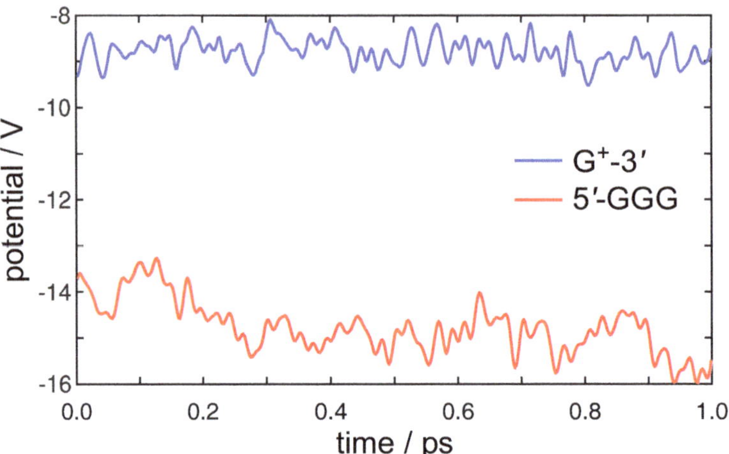

Figure 6. Electrostatic potential fluctuations at the G-3′ site (blue line) and 5′-GGG nucleobase stack (red line) for the initial 100 fs dynamics for ds-5′-GGGTG-3′ B-DNA sequence. Averages over 3 simulations are reported. The hole is localized at G-3′.

The HOMO energies of the GGG and G moieties retrieved by ONIOM computations (carried out for each snapshot) are plotted in Figure 7 as a function of time, a similar figure, also including statistical errors is provided in the Supplementary Materials.

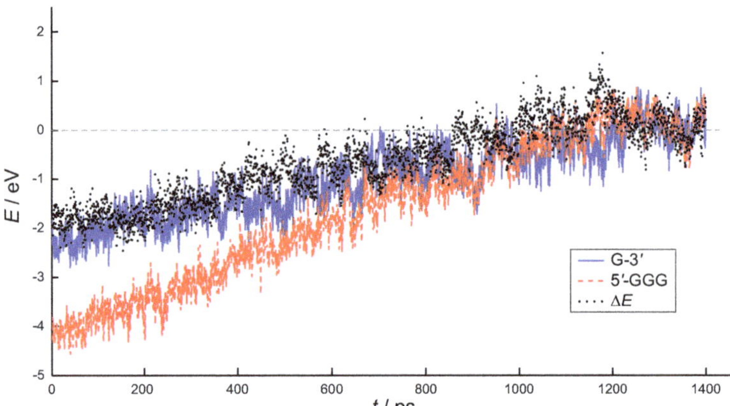

Figure 7. HOMO energy (eV) of 5′-GGG nucleobase stack (red dashed line) and G-3′ nucleobase (blue line) together with their difference (ΔE, black dots) for the 5′-GGGTG-3′ double strand immersed in a box consisting of *ca.* 15,000 water molecules and including counterions. Average values over 3 simulations are reported. The hole is initially ($t = 0$) localized on G.

At the starting time, the vertical ionization potential of G is predicted to lie ≈2 eV below the one of GGG, as expected, because the positive charge has been initially injected on G-3′. An almost linear increase in the HOMO energy of both the G and the GGG moieties is predicted for the first 800 ps ($\Delta E \approx 0$, dotted black curve, Figure 7), with the hole site energy of GGG decreasing faster, in line with the results of Figure 6. Equalization of HOMO levels of G and GGG is found to occur at *ca.* 800 ps and then the G and GGG HOMO levels remain almost degenerate up to 2 ns, i.e., the whole simulation time.

The autocorrelation function of $\Delta E(t)$, defined as

$$\text{ACF} = \frac{\langle \delta E(0) \delta E(t) \rangle}{\langle (\delta E)^2 \rangle},$$

with $\delta E(t) = \Delta E(t) - \langle \Delta E \rangle$ denoteing the instantaneous fluctuation of $\Delta E(t)$ from its equilibrium value, is diagrammed in Figure 8.

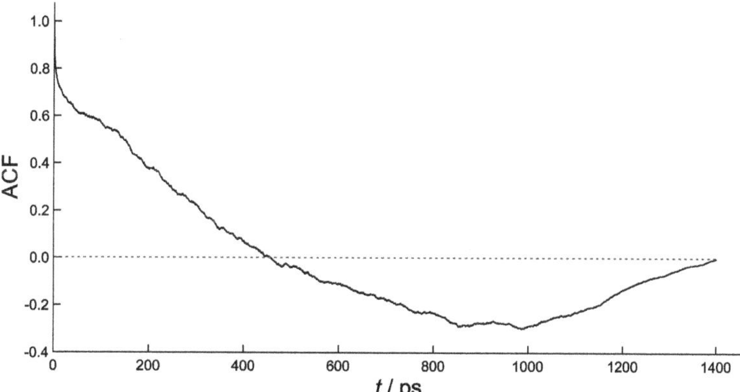

Figure 8. Autocorrelation function ACF $= \langle \delta E(0) \delta E(t) \rangle / \langle (\delta E)^2 \rangle$ averaged over 3 simulations.

This correlation function, that measures the loss of memory of random fluctuations in the hole energy difference, does not exhibit the simple exponential decay typical of linear processes. Although the sampling time (0.5 ps) does not permit the evaluation of the initial decay, it is clearly seen that ACF reverses its direction at *ca.* 450 ps with a half period of *ca.* 900 ps, roughly the same time needed to annihilate $\Delta E(t)$. That confirms that the initial state is indeed at equilibrium, thus the half period $\tau_{act} = 900$ ps can safely be taken as the time needed to bring G^+-3′ and 5′-GGG^+ into electronic resonance. Present estimate thus appears to be *ca.* one order of magnitude larger than the values, 50–100 ps, inferred in previous work [34,35].

In order to assess the reliability of the present estimate of the activation time, we have computed the P_{GGG}/P_G yield ratios of oxidative damage at GGG and G sites for the 5′-GGG(T)$_n$G-3′ sequences in which the charge is initially injected on G-3′, by using $\tau_{act} = 900$ ps, i.e., $k_{act} = 1/\tau_{act} = 1.1 \times 10^9$ s^{-1} in the kinetic scheme (Figure 1), together with the k_{HT}'s predicted by the quantum dynamics simulations of [34]. Current estimates of the oxidative damage yields are reported in Table 1 together with the results of [34] (obtained by adopting $\tau_{act} = 100$ ps) and their experimental counterparts from [26].

Table 1. P_{GGG}/P_G yield ratios of oxidative damage at GGG and G sites for the oxidized 5′-GGG(T)$_n$G-3′ DNA sequences predicted by using the kinetic model of Figure 1, together with $\tau_{act} = 900$ ps (first column, present estimate), and $\tau_{act} = 100$ ps (second column, ref [34]). Experimental values ([26]) are reported in the third column.

n	$\tau_{act} = 900$ ps	$\tau_{act} = 100$ ps	exper.
1	31	273	250
2	4.2	42	30
3	0.5	5.2	4.0
4	0.35	3.0	3.5
5	0.28	2.5	3.0
6	0.23	2.2	2.6
7	0.20	2.0	2.5

The relative yield ratios predicted upon enlarging the size of the $(T)_n$ bridge in 5′-GGG$(T)_n$G-3′ sequences are well reproduced also by using an activation time amounting to *ca.* 1000 ps (Table 1), i.e., one order of magnitude larger than previous estimates. However, absolute yield ratios turn out to be underestimated by a factor of ten. Therefore, it appears that faster kinetics than those presently inferred for the activation step are needed to achieve quantitative predictions. That suggests that the approximations introduced in the theoretical model, in particular the restraints imposed to G^+-3′, result in a too slow process possibly because of the lack of nuclear degrees of freedom of G^+-3′ in simulation.

However, one should not put too much reliance to the discrepancy of the different values, Indeed, lacking a reliable force field for oxidized guanine, present preliminary simulations do not allow for the vibrational relaxation of the G-3′ moiety occurring before the removal of the positive charge. That, in turn, gives rise to a larger equilibration time. Therefore, the presently estimated activation time should be considered as an upper limit.

5. Conclusions

The activation time, i.e., the time needed for bringing the donor (G) and the acceptor site (GGG) into resonance and activating the hole transfer in the 5′-GGGTG^+-3′ double stranded oxidized DNA sequence has been predicted by a theoretical protocol consisting in molecular dynamics simulations including the whole DNA strand, water and counterions at realistic concentration, followed by QM/MM computations. Activation appears to be modulated by both the vibrational modes of DNA and solvent response. The results of this preliminary study are consistent with an activation occurring in $\approx$0.9 ns. The time constant obtained by present computations is *ca.* one order of magnitude larger than previous estimates, $\approx$50–100 ps, inferred by a kinetic model capable of reproducing almost quantitatively the observed yield ratios of damaged DNA products, the most significant among observable quantities, at different sites for several diverse oxidized DNA sequences.

Of course, the computational strategy herein developed can be further improved. Work is in progress along that line.

Supplementary Materials: The following are available online. Cartesian coordinates of the whole system optimized at the AMBER OL15/TIP3P MM level of theory. Figure S1, HOMO energy of 5′-GGG stack and G-3′ nucleobase as a function of time, also including error bars.

Author Contributions: Conceptualization, A.P. and A.C.; funding acquisition, A.P. and A.C.; investigation, A.P., A.L. and A.C.; methodology, A.P., A.L. and A.C.; project administration and resources, A.C. and A.P.; software, A.L. and A.C.; writing, A.C., A.L. and A.P. All authors have read and agreed to the published version of the manuscript.

Funding: This research was funded by the Università di Salerno, grants: FARB 2018 and FARB 2019.

Conflicts of Interest: The authors declare no conflict of interest.

References

1. Kanvah, S.; Joseph, J.; Schuster, G.B.; Barnett, R.N.; Cleveland, C.L.; Landman, U. Oxidation of DNA: Damage to Nucleobases. *Acc. Chem. Res.* **2010**, *43*, 280–287. [CrossRef]
2. Genereux, J.C.; Barton, J.K. Mechanisms for DNA Charge Transport. *Chem. Rev.* **2010**, *110*, 1642–1662. [CrossRef]
3. Kumar, A.; Sevilla, M.D. Proton-Coupled Electron Transfer in DNA on Formation of Radiation-Produced Ion Radicals. *Chem. Rev.* **2010**, *110*, 7002–7023. [CrossRef]
4. Kawai, K.; Majima, T. Hole Transfer Kinetics of DNA. *Acc. Chem. Res.* **2013**, *46*, 2616–2625. [CrossRef]
5. Peluso, A.; Caruso, T.; Landi, A.; Capobianco, A. The Dynamics of Hole Transfer in DNA. *Molecules* **2019**, *24*, 4044. [CrossRef]
6. Heller, A. Spiers Memorial Lecture. On the Hypothesis of Cathodic Protection of Genes. *Faraday Discuss.* **2000**, *116*, 1–13. [CrossRef] [PubMed]
7. Rokhlenko, Y.; Cadet, J.; Geacintov, N.E.; Shafirovich, V. Mechanistic Aspects of Hydration of Guanine Radical Cations in DNA. *J. Am. Chem. Soc.* **2014**, *136*, 5956–5962. [CrossRef] [PubMed]
8. Shi, W.; Han, S.; Huang, W.; Yu, J. High Mobility Organic Field-Effect Transistor Based on Water-Soluble Deoxyribonucleic Acid via Spray Coating. *Appl. Phys. Lett.* **2015**, *106*, 043303. [CrossRef]
9. Zhang, Y.; Zalar, P.; Kim, C.; Collins, S.; Bazan, G.C.; Nguyen, T.Q. DNA Interlayers Enhance Charge Injection in Organic Field-Effect Transistors. *Adv. Mater.* **2012**, *24*, 4255–4260. [CrossRef] [PubMed]

10. Zalar, P.; Kamkar, D.; Naik, R.; Ouchen, F.; Grote, J.G.; Bazan, G.C.; Nguyen, T.Q. DNA Electron Injection Interlayers for Polymer Light-Emitting Diodes. *J. Am. Chem. Soc.* **2011**, *133*, 11010–11013. [CrossRef] [PubMed]
11. Gomez, E.F.; Venkatraman, V.; Grote, J.G.; Steckl, A.J. Exploring the Potential of Nucleic Acid Bases in Organic Light Emitting Diodes. *Adv. Mater.* **2015**, *27*, 7552–7562. [CrossRef]
12. Liedl, T.; Sobey, T.L.; Simmel, F.C. DNA-Based Nanodevices. *Nano Today* **2007**, *2*, 36–41. [CrossRef]
13. Liang, L.; Fu, Y.; Wang, D.; Wei, Y.; Kobayashi, N.; Minari, T. DNA as Functional Material in Organic-Based Electronics. *Appl. Sci.* **2018**, *8*, 90. [CrossRef]
14. Yoshioka, Y.; Kitagawa, Y.; Takano, Y.; Yamaguchi, K.; Nakamura, T.; Saito, I. Experimental and Theoretical Studies on the Selectivity of GGG Triplets toward One-Electron Oxidation in B-Form DNA. *J. Am. Chem. Soc.* **1999**, *121*, 8712–8719. [CrossRef]
15. Capobianco, A.; Caruso, T.; D'Ursi, A.M.; Fusco, S.; Masi, A.; Scrima, M.; Chatgilialoglu, C.; Peluso, A. Delocalized Hole Domains in Guanine-Rich DNA Oligonucleotides. *J. Phys. Chem. B* **2015**, *119*, 5462–5466. [CrossRef]
16. Capobianco, A.; Landi, A.; Peluso, A. Modeling DNA Oxidation in Water. *Phys. Chem. Chem. Phys.* **2017**, *19*, 13571–13578. [CrossRef] [PubMed]
17. Capobianco, A.; Peluso, A. The Oxidization Potential of AA Steps in Single Strand DNA Oligomers. *RSC Adv.* **2014**, *4*, 47887–47893. [CrossRef]
18. Capobianco, A.; Caruso, T.; Peluso, A. Hole Delocalization over Adenine Tracts in Single Stranded DNA Oligonucleotides. *Phys. Chem. Chem. Phys.* **2015**, *17*, 4750–4756. [CrossRef]
19. Capobianco, A.; Caruso, T.; Celentano, M.; D'Ursi, A.M.; Scrima, M.; Peluso, A. Stacking Interactions between Adenines in Oxidized Oligonucleotides. *J. Phys. Chem. B* **2013**, *117*, 8947–8953. [CrossRef]
20. Capobianco, A.; Velardo, A.; Peluso, A. Single-Stranded DNA Oligonucleotides Retain Rise Coordinates Characteristic of Double Helices. *J. Phys. Chem. B* **2018**, *122*, 7978–7989. [CrossRef]
21. Harris, M.A.; Mishra, A.K.; Young, R.M.; Brown, K.E.; Wasielewski, M.R.; Lewis, F.D. Direct Observation of the Hole Carriers in DNA Photoinduced Charge Transport. *J. Am. Chem. Soc.* **2016**, *138*, 5491–5494. [CrossRef] [PubMed]
22. Schuster, G.B. Long-Range Charge Transfer in DNA: Transient Structural Distortions Control the Distance Dependence. *Acc. Chem. Res.* **2000**, *33*, 253–260. [CrossRef] [PubMed]
23. Caruso, T.; Carotenuto, M.; Vasca, E.; Peluso, A. Direct Experimental Observation of the Effect of the Base Pairing on the Oxidation Potential of Guanine. *J. Am. Chem. Soc.* **2005**, *127*, 15040–15041. [CrossRef] [PubMed]
24. Caruso, T.; Capobianco, A.; Peluso, A. The Oxidation Potential of Adenosine and Adenosine-Thymidine Base-Pair in Chloroform Solution. *J. Am. Chem. Soc.* **2007**, *129*, 15347–15353. [CrossRef] [PubMed]
25. Capobianco, A.; Carotenuto, M.; Caruso, T.; Peluso, A. The Charge-Transfer Band of an Oxidized Watson-Crick Guanosine-Cytidine Complex. *Angew. Chem. Int. Ed.* **2009**, *48*, 9526–9528. [CrossRef]
26. Giese, B.; Amaudrut, J.; Köhler, A.K.; Spormann, M.; Wessely, S. Direct Observation of Hole Transfer through DNA by Hopping between Adenine Bases and by Tunneling. *Nature* **2001**, *412*, 318–320. [CrossRef]
27. Bixon, M.; Giese, B.; Wessely, S.; Langenbacher, T.; Michel-Beyerle, M.E.; Jortner, J. Long-Range Charge Hopping in DNA. *Proc. Natl. Acad. Sci. USA* **1999**, *96*, 11713–11716. [CrossRef]
28. Borrelli, R.; Capobianco, A.; Landi, A.; Peluso, A. Vibronic Couplings and Coherent Electron Transfer in Bridged Systems. *Phys. Chem. Chem. Phys.* **2015**, *17*, 30937–30945. [CrossRef]
29. Renger, T.; Marcus, R.A. Variable Range Hopping Electron Transfer Through Disordered Bridge States: Application to DNA. *J. Phys. Chem. A* **2003**, *107*, 8404–8419. [CrossRef]
30. Bixon, M.; Jortner, J. Incoherent Charge Hopping and Conduction in DNA and Long Molecular Chains. *Chem. Phys.* **2005**, *319*, 273–282. [CrossRef]
31. Grozema, F.C.; Tonzani, S.; Berlin, Y.A.; Schatz, G.C.; Siebbeles, L.D.A.; Ratner, M.A. Effect of Structural Dynamics on Charge Transfer in DNA Hairpins. *J. Am. Chem. Soc.* **2008**, *130*, 5157–5166. [CrossRef]
32. Zhang, Y.; Liu, C.; Balaeff, A.; Skourtis, S.S.; Beratan, D.N. Biological Charge Transfer via Flickering Resonance. *Proc. Natl. Acad. Sci. USA* **2014**, *111*, 10049–10054. [CrossRef]
33. Levine, A.D.; Iv, M.; Peskin, U. Length-Independent Transport Rates in Biomolecules by Quantum Mechanical Unfurling. *Chem. Sci.* **2016**, *7*, 1535–1542. [CrossRef]
34. Landi, A.; Borrelli, R.; Capobianco, A.; Peluso, A. Transient and Enduring Electronic Resonances Drive Coherent Long Distance Charge Transport in Molecular Wires. *J. Phys. Chem. Lett.* **2019**, *10*, 1845–1851. [CrossRef]
35. Landi, A.; Capobianco, A.; Peluso, A. Coherent Effects in Charge Transport in Molecular Wires: Toward a Unifying Picture of Long-Range Hole Transfer in DNA. *J. Phys. Chem. Lett.* **2020**, *11*, 7769–7775. [CrossRef]
36. Jimenez, R.; Fleming, G.R.; Kumar, P.V.; Maroncelli, M. Femtosecond Solvation Dynamics of Water. *Nature* **1994**, *369*, 471–473. [CrossRef]
37. Fleming, G.R.; Cho, M. Chromophore-Solvent Dynamics. *Annu. Rev. Phys. Chem.* **1996**, *47*, 109–134. [CrossRef]
38. Kobayashi, K.; Tagawa, S. Direct Observation of Guanine Radical Cation Deprotonation in Duplex DNA Using Pulse Radiolysis. *J. Am. Chem. Soc.* **2003**, *125*, 10213–10218. [CrossRef] [PubMed]
39. Steenken, S. Purine Bases, Nucleosides, and Nucleotides: Aqueous Solution Redox Chemistry and Transformation Reactions of Their Radical Cations and e^- and OH Adducts. *Chem. Rev.* **1989**, *89*, 503–520. [CrossRef]

40. Capobianco, A.; Caruso, T.; Celentano, M.; La Rocca, M.V.; Peluso, A. Proton Transfer in Oxidized Adenosine Self-Aggregates. *J. Chem. Phys.* **2013**, *139*, 145101-4. [CrossRef] [PubMed]
41. Candeias, L.P.; Steenken, S. Structure and Acid-Base Properties of One-Electron-Oxidized Deoxyguanosine, Guanosine, and 1-Methylguanosine. *J. Am. Chem. Soc.* **1989**, *111*, 1094–1099. [CrossRef]
42. Chatgilialoglu, C.; Caminal, C.; Guerra, M.; Mulazzani, Q.G. Tautomers of One-Electron-Oxidized Guanosine. *Angew. Chem. Int. Ed.* **2005**, *44*, 6030–6032. [CrossRef] [PubMed]
43. Chatgilialoglu, C.; Caminal, C.; Altieri, A.; Vougioukalakis, G.C.; Mulazzani, Q.G.; Gimisis, T.; Guerra, M. Tautomerism in the Guanyl Radical. *J. Am. Chem. Soc.* **2006**, *128*, 13796–13805. [CrossRef]
44. Joseph, J.; Schuster, G.B. Emergent Functionality of Nucleobase Radical Cations in Duplex DNA: Prediction of Reactivity Using Qualitative Potential Energy Landscapes. *J. Am. Chem. Soc.* **2006**, *128*, 6070–6074. [CrossRef]
45. Lu, X.J.; Olson, W.K. 3DNA: A Versatile, Integrated Software System for the Analysis, Rebuilding and Visualization of Three-dimensional Nucleic-Acid Structures. *Nat. Protoc.* **2008**, *3*, 1213–1227. [CrossRef]
46. Tomasi, J.; Mennucci, B.; Cammi, R. Quantum Mechanical Continuum Solvation Models. *Chem. Rev.* **2005**, *105*, 2999–3094. [CrossRef] [PubMed]
47. Case, D.A.; Ben-Shalom, I.Y.; Brozell, S.R.; Cerutti, D.S.; Cheatham, T.E., III; Cruzeiro, V.W.D.; Darden, T.A.; Duke, R.E.; Ghoreishi, D.; Gilson, M.K.; et al. *AMBER 18*; University of California: San Francisco, CA, USA, 2018.
48. Galindo-Murillo, R.; Robertson, J.C.; Zgarbovic, M.; Šponer, J.; Otyepka, M.; Jurečka, P.; Cheatham, T.E., III. Assessing the Current State of Amber Force Field Modifications for DNA. *J. Chem. Theory Comput.* **2016**, *12*, 4114–4127. [CrossRef] [PubMed]
49. Price, D.J.; Brooks, C.L. A modified TIP3P Water Potential for Simulation with Ewald Summation. *J. Chem. Phys.* **2004**, *121*, 10096–10103. [CrossRef]
50. Joung, I.S.; Cheatham, T.E. Determination of Alkali and Halide Monovalent Ion Parameters for Use in Explicitly Solvated Biomolecular Simulations. *J. Phys. Chem. B* **2008**, *112*, 9020–9041. [CrossRef]
51. Steinbrecher, T.; Koslowski, T.; Case, D.A. Direct Simulation of Electron Transfer Reactions in DNA Radical Cations. *J. Phys. Chem. B* **2008**, *112*, 16935–16944. [CrossRef]
52. Li, P.; Merz, K.M. Metal Ion Modeling Using Classical Mechanics. *Chem. Rev.* **2017**, *117*, 1564–1686. [CrossRef]
53. Šponer, J.; Bussi, G.; Krepl, M.; Banáš, P.; Bottaro, S.; Cunha, R.A.; Gil-Ley, A.; Pinamonti, G.; Poblete, S.; Jurečka, P.; et al. RNA Structural Dynamics As Captured by Molecular Simulations: A Comprehensive Overview. *Chem. Rev.* **2018**, *118*, 4177–4338. [CrossRef]
54. Chen, A.A.; Draper, D.E.; Pappu, R.V. Molecular Simulation Studies of Monovalent Counterion-Mediated Interactions in a Model RNA Kissing Loop. *J. Mol. Biol.* **2009**, *390*, 805–819. [CrossRef]
55. Chen, A.A.; Marucho, N.; Baker, N.A.; Pappu, R.V. Simulations of RNA Interactions with Monovalent Ions. In *Methods in Enzymology*; Academic Press: New York, NY, USA, 2009; Volume 469, pp. 411–432.
56. Várnai, P.; Zakrzewska, K. DNA and Its Counterions: A Molecular Dynamics Study. *Nucleic Acids Res.* **2004**, *32*, 4269–4280. [CrossRef]
57. Cornell, W.D.; Cieplak, P.; Bayly, C.I.; Gould, I.R.; Merz, K.M.; Ferguson, D.M.; Spellmeyer, D.C.; Fox, T.; Caldwell, J.W.; Kollman, P.A. A Second Generation Force Field for the Simulation of Proteins, Nucleic Acids, and Organic Molecules. *J. Am. Chem. Soc.* **1995**, *117*, 5179–5197. [CrossRef]
58. Bayly, C.I.; Cieplak, P.; Cornell, W.; Kollman, P.A. A Well-Behaved Electrostatic Potential Based Method Using Charge Restraints for Deriving Atomic Charges: The RESP Model. *J. Phys. Chem.* **1993**, *97*, 10269–10280. [CrossRef]
59. Miyamoto, S.; Kollman, P.A. SETTLE: An Analytical Version of the SHAKE and RATTLE Algorithm for Rigid Water Models. *J. Comput. Chem.* **1992**, *13*, 952–962. [CrossRef]
60. Crowley, M.F.; Darden, T.A.; Cheatham, T.E., III; Deerfield, D.W., II. Adventures in Improving the Scaling and Accuracy of a Parallel Molecular Dynamics Program. *J. Supercomput.* **1997**, *11*, 255–278. [CrossRef]
61. Galindo-Murillo, R.; Roe, D.R.; Cheatham, T.E., III. Convergence and Reproducibility in Molecular Dynamics Simulations of the DNA Duplex d(GCACGAACGAACGAACGC). *Biochim. Biophys. Acta Gen. Subj.* **2015**, *1850*, 1041–1058. [CrossRef]
62. Chung, L.W.; Sameera, W.M.C.; Ramozzi, R.; Page, A.J.; Hatanaka, M.; Petrova, G.P.; Harris, T.V.; Li, X.; Ke, Z.; Liu, F.; et al. The ONIOM Method and Its Applications. *Chem. Rev.* **2015**, *115*, 5678–5796. [CrossRef] [PubMed]
63. Stowasser, R.; Hoffmann, R. What Do the Kohn-Sham Orbitals and Eigenvalues Mean? *J. Am. Chem Soc.* **1999**, *121*, 3414–3420. [CrossRef]
64. Clemente, F.R.; Vreven, T.; Frisch, M.J. Getting the Most out of ONIOM: Guidelines and Pitfalls. In *Quantum Biochemistry*; John Wiley & Sons, Ltd.: Hoboken, NJ, USA, 2010; Chapter 2, pp. 61–83.
65. Austin, A.; Petersson, G.A.; Frisch, M.J.; Dobek, F.J.; Scalmani, G.; Throssell, K. A Density Functional with Spherical Atom Dispersion Terms. *J. Chem. Theory Comput.* **2012**, *8*, 4989–5007. [CrossRef]
66. Frisch, M.J.; Trucks, G.W.; Schlegel, H.B.; Scuseria, G.E.; Robb, M.A.; Cheeseman, J.R.; Scalmani, G.; Barone, V.; Mennucci, B.; Petersson, G.A.; et al. *Gaussian 09 Revision D.01*; Gaussian Inc.: Wallingford, CT, USA, 2009.
67. Olson, W.K.; Bansal, M.; Burley, S.K.; Dickerson, R.E.; Gerstein, M.; Harvey, S.C.; Heinemann, U.; Lu, X.J.; Neidle, S.; Shakked Z.; et al. A Standard Reference Frame for the Description of Nucleic Acid Base-pair Geometry. *J. Mol. Biol.* **2001**, *313*, 229–237. [CrossRef] [PubMed]

68. Lu, X.J.; Olson, W.K. 3DNA: A Software Package for the Analysis, Rebuilding and Visualization of Three-Dimensional Nucleic Acid Structures. *Nucleic Acids Res.* **2003**, *31*, 5108–5121. [CrossRef] [PubMed]
69. Voityuk, A.A.; Jortner, J.; Bixon, M.; Rösch, N. Energetic of Hole Transfer in DNA. *Chem. Phys. Lett.* **2000**, *324*, 430–434. [CrossRef]
70. Troisi, A.; Orlandi, G. The Hole Transfer in DNA: Calculation of Electron Coupling between Close Bases. *Chem. Phys. Lett.* **2001**, *344*, 509–518. [CrossRef]
71. Voityuk, A.A.; Jortner, J.; Bixon, M.; Rösch, N. Electronic Coupling between Watson-Crick Pairs for Hole Transfer and Transport in Desoxyribonucleic Acid. *J. Chem. Phys.* **2001**, *114*, 5614–5620. [CrossRef]
72. Voityuk, A.A.; Siriwong, K.; Rösch, N. Environmental Fluctuations Facilitate Electron-Hole Transfer from Guanine to Adenine in DNA π Stacks. *Angew. Chem. Int. Ed.* **2004**, *43*, 624–627. [CrossRef]
73. Denisov, V.P.; Carlström, G.; Venu, K.; Halle, B. Kinetics of DNA Hydration. *J. Mol. Biol.* **1997**, *268*, 118–136. [CrossRef]
74. Phan, A.T.; Leroy, J.L.; Guéron, M. Determination of the Residence Time of Water Molecules Hydrating B′-DNA and B-DNA, by One-Dimensional Zero-Enhancement Nuclear Overhauser Effect Spectroscopy. *J. Mol. Biol.* **1999**, *286*, 505–519. [CrossRef]
75. Marcus, R.A. Solvent Dynamics-Modified RRKM Theory in Clusters. *Chem. Phys. Lett.* **1995**, *244*, 10–18. [CrossRef]
76. El Hassan, M.A.; Calladine, C.R. Conformational Characteristics of DNA: Empirical Classifications and a Hypothesis for the Conformational Behaviour of Dinucleotide Steps. *Philos. Trans. R. Soc. A* **1997**, *355*, 43–100. [CrossRef]
77. Heddi, B.; Oguey, C.; Lavelle, C.; Foloppe, N.; Hartmann, B. Intrinsic Flexibility of B-DNA: The Experimental TRX Scale. *Nucleic Acids Res.* **2010**, *38*, 1034–1047. [CrossRef] [PubMed]
78. Zgarbová, M.; Otyepka, M.; Šponer, J.; Lankaš, F.; Jurečka, P. Base Pair Fraying in Molecular Dynamics Simulations of DNA and RNA. *J. Chem. Theory Comput.* **2014**, *10*, 3177–3189. [CrossRef]
79. Sugiyama, H.; Saito, I. Theoretical Studies of GG-Specific Photocleavage of DNA via Electron Transfer: Significant Lowering of Ionization Potential and 5′-Localization of HOMO of Stacked GG Bases in B-form DNA. *J. Am. Chem. Soc.* **1996**, *118*, 7063–7068. [CrossRef]
80. Senthilkumar, K.; Grozema, F.C.; Fonseca Guerra, C.; Bickelhaupt, F.M.; Lewis, F.D.; Berlin, Y.A.; Ratner, M.A.; Siebbeles, L.D.A. Absolute Rates of Hole Transfer in DNA. *J. Am. Chem. Soc.* **2005**, *127*, 14894–14903. [CrossRef]
81. Persson, R.A.X.; Pattni, V.; Singh, A.; Kast, S.M.; Heyden, M. Signatures of Solvation Thermodynamics in Spectra of Intermolecular Vibrations. *J. Chem. Theory Comput.* **2017**, *13*, 4467–4481. [CrossRef] [PubMed]

Article

Mapping the DNA Damaging Effects of Polypyridyl Copper Complexes with DNA Electrochemical Biosensors

Anna Banasiak [1], Nicolò Zuin Fantoni [2,3], Andrew Kellett [3,4,*] and John Colleran [1,5,*]

1 Applied Electrochemistry Group, FOCAS Institute, Technological University Dublin, Camden Row, Dublin 8, D08 CKP1 Dublin, Ireland; annabanasiak89@gmail.com
2 Department of Chemistry, University of Oxford, Oxford OX1 3TA, UK; nicolo.fantonizuin@chem.ox.ac.uk
3 School of Chemical Sciences and National Institute for Cellular Biotechnology, Dublin City University, Glasnevin, Dublin 9, D09 NR58 Dublin, Ireland
4 Synthesis and Solid-State Pharmaceutical Centre, School of Chemical Sciences, Dublin City University, Glasnevin, Dublin 9, D09 NR58 Dublin, Ireland
5 Central Quad Grangegorman, School of Chemical and Pharmaceutical Sciences, Technological University Dublin, Dublin 7, D07 H6K8 Dublin, Ireland
* Correspondence: andrew.kellett@dcu.ie (A.K.); john.colleran@tudublin.ie (J.C.); Tel.: +353-1-700-5461 (A.K.); +353-1-220-5562 (J.C.)

Abstract: Several classes of copper complexes are known to induce oxidative DNA damage that mediates cell death. These compounds are potentially useful anticancer agents and detailed investigation can reveal the mode of DNA interaction, binding strength, and type of oxidative lesion formed. We recently reported the development of a DNA electrochemical biosensor employed to quantify the DNA cleavage activity of the well-studied $[Cu(phen)_2]^{2+}$ chemical nuclease. However, to validate the broader compatibility of this sensor for use with more diverse—and biologically compatible—copper complexes, and to probe its use from a drug discovery perspective, analysis involving new compound libraries is required. Here, we report on the DNA binding and quantitative cleavage activity of the $[Cu(TPMA)(N,N)]^{2+}$ class (where TPMA = tris-2-pyridylmethylamine) using a DNA electrochemical biosensor. TPMA is a tripodal copper caging ligand, while *N,N* represents a bidentate planar phenanthrene ligand capable of enhancing DNA interactions through intercalation. All complexes exhibited electroactivity and interact with DNA through partial (or semi-) intercalation but predominantly through electrostatic attraction. Although TPMA provides excellent solution stability, the bulky ligand enforces a non-planar geometry on the complex, which sterically impedes full interaction. $[Cu(TPMA)(phen)]^{2+}$ and $[Cu(TPMA)(DPQ)]^{2+}$ cleaved 39% and 48% of the DNA strands from the biosensor surface, respectively, while complexes $[Cu(TPMA)(bipy)]^{2+}$ and $[Cu(TPMA)(PD)]^{2+}$ exhibit comparatively moderate nuclease efficacy (ca. 26%). Comparing the nuclease activities of $[Cu(TPMA)(phen)]^{2+}$ and $[Cu(phen)_2]^{2+}$ (ca. 23%) confirms the presence of TPMA significantly enhances chemical nuclease activity. Therefore, the use of this DNA electrochemical biosensor is compatible with copper(II) polypyridyl complexes and reveals TPMA complexes as a promising class of DNA damaging agent with tuneable activity due to coordinated ancillary phenanthrene ligands.

Keywords: DNA biosensor; chemical nucleases; DNA-drug interaction; copper complexes; metallodrugs

Citation: Banasiak, A.; Zuin Fantoni, N.; Kellett, A.; Colleran, J. Mapping the DNA Damaging Effects of Polypyridyl Copper Complexes with DNA Electrochemical Biosensors. *Molecules* **2022**, *27*, 645. https://doi.org/10.3390/molecules27030645

Academic Editors: Ivo Piantanida and Michael Smietana

Received: 28 November 2021
Accepted: 1 January 2022
Published: 19 January 2022

1. Introduction

Life expectancy has risen significantly during the last few decades due to better accessibility to health care, education, clean water, and sufficient food. During an average lifetime people therefore experience greater exposure to mutagenic substances and suffer age-related losses to cellular function, increasing the chances of permanent DNA damage [1]. It is not surprising then that the number of people diagnosed with cancer is incrementing year-on-year [2]. This trend has emphasised the need for developing effective new anticancer drugs with high specificity, tailored to the individual [3].

Many bioinorganic compounds containing transition metals such as Pt, Ru, Ag, Cu, and Mn exhibit anticancer properties [4–8] with a number undergoing clinical trial [9–11]. Among these, copper complexes have proved interesting as their biological activity is closely linked to redox cycling through several oxidation states—Cu(I) and Cu(II) [4]. Their biological and physicochemical properties are also highly dependent on the chelating ligands used in their construction [12,13]. The type and shape of the Cu-coordinated organic scaffold defines the DNA-interaction of the metal complex, which can bind through intercalation, groove binding, or electrostatic interactions [12]. Upon DNA-binding, the metal complex mediates DNA cleavage by hydrolysis of the DNA backbone or by oxidation of DNA sugars or bases [14]. Many copper complexes, in the presence of reductant (ascorbate, glutathione, or NADH) and an oxidant (H_2O_2 or O_2), generate reactive oxygen species (ROS) through Fenton-like [15,16] or other radical-generating mechanisms [17]. These ROS induce damage to DNA, such as the oxidation of bases (e.g., 8-oxoguanine) or to the sugar unit through C–H activation, that result in single- or double-strand DNA breaks [18]. Since the type of DNA damage induced here differs from classical platinum(II) therapy, copper complexes might therefore circumvent existing clinical treatment limitations—especially for recalcitrant cancers of the breast, brain, and pancreas [19,20].

The first reported copper-based complex to exhibit oxidative chemical nuclease activity was $[Cu(phen)_2]^{2+}$ [21]. In recent years, many related copper complexes have been prepared and their chemical nuclease activity explored both biologically and chemically not only in the context of anticancer drug discovery [5,14,22,23], but also for applications in gene-directed cleavage [24–27] and protein engineering [28]. Chemical nuclease activity is usually examined using gel electrophoresis techniques [29–31], further advanced through lab-on-a-chip technology [32]. These methodologies yield valuable insight into the ligand-directed DNA nuclease mechanisms of copper complexes [33]. Attempts to create electrochemical methods to detect DNA cleavage events induced by bioinorganic complexes, have also been reported [34–36]. However, these techniques do not provide quantitative data regarding nuclease efficacy and site-specific cleavage. In the current work, the chemical nuclease activities of a new class of copper complex was examined using a quantitative electrochemical method described recently by us [37]. Here, the DNA surface coverage of the biosensor is measured before and after its exposure to the copper nuclease in the presence of reductant and oxidant. Upon DNA cleavage, nucleic acid fragments are released from the immobilised DNA on the electrode surface. This decrease in the DNA surface coverage can be determined electrochemically and presented as the percentage in nuclease efficacy of the complex.

The copper complex class explored herein contain a tris-2-pyridylmethylamine (TPMA) caging ligand and one bidentate ligand selected from: 1,10-phenanthroline (phen); 2,2′-bipyridine (bipy); dipyridoquinoxaline (DPQ); or 1,10-phenanthroline-5,6-dione (PD). TPMA is a tripodal ligand that alone, and as part of a complex with copper, exhibits cytotoxicity against several cancer cell lines [38]. Planar bidentate phenanthrene ligands, on the other hand, are known to interact with nucleic acids by intercalation or semi-intercalation and coordinate a copper ion via *N,N* ligation. Importantly, the generation of radical species at the DNA interface by this class leads to strand excision [32,39]. Although the DNA binding and damaging properties of copper, cobalt, and ruthenium complexes of TPMA have been reported [40–43], these interactions have not, as yet, been characterised using DNA electrochemical biosensors.

In this work, the electrochemical and chemical nuclease properties of copper TPMA complexes were characterised using DNA electrochemical biosensor, recently employed in the analysis of $[Cu(phen)_2]^{2+}$—an agent which semi-intercalates and cleaves DNA from the minor groove [37]. Herein, the DNA binding and quantitative cleavage activity of the $[Cu(TPMA)(N,N)]^{2+}$ class is reported using a DNA electrochemical biosensor. These results not only validate the broader compatibility of our DNA sensor for use with polypyridyl complexes, but they also provide valuable drug discovery information related to target identification and selection.

2. Results and Discussion

2.1. Electrochemical Characterisation of Copper Complexes at Gold Electrodes

Four copper complexes were examined in this work [44]: $[Cu(TPMA)(phen)](ClO_4)_2$; $[Cu(TPMA)(bipy)](ClO_4)_2$; $[Cu(TPMA)(DPQ)](ClO_4)_2$; and $[Cu(TPMA)(PD)](ClO_4)_2$ (Scheme 1). All complexes were previously characterised using single crystal X-ray analysis and their solution stabilities identified using cw-EPR, HYSCORE, and ENDOR spectroscopies. As shown, the copper centre is bound to four nitrogen atoms of the TPMA ligand and two nitrogen atoms of the bidentate ligand, creating a six-coordinated complex of distorted octahedral geometry [17].

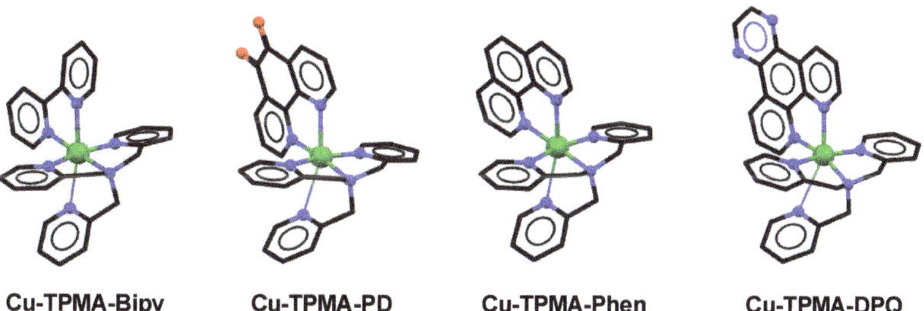

Scheme 1. The structures of $[Cu(TPMA)(bipy)](ClO_4)_2$, $[Cu(TPMA)(PD)](ClO_4)_2$, $[Cu(TPMA)(phen)](ClO_4)_2$ and $[Cu(TPMA)(DPQ)](ClO_4)_2$.

All copper complexes are electroactive and exhibit an ill-defined electrochemical response at bare gold electrodes in O_2-free solutions (argon degassed) aqueous solution. The electrochemical profiles obtained for $[Cu(TPMA)(phen)]^{2+}$, $[Cu(TPMA)(bipy)]^{2+}$, and $[Cu(TPMA)(DPQ)]^{2+}$ are very similar (Figure 1a) in that two reduction peaks (C_1 and C_2) and one oxidation peak (A_1) were observed. The electrochemical behaviour of $[Cu(TPMA)(PD)]^{2+}$ is more complicated and several additional oxidation and reduction peaks appear at the bare gold electrode (Figure 1b).

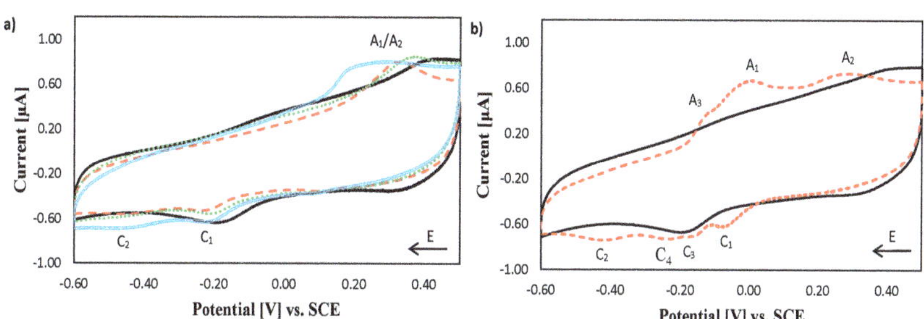

Figure 1. Typical cyclic voltammograms registered in O_2-free solutions at the bare gold electrode, scan rate 100 mV s^{-1}, in 0.1 M PB, pH 7.0 (black solid trace): (**a**) 20 µM of $[Cu(TPMA)(phen)]^{2+}$ (red dashed trace), $[Cu(TPMA)(bipy)]^{2+}$ (blue double trace) and $[Cu(TPMA)(DPQ)]^{2+}$ (green dotted trace), (**b**) 20 µM of $[Cu(TPMA)(PD)]^{2+}$ (red dashed trace).

Complications due to the electrochemistry of gold oxide monolayer formation were prevented by removing dissolved oxygen from the solutions. To determine how the individual peaks are correlated with one another, additional measurements in narrower potential windows were performed for $[Cu(TPMA)(phen)]^{2+}$ and $[Cu(TPMA)(PD)]^{2+}$ (S-3, Figures S2–S9, SM).

In measurements performed for $[Cu(TPMA)(phen)]^{2+}$, it was determined that the C_1/A_1 peaks are coupled. The C_2 redox process occurs only when it is preceded by an A_2 oxidation event and, thus, the C_2/A_2 are also coupled. The C_2 peak was not observed in our studies of $[Cu(phen)_2]^{2+}$ [37], hence, its appearance is linked to the presence of the TPMA ligand in $[Cu(TPMA)(phen)]^{2+}$. This was verified by electrochemical characterisation of $[Cu(TPMA)]^{2+}$ under the same conditions which returned redox potential comparable to those reported in polyelectrolyte [45] (S-3, Figure S4, SM). A plausible justification is that after reduction from $[Cu(TPMA)(phen)]^{2+}$ to $[Cu(TPMA)(phen)]^{+}$, C_1, the complex undergoes structural reorganisation, and subsequent electron transfer, described by the reduction event, C_2. Elucidation of the mechanism involved here requires further study but may be indicative of dissociation of one of the TPMA chelating N atoms to generate a five-coordinated species [17] (structurally—and possibly chemically—distinct from the initial complex). A similar interpretation can be applied to the C_1/A_1 and C_2/A_2 redox processes of $[Cu(TPMA)(bipy)]^{2+}$ and $[Cu(TPMA)(DPQ)]^{2+}$. The proposed electrochemical behaviour in aqueous solution justifies the two observed redox couples, referred to hereafter as 'Cu(TPMA)' and 'Cu(*N*,*N*′)' character, and DNA nuclease properties observed at the DNA modified electrodes, for these complexes.

In measurements performed on $[Cu(TPMA)(PD)]^{2+}$, it was determined that the C_1/A_1 redox coupled is ascribed to the 'Cu(PD) character' of the complex and C_3/A_3 can be associated with the reversible oxidation of the 1,10-phen-5,6-dione (PD) ligand. Transition metal PD complexes can undergo a two-step 2 electron reduction [46–48], while the reduction of the free PD ligand in aqueous solution, from quinone to semiquinone and further to hydroquinone, has also been reported [49,50] (S-3, Figure S7, SM). Moreover, the C_2 peak, also common to the other complexes at −0.4 V, is linked to the A_2 oxidation processes and describes the electroactivity of the reorganised complex ('Cu(TPMA) character').

In general, at both bare and DNA modified gold electrodes, one redox event (C_1/A_1), Cu(*N*,*N*′) character', was recorded at similar potentials for all copper complexes and is comparable to the C_1/A_1 data obtained for $[Cu(phen)_2]^{2+}$ (Table 1), while the C_2/A_2 couples describe the 'Cu(TPMA) character' redox behaviour of the complexes:

$$[\mathrm{Cu(TPMA)}(N,N)]^{2+} \rightleftharpoons [\mathrm{Cu(TPMA)}(N,N)]^{+} + e^{-}$$

Table 1. The values of C_1/A_1 and C_2/A_2 redox couple anodic and cathodic peak potentials at bare gold electrodes.

Complex	$Ep_{,C1}/Ep_{,A1}$ [V]	$\Delta Ep_{,C1/A1}$ [mV]	$Ep_{,C2}/Ep_{,A2}$ [V]	$\Delta Ep_{,C2/A2}$ [mV]
$[Cu(TPMA)(Phen)]^{2+}$	−0.18/+0.30	480	−0.46/+0.30	760
$[Cu(TPMA)(DPQ)]^{2+}$	−0.21/+0.23	440	−0.46/+0.23	590
$[Cu(TPMA)(Bipy)]^{2+}$	−0.19/+0.35	540	−0.45/+0.35	800
$[Cu(TPMA)(PD)]^{2+}$	−0.07/−0.04	30	−0.43/+0.27	700
$[Cu(TPMA)]^{2+}$	−0.21/+0.01	220	−0.42/+0.22	640
*$[Cu(Phen)_2]^{2+}$	−0.20/−0.08	120	NA	NA

Scan rate 100 mV s^{-1}, in 0.1 M PB, and the separation between these potentials, ΔEp. * From reference [37].

With the exception of $[Cu(TPMA)(PD)]^{2+}$, where reversible redox behaviour is observed in the C_1/A_1 couple ($\Delta E_{p,\,C1/A1}$ = 30 mV), large peak separation values indicate that both C_1/A_1 and C_2/A_2 redox couples were electrochemically irreversible ($\Delta Ep_{,\,C1/A1}$ > 440 mV and $\Delta Ep_{,\,C2/A2}$ > 590 mV) at the bare gold electrode. The C_2 reduction peaks for all copper complexes were registered at similar potentials, whereas the oxidation peaks were observed at positive potential values (>0.23 V). When compared to that of $[Cu(phen)_2]^{2+}$, it is apparent that the C_1/A_1 couple potentials are dictated by the presence of the TPMA ligand. The sluggish A_2 oxidation processes may indicate that, in terms of complex stability, the reduced state, Cu(I), is favoured, i.e., once reduced, re-oxidation of the complexes is kinetically difficult. These potential values show that differences in the structures of the complexes have a significant bearing on the respective redox properties.

2.2. Electrochemical Characterisation of $[Cu(phen)_2]^{2+}$, $[Cu(TPMA)]^{2+}$ and the TPMA Ligand at the DNA Biosensor

To interpret the data obtained for $[Cu(TPMA)(N,N')]^{2+}$, electrochemical analysis of $[Cu(phen)_2]^{2+}$, $[Cu(TPMA)]^{2+}$ and the TPMA ligand at the DNA biosensor was carried out (Figure 2). The redox peaks for $[Cu(phen)_2]^{2+}$ were clearly visible at −0.12 V (C_1) and −0.07 V (A_1) vs. SCE (Figure 2a). This redox process can be associated with the reduction and oxidation of the complex, $[Cu(phen)_2]^{2+}/[Cu(phen)_2]^{+}$, mediated by the immobilised DNA layer via long-range electron transfer. A peak potential separation, ΔEp, of 50 mV, and the ratio between the oxidation and reduction peak currents, $I_{p,a}/I_{p,c} > 2$, indicate that the redox reaction is quasi-reversible. The electrochemistry of this compound is discussed in detail elsewhere. An ill-defined C_1/A_1 redox couple was evident for $[Cu (TPMA)]^{2+}$ (Figure 2b). The small reduction C_1 peak appears at a potential similar to that for $[Cu(phen)_2]^{2+}$ but the oxidation A_1 peak is shifted to a much more positive potential value. The differences in the redox potentials between the complexes are directed by the variations in complex structure and ligand type. The planar aromatic ligand of $[Cu(phen)_2]^{2+}$ facilitates the partial intercalation of the complex within DNA bases [51] and the complex redox reaction is then mediated via long-range electron transfer.

The bulky non-planar TPMA ligand impedes intercalation between DNA bases, precluding close contact between the complex and DNA; hence, the electrochemical response of this complex is less well-defined than that observed for $[Cu(phen)_2]^{2+}$. An additional reduction peak C_2, at −0.42 V vs. SCE, (not observed in $[Cu(phen)_2]^{2+}$) can be linked to the presence of TPMA ligand in the complex, confirmed by the single small reduction peak (C_2) at −0.40 V vs. SCE for the TPMA ligand alone (Figure 2c). The presence of this peak indicates that the TPMA ligand can be reduced via DNA-mediated electron transfer.

The TPMA ligand consists of three pyridine rings that can, in theory, be reduced when free in solution [52]. The pyridine ring (pKa 5.2 [53]) after protonation in PB, pH 7.0, can undergo a one electron reduction to the pyridinyl radical (S-4, Figure S10, SM). Finally, the C_2 peak current was ca. 15 times larger for $[Cu(TPMA)]^{2+}$ than for the uncharged TPMA ligand alone, for the same concentration of compounds—evidence of a synergistic effect between the redox active copper and TPMA ligand, coupled to electrostatically-facilitated complex interaction with the immobilised DNA strands.

2.3. Electrochemical Characterisation of $[Cu(TPMA)(N,N')]^{2+}$ Complexes at the DNA Biosensor

Two prominent redox couples (C_1/A_1 and C_2/A_2) were observed for $[Cu(TPMA)(phen)]^{2+}$ and $[Cu(TPMA)(DPQ)]^{2+}$ at the DNA biosensor (Figure 3). Analysis of the peak potentials (S-5, Table S1, SM) indicate that the C_1/A_1 redox processes are quasi- reversible, while the C_2/A_2 processes tend toward irreversibility [54]. The C_1/A_1 redox waves, observed for $[Cu(TPMA)(phen)]^{2+}$ and $[Cu(phen)_2]^{2+}$, occur at similar potentials indicating that both are related to the same redox event, i.e., representing the '$[Cu(phen)_2]^{2+}$ electronic character' of the $[Cu(TPMA)(phen)]^{2+}$ complex. The C_2/A_2 wave was observed for all copper complexes except for $[Cu(phen)_2]^{2+}$, which suggests that the redox wave observed at these potentials is directed by the presence of the TPMA ligand. In support of this assignment, the C_2 reduction peak appears at very similar potentials for $[Cu(TPMA)]^{2+}$ and $[Cu(TPMA)(N,N')]^{2+}$ at the DNA biosensors. Thus, the C_2/A_2 redox wave is associated with the '$[Cu(TPMA)]^{2+}$ electronic character' of the complex. Copper complexes containing TPMA have shown solvent-dependent coordination spheres—five-coordinate in H_2O (no salts present) and mixed five and six-coordinate in organic solvent [17]. Data generated at the DNA modified electrodes seem to indicate that the complexes are six-coordinate in phosphate buffer and the dissociation of one Cu–N dative bond in aqueous solution generates a single unoccupied catalytic site within the complexes. This structural reorganisation would seem to justify the distinct mixed redox behaviour observed in these complexes at DNA modified electrodes.

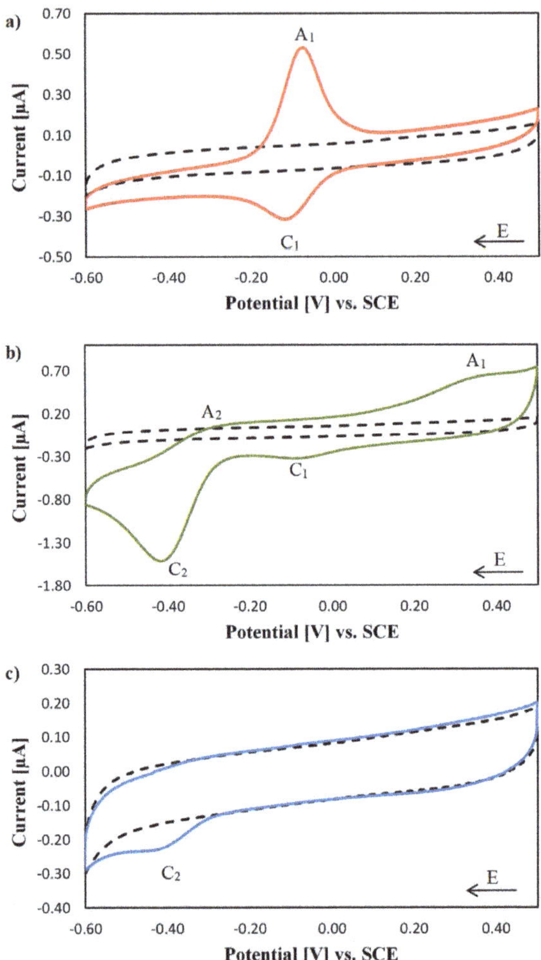

Figure 2. Typical cyclic voltammograms, scan rate 100 mV s^{-1}, registered at the DNA biosensor in 0.1 M PB, pH 7.0 (black dashed trace), and in 20 μM solutions of (**a**) $[Cu(phen)_2]^{2+}$ (red solid trace), (**b**) $[Cu(TPMA)]^{2+}$ (green solid trace) and (**c**) TPMA ligand (blue solid trace). Scan rates, 100 mV s^{-1}.

The interactions between DNA and the complexes were next evaluated through the changes in the $E^{0'}$ values for the C_1/A_1 and C_2/A_2 redox couples at the bare and DNA modified electrodes (S-5, Table S2, SM). Similar analysis using the C_2/A_2 redox couple could not be conducted since the A_2 peak was not evident at bare gold electrodes and a formal potential could not be estimated. The negative shift in the C_1/A_1 formal potential, observed in both $[Cu(TPMA)(phen)]^{2+}$ and $[Cu(TPMA)(DPQ)]^{2+}$, indicate that these complexes interact with DNA predominantly through electrostatic interactions in these conditions [55]. However, the presence of the C_1/A_1 redox couple observed at similar potentials for $[Cu(TPMA)(phen)]^{2+}$ and $[Cu(phen)_2]^{2+}$ indicates that, despite steric hindrance impeding intercalation, $[Cu(TPMA)(phen)]^{2+}$ is positioned in close enough proximity to the DNA strands to undergo DNA mediated redox reactions.

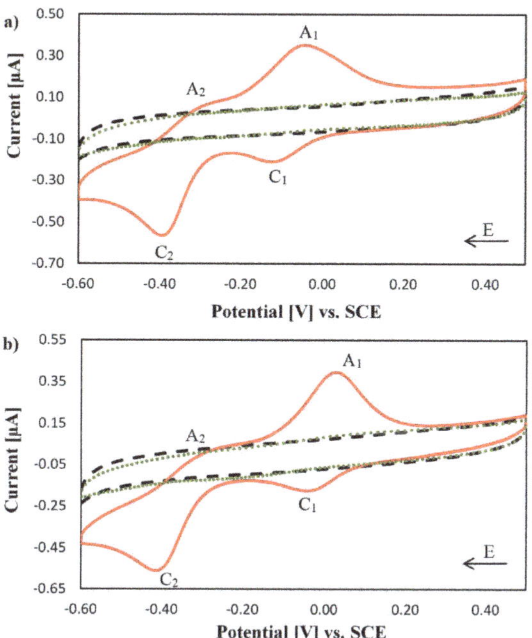

Figure 3. Typical cyclic voltammograms registered at the DNA biosensors, in 0.1 M PB, pH 7.0, scan rate 100 mV s^{-1} (black dashed trace), in 20 µM of (**a**) $[Cu(TPMA)(phen)]^{2+}$ (**b**) $[Cu(TPMA)(DPQ)]^{2+}$ (red solid trace) and after washing the complex from the DNA layer for 1 h (green dotted trace).

The $[Cu(TPMA)(DPQ)]^{2+}$ C_1/A_1 redox couple is shifted to less negative potentials compared to those observed with $[Cu(TPMA)(phen)]^{2+}$, and $[Cu(TPMA)(PD)]^{2+}$, inferring a more facile redox reaction than for $[Cu(TPMA)(phen)]^{2+}$. The planar aromatic ligand, DPQ, is of higher aromatic surface area than phenanthroline, phendione, and bipyridine; hence, despite the bulky TPMA ligand, the complex can intercalate between DNA bases to a greater extent than the phenanthroline, phendione, and bipyridine analogues, placing the copper centre in closer proximity to the DNA strands. This effect corroborates earlier topoisomerase unwinding experiments where the DPQ complex unwound superhelical plasmid DNA in manner similar to the classical DNA intercalator ethidium bromide while the phenanthroline complex was not effective [17].

The disappearance of $[Cu(TPMA)(phen)]^{2+}$ and $[Cu(TPMA)(DPQ)]^{2+}$ redox waves, after washing the DNA layer for one hour in 0.1 M PB (0.5% ACN), pH 7.0, confirms that the interactions between the complexes and DNA are non-covalent in character (Figure 3; green dotted traces). The 2,2′-bipyridine ligand (bipy) is known to interact with DNA through minor groove binding [56,57]; however, if more than one bipy ligand is present in the complex, the complex tends to interact with DNA through electrostatic attraction, presumably due to steric hindrance caused by the size of the complex [55,58]. The electrochemical profile for $[Cu(TPMA)(bipy)]^{2+}$ was expected to be then of a slightly different character than complexes presented above. The electrochemical response registered at the DNA biosensor for $[Cu(TPMA)(bipy)]^{2+}$ reveals the presence of C_1/A_1 and C_2/A_2 redox couples and an additional oxidation peak A_3 (Figure 4). The C_1 peak is ill-defined compared to that observed for $[Cu(TPMA)(phen)]^{2+}$ and $[Cu(TPMA)(DPQ)]^{2+}$ indicating a less facile reduction reaction occurs in $[Cu(TPMA)(bipy)]^{2+}$. Moreover, the C_1/A_1 redox process appeared to be irreversible in contrast to the results obtained for $[Cu(TPMA)(phen)]^{2+}$ and $[Cu(TPMA)(DPQ)]^{2+}$ (Table S2).

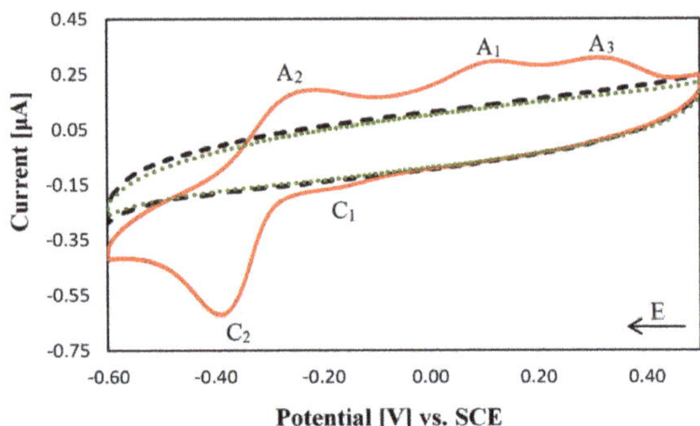

Figure 4. Typical cyclic voltammograms registered at the DNA biosensor in 0.1 M PB, pH 7.0 (black dashed trace), in 20 μM of $[Cu(TPMA)(bipy)]^{2+}$ (red solid trace) and after washing $[Cu(TPMA)(bipy)]^{2+}$ from the DNA layer for 1 h (green dotted trace). Scan rates, 100 mV s^{-1}.

Therefore, it appears that the 2,2′-bipyridine ligand cannot position the copper metal centre in close proximity to DNA, while 1,10-phenanthroline and DPQ—even if only through partial intercalation—can position the metal centre close enough to facilitate redox reactions. The additional A_3 oxidation peak appeared at +0.30 V vs. SCE, at a similar potential to the A_1 peak of $[Cu(TPMA)]^{2+}$ at the DNA biosensor (S-5, Table S1, SM). This result suggests that the A_3 redox peak is of '$[Cu(TPMA)]^{2+}$ character'. Moreover, the A_2 peak, although still irreversible, was much more prominent in $[Cu(TPMA)(bipy)]^{2+}$ than in $[Cu(TPMA)(phen)]^{2+}$ and $[Cu(TPMA)(DPQ)]^{2+}$, indicating again that the '$[Cu(TPMA)]^{2+}$ character' presents the dominant redox process in $[Cu(TPMA)(bipy)]^{2+}$. The negative shift in the $E^{0'}$ in the presence of DNA compared to the $E^{0'}$ obtained in the absence of DNA (S-5, Table S2, SM) suggests that $[Cu(TPMA)(bipy)]^{2+}$ interacts with DNA predominantly through electrostatic attraction. Additionally, the lack of any faradaic processes at the DNA biosensor after washing indicates that the complex interacts with DNA in a non-covalent manner.

Analysis of the electrochemical profile of $[Cu(TPMA)(PD)]^{2+}$ revealed time-dependent changes. The redox couples C_1/A_1 and C_2/A_2 observed for $[Cu(TPMA)(PD)]^{2+}$ at the DNA biosensor were initially much smaller in magnitude than redox peaks observed with the other complexes (Figure 5a). The C_1/A_1 redox wave was recorded at similar potentials to those at the bare gold electrode (C_1 at −0.08 V and A_1 at −0.04 V vs. SCE) and can be associated with the electrochemical reduction of the complex [48], directed by PD electronic characteristics.

The C_2/A_2 redox couple, the '$[Cu(TPMA)]^{2+}$ character' of the complex, was evident −0.42/−0.33 V vs. SCE. After 60 min, at the same DNA sensor, the peaks expanded and shifted over time (Figure 5b). Moreover, an additional peak (A_3) at +0.14 V became evident. The changes in the electrochemical behaviour of the complex could be due to disruption of the DNA layer and further insertion of the complex within the DNA strands.

The $[Cu(TPMA)(PD)]^{2+}$, in contrast to the other studied complexes, was not easily removed by washing from the DNA layer. Some $[Cu(TPMA)(PD)]^{2+}$ redox activity was recorded at the DNA biosensor in pure PB up to two days after exposure of the biosensor to the $[Cu(TPMA)(PD)]^{2+}$ solution (S-6, Figure S11, SM). Despite washing, the presence of the complex in the layer for an extended period of time suggests that the complex interacts with DNA in a covalent manner. According to the literature, some quinones can covalently bind to DNA bases, creating DNA adducts [59,60].

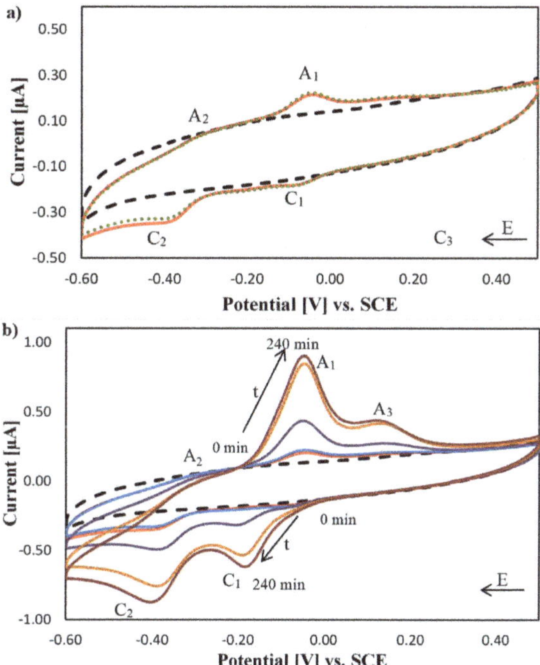

Figure 5. Typical cyclic voltammograms, obtained at 100 mV s^{-1} scan rates, registered at the DNA biosensors in 0.1 M PB, pH 7.0 (black dashed trace), and in 20 μM of $[Cu(TPMA)(PD)]^{2+}$ (**a**) just after addition (red solid trace) and after 60 min (green dotted trace); (**b**) over time using a single DNA sensor: 0 min (red trace); 60 min (blue trace); 120 min (violet trace); 180 min (orange trace); and 240 min (brown trace).

While fresh DNA sensors were prepared for the characterisation of each complex, scan rate experiments were performed at single DNA sensors and presented results are representative of typical analyses. These data were repeatable between freshly made DNA sensors. The relationship between the oxidation and reduction peak currents and the square root of scan rate, for each copper complex at the DNA biosensor, was linear while the relationship between peak current and scan rate was not (S-7, Figures S12–S15, SM). The redox reactions of the presented copper complexes were then deemed to be under diffusion control. In contrast, the redox reaction of $[Cu(phen)_2]^{2+}$, measured under the same conditions at the DNA biosensors, is under adsorption control [37]. This indicates that the presence of the bulky cage ligand, TPMA, directs a diffusion-controlled process and confirms that intercalation within the immobilised DNA strands is effectively prevented. The shift in estimated formal potentials are all negative for the complexes at the DNA biosensors, indicative of electrostatic interaction regimes (S-5, Table S2, SM). The negative shift in the C_2/A_2 formal potentials follow the trend $Cu(TPMA)(DPQ)]^{2+}$ < $[Cu(TPMA)(phen)]^{2+}$ < $[Cu(TPMA)(PD)]^{2+}$ < $[Cu(TPMA)(bipy)]^{2+}$ which would seem to reflect expected DNA intercalative properties of the ancillary *N,N′* ligands.

The electrochemical profiles for the copper complexes were also registered using square wave voltammetry. These data were used to check the stability of copper complex interaction with DNA. Prolonged exposure of the complexes $[Cu(TPMA)(phen)]^{2+}$, $[Cu(TPMA)(DPQ)]^{2+}$ and $[Cu(TPMA)(bipy)]^{2+}$, to the DNA biosensors revealed that the redox peaks were stable indefinitely, while a marked changed was observed for $[Cu(TPMA)(PD)]^{2+}$ after one hour (S-8, Figures S16–S19, SM).

2.4. DNA Nuclease Efficacy of Copper Complexes

Double-strand DNA breaks (DSBs) mediated by the copper complexes in the presence of an exogenous oxidant and reductant were monitored electrochemically using the DNA biosensors. The biosensors do not distinguish independent DSBs from proximate single strand breaks (SSBs) leading to double strand cleavage and in this assay, detection of the latter is likely given our earlier nicking observations by $[Cu(TPMA)(N,N)]^{2+}$ complexes with supercoiled pUC19 DNA [17]. The DNA surface coverages of the DNA biosensors were measured, prior to exposure to the copper complex nuclease assays, in 10 mM Tris-HCl, pH 7.4, using RuHex as a redox probe [37,61]. The DNA biosensors were then exposed to a nuclease assay containing 10, 20, and 50 μM of the complex, 1 mM ascorbic acid (AA) as the reductant and 1 mM H_2O_2 as the oxidant in 0.1 M PB, pH 7.0, at 37 °C for two hours. The DNA surface coverage was measured again after cooling and washing the DNA biosensor. The DNA layer washing is essential to ensure that the components of the nuclease assay are removed from the DNA layer and do not affect the subsequent interactions between DNA and RuHex. The changes in the DNA surface coverage values are then the consequence of DNA cleavage from the electrode surface. The cleavage efficacy (*C.E.*) can be then calculated as shown in Equation (1):

$$C.E. = 100\,\% - \left(\frac{\Gamma_{DNA\ after\ nuclease\ assay}}{\Gamma_{DNA\ before\ nuclease\ assay}} \times 100\% \right) \tag{1}$$

Since the Oligo DNA was 30 base pairs long, the number of cleaved base pairs was calculated as follows, Equation (2):

$$BP\ Cleaved = 30\ \text{bp} - \left(\frac{\Gamma_{DNA\ after\ nuclease\ assay}}{\Gamma_{DNA\ before\ nuclease\ assay}} \times 30\ \text{bp} \right) \tag{2}$$

The nuclease activities of all copper complexes at concentrations of 10, 20, and 50 μM are presented in Figure 6. Individual data sets for the nuclease and control experiments, with %RSD values, are available in the supporting information (S-9, Tables S3–S6, SM). All measurements were performed in triplicate using DNA biosensors freshly prepared for each measurement. The control measurements for this method, namely, the oxidant alone, the reductant alone and a mixture of the reductant and oxidant have no effect on the DNA surface coverage and are published elsewhere [37]. It was determined that the presence of a small amount of ACN (up to 1% v/v checked, S-2, Figure S1, SM) in the solution did not affect the DNA surface coverage values. Moreover, the complexes alone, in the absence of an exogenous reductant and oxidant, did not cause DNA cleavage.

$[Cu(TPMA)]^{2+}$ was found to cause insignificant DNA cleavage, 5.36% (%RSD 7.50%)—estimated as one base pair, in the presence of an exogenous reductant and oxidant. The non-planar geometry of TPMA dictates the proximity of the copper centre of the complex to the DNA strands. A relatively large distance between the copper centre and the DNA strands limits the ability to cleave DNA as the potency and frequency of radical species generated by the complex are greatly diminished. This result is in good agreement with results already published for this complex. Humphreys et al. first examined the nuclease activity of $[Cu(TPMA)]^{2+}$ using plasmid DNA and radiolabelled sequences [40,41]. In that work, the complex did not cause double-strand breaks in any of the tested DNA types; however, $[Cu(TPMA)]^{2+}$ did cause single-strand breaks in the plasmid. The same effect was observed by us on plasmid DNA [17] and in the current study it is likely proximate single strand breaks by $[Cu(TPMA)]^{2+}$ gives rise to the observed base pair cleaved. In contrast to this finding, the $[Cu(TPMA)(N,N)]^{2+}$ complexes exhibited high cleavage activity under the same conditions. Planar aromatic ligands in complexes enhance the binding activity between DNA and complex [17] and the redox properties of the complex (different $E^{0'}$ values for different complexes), having a marked effect on the observed nuclease efficacies. Previous studies have shown that superoxide radical and hydrogen peroxide generation play main roles in the DNA damage induced by the $[Cu(TPMA)(N,N)]^{2+}$ family. Moreover, some evidence of DNA damage associated with deamination of adenine (via the creation of deoxyinosine) was evident [17].

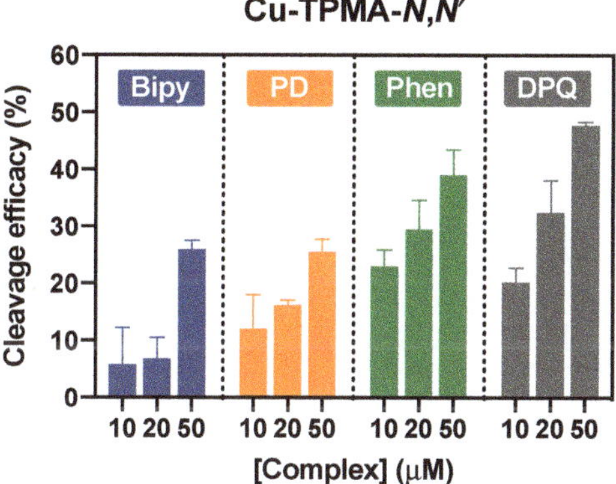

Figure 6. Average DNA nuclease efficacies determined at DNA modified electrodes for the copper complexes in 10, 20, and 50 μM concentrations; $[Cu(TPMA)(bipy)]^{2+}$ (blue bars), $[Cu(TPMA)(PD)]^{2+}$ (orange bars), $[Cu(TPMA)(phen)]^{2+}$ (green bars), and $[Cu(TPMA)(DPQ)]^{2+}$ (grey bars).

In this work, $[Cu(TPMA)(phen)]^{2+}$ exhibited high nuclease activity, even in small concentrations (S-9, Table S3, SM). Comparatively, 10 μM $[Cu(TPMA)(phen)]^{2+}$ cleaved as much DNA from the electrodes as 50 μM of $[Cu(phen)_2]^{2+}$ (23%) under the same experimental conditions [37]. The higher nuclease activity of $[Cu(TPMA)(phen)]^{2+}$ is likely due to the presence of the TPMA ligand enhancing the complex solution stability. The ligand also dictates the oxidation potential of the complex through stabilisation of the Cu(I) oxidation state (Section 3.1), serving to promote the generation of ROS and, ultimately, enhanced nuclease efficacy in comparison to $[Cu(phen)_2]^{2+}$. This effect was evident for the family of $[Cu(TPMA)(N,N')]^{2+}$ analogues presented here.

$[Cu(TPMA)(DPQ)]^{2+}$ in the presence of the reductant and oxidant exhibited the highest nuclease efficacy of all complexes (S-9, Table S4, SM). A concentration of 50 μM $[Cu(TPMA)(DPQ)]^{2+}$ cleaves almost half of the DNA strands present on the electrode surface. The nuclease activity of a copper complex containing a DPQ ligand, $[Cu(DPQ)_2(H_2O)]^{2+}$, was also reported by Santra et al. [62], with the cleavage of plasmid DNA observed on an agarose gel. The DNA degradation was more extensive for $[Cu(DPQ)_2(H_2O)]^{2+}$ than for the same concentration of $[Cu(phen)_2(H_2O)]^{2+}$ (non-quantitative data). Following the logic attributed to $[Cu(TPMA)(phen)]^{2+}$, the extended phenanthrene DPQ ligand should penetrate further into the DNA base pairs placing the complex in closer proximity to the DNA strands, facilitating more efficient DNA mediated charge transfer. Enhanced redox cycling leading to efficient in situ generation of ROS results in a complex with potent DNA cleavage properties.

The $[Cu(TPMA)(bipy)]^{2+}$ complex did not exhibit significant DNA cleavage at concentrations of 10 μM and 20 μM (S-9, Table S5, SM). Moderate cleavage was observed only when a 50 μM concentration was used. These results suggest that an excess of the complex is necessary to cause double-strand breaks (again resulting from proximate SSBs). According to the literature, copper complexes with 2,2′-bipyridine do not possess significant DNA nuclease properties [15,63]. The presence of the TPMA ligand appears to enhance complex cleavage efficacy to some degree, likely due enhanced solution stability leading to the creation of a favourable redox environment, promoting radical formation.

$[Cu(TPMA)(PD)]^{2+}$ did not exhibit nuclease activity in the absence of exogenous oxidant and reductant (S-9, Table S6, SM). However, the electrochemical profile of $[Cu(TPMA)(PD)]^{2+}$ registered at the DNA biosensors (in the absence of exogenous oxidant and reductant) reveals that the complex disrupts the DNA duplex significantly over time (Figure 5). The observed changes in the oxidation and reduction peaks then are not associated with DNA ablation from the electrode surface but are likely representative of a covalent DNA binding process. Conversely, in the presence of the exogenous reductant and oxidant, $[Cu(TPMA)(PD)]^{2+}$ exhibits moderate and concentration dependent DNA cleavage.

3. Materials and Methods

3.1. Materials

$[Cu(phen)_2](NO_3)_2$ and the copper complexes of the general formula: $[Cu(TPMA)(N,N')]^{2+}$ were synthesised as reported [17,44]. The $[Cu(TPMA)(N,N')]^{2+}$ complexes were dissolved in acetonitrile (ACN) prior to further dilution in phosphate buffer (PB). The final concentration of ACN in the PB solutions of the complexes did not exceed 0.2% (v/v). Ruthenium(III) hexaamine trichloride 98% and 4-mercaptotoluene 98% were purchased from Fisher. All other chemicals were of analytical grade and were purchased from Sigma–Aldrich. DNA oligomers containing 30 nucleotides were also purchased from Sigma–Aldrich and had the following sequences:

Thiol-modified strand: SH-$(CH_2)_6$- 5′-AGTACAGTCATCGCTTAATTATCGTACGTA-3′
Complementary strand: 3′-TCATGTCAGTAGCGAATTAATAGCATGCAT-5′.

DNA strands were hybridised prior to use according to the procedure provided by Sigma–Aldrich, Arklow, Ireland. Argon gas of technical grade was purchased from Air Products. Nuclease-free water was used for the preparation of the oligo DNA solutions, while Milli-Q® water was used to prepare all other solutions.

3.2. Equipment

The electrochemical measurements were performed with a CH Instruments, Inc. (IJ Cambria Scientific Ltd., Llanelli, UK) potentiostat, model 620A. DNA-modified gold electrodes (2 mm diameter, from CH Instruments, Inc.), a saturated calomel electrode (SCE, from BAS Inc., West Lafayette, IN, USA) and a platinum wire (from Surepure Chemetals, Florham Park, NJ, USA) were used as the working, reference, and counter electrodes, respectively. The preparation of the DNA-modified gold electrodes (DNA biosensors) is described in S-1, Supplementary Materials (SM). All glassware in contact with DNA was silanised to offset adsorption of DNA from solutions [64]. Prior to DNA immobilisation, bare gold electrodes were cleaned by manual polishing with a 0.05 μm Al_2O_3 slurry on Buehler microcloth for 5 min, followed by electrochemical cycling in 0.5 M deaerated H_2SO_4, over the potential range −0.2 V to +1.5 V vs. SCE at a scan rate of 100 mV s^{-1}, until steady-state current was attained. The thiol-modified gold electrodes were cleaned by electrochemical desorption of thiols through cycling in 0.5 M KOH, over the potential range +0.1 V to −1.4 V vs. SCE at a scan rate of 50 mV s^{-1}, followed by manual polishing on microcloth with 0.05 μm Al_2O_3 for 5 min. The supporting electrolyte was deoxygenated with argon before measurements and a blanket of argon was maintained above the solutions during measurements. The electrochemical measurements were performed at room temperature, while the chemical nuclease assays were carried out at 37 °C.

3.3. Interactions between Immobilised DNA and Bioinorganic Compounds

The interactions between the complexes and DNA were deciphered through analysis of the electrochemical profile of the compound in the absence and presence of DNA [55]. If the $E^{0'}$ value, estimated as the mean peak potential for the redox couple of the complex, shifts to more positive potentials in the presence of DNA, the complex was interpreted to interact with DNA predominantly through intercalation. If the $E^{0'}$ value shifts to more negative potentials, the interactions between DNA and the compound was interpreted as predominantly electrostatic.

3.4. Determination of DNA Surface Coverage

The DNA surface coverage was measured electrochemically using ruthenium(III) hexaamine (RuHex) as a redox probe [37,61,65]. In low ionic strength electrolyte (10 mM Tris-HCl), the positively charged RuHex interacts with DNA electrostatically, as an external binder, replacing the existing phosphate backbone counterions (Na^+ or K^+). The amount of RuHex molecules adsorbed in the DNA strands can be determined electrochemically using chronocoulometry. Upon DNA layer saturation, the number of adsorbed RuHex molecules is proportional to the number of phosphate groups in the DNA strands. Hence, knowing the number of phosphate groups present in the custom DNA strands (one nucleotide contains one phosphate group), the amount of adsorbed RuHex can be related to the number of DNA molecules immobilised at the electrode surface. After each aliquot addition of RuHex, and prior to data acquisition, the solution was allowed to equilibrate with the DNA sensor for two minutes.

3.5. Electrochemical Measurements and Nuclease Assay Conditions

Each measurement was performed independently on freshly prepared electrodes in 0.1 M phosphate buffer (PB), pH 7.0, containing 20 μM of the complex. Cyclic voltammetry was carried out over the potential range +0.5 V to −0.6 V vs. SCE, at a scan rate of 0.1 V s^{-1}. The $[Cu(TPMA)(N,N')]^{2+}$ solutions were allowed to equilibrate with the DNA sensors for 20 min prior to characterisation using cyclic voltammetry. To promote solubility in PB, the complexes were first dissolved in ACN, and aliquots were then added to PB (amount of ACN in PB was 0.2% v/v). The addition of ACN to PB does not affect the integrity of the DNA layer (S-2, Figure S1, SM). Chronocoulometry was then performed over the potential range +0.1 V to −0.6 V vs. SCE with a pulse width of 0.5 s. The DNA biosensor was exposed to the DNA nuclease assay, comprising of 10 μM, 20 μM, or 50 μM of the chosen complex, 1 mM of sodium-*L*-ascorbate (AA) and 1 mM of H_2O_2 for two hours at 37 °C.

4. Conclusions

The copper complexes investigated in this work were all found to be electroactive and interact with DNA predominantly through electrostatic attraction. The intercalation of planar aromatic bidentate ligands, such as 1,10-phenanthroline and DPQ, was hindered due to the bulky TPMA scaffold. The electrochemical response of $[Cu(TPMA)(bipy)]^{2+}$ was dominated by the '$[Cu(TPMA)]^{2+}$ character' of the complex. It was also found that $[Cu(TPMA)(PD)]^{2+}$ caused strong disruption of the DNA layer over time and remains within the DNA layer for an extended period of time even after washing. Dissociation was not observed in any of studied complexes under our experimental conditions, which corroborates earlier EPR results. Detailed investigations into the nuclease activity toward immobilised DNA strands provided an interesting array of results. Firstly, in the presence of an exogenous reductant and oxidant, $[Cu(TPMA)(phen)]^{2+}$ and $[Cu(TPMA)(DPQ)]^{2+}$ displayed relatively high nuclease activity by cleaving 39% and 48% of the DNA strands at 50 μM drug-loading, respectively. Under the same conditions, $[Cu(TPMA)(bipy)]^{2+}$ and $[Cu(TPMA)(PD)]^{2+}$ exhibit moderate nuclease efficacy with ca. 26% cleavage activity detected for both agents. Although the $[Cu(TPMA)]^{2+}$ complex alone displays poor chemical nuclease activity, coordination of phen or DPQ promotes excellent cleavage ability. The type of ancillary ligand present also significantly affects both the DNA binding activity and oxidative DNA damage induced. Although electrostatic forces dominate DNA-$[Cu(TPMA)(N,N)]^{2+}$ interactions, the intercalative properties of phen and DPQ facilitate greater accessibility to DNA bases. Consequently, greater electronic communication to the DNA base pairs grant $[Cu(TPMA)(phen)]^{2+}$ and $[Cu(TPMA)(DPQ)]^{2+}$ enhanced DNA cleavage abilities, through more efficient DNA mediated redox cycling, and concomitant in situ generation of reactive oxygen species. This assertion may seem to contradict data gleaned for $[Cu(TPMA)(PD)]^{2+}$, but based on electrochemical behaviour observed at the DNA biosensors in the absence of exogenous reagents, we postulate that PD-initiated

DNA coupling reactions occur that compete with the DNA oxidative damage pathway, limiting the nuclease efficacy observed for this complex. Overall, DNA electrochemical biosensors appear highly suitable for analysing copper polypyridyl complexes and can provide valuable structure-activity-relationship (SAR) data from both DNA damage and drug discovery perspectives. It appears TPMA provides strong solution stability that stabilises the Cu(I) oxidation state of this complex series thereby facilitating the generation of reactive oxygen species. Despite the predominant electrostatic interaction of $[Cu(TPMA)(phen)]^{2+}$ with DNA, its nuclease activity was higher than the known DNA semi-intercalator, $[Cu(phen)_2]^{2+}$. The results obtained here clearly show how the ligand scaffold and ancillary *N,N* ligands potentiate both the redox and chemical nuclease properties of bioinorganic copper complexes.

Supplementary Materials: The following are available online. Figure S1: The effect of acetonitrile on the DNA layer of the DNA biosensor; Figures S2–S6, S8, S9: Electrochemical response of $[Cu(TPMA)(phen)]^{2+}$, $[Cu(TPMA)]^{2+}$, and $[Cu(TPMA)(PD)]^{2+}$ at the gold electrode; Figure S7: Scheme of quinone electrochemical reduction; Figure S10: Reduction of the pyridine ring; Tables S1 and S2: Electrochemical parameters for the copper complexes obtained at the DNA biosensors; Figure S11: Washing of the DNA layer after interaction with $[Cu(TPMA)(PD)]^{2+}$; Figures S12–S15: The relationship between the oxidation and reduction wave peak currents versus the scan rate, and the square root of the scan rate, at the DNA biosensor; Figures S16–S19: The stability of the complexes at the DNA biosensor; Tables S3–S6: DNA nuclease efficacy of copper complexes. References [17,37,45–50,52,61,64,65] have been cited in the Supplementary Materials.

Author Contributions: Conceptualization, J.C.; methodology, A.B. and J.C.; validation, A.B. and J.C.; formal analysis, A.B., N.Z.F., A.K. and J.C.; investigation, A.B.; resources, A.K. and J.C.; data curation, A.B., N.Z.F., A.K. and J.C.; writing—original draft preparation, A.B.; writing—review and editing, A.B., N.Z.F., A.K. and J.C.; visualization, A.B. and N.Z.F.; supervision, A.K. and J.C.; project administration, A.K. and J.C.; funding acquisition, A.K. and J.C. All authors have read and agreed to the published version of the manuscript.

Funding: A.B. and J.C. acknowledge PhD scholarship funding from TU Dublin (Fiosraigh Award, Grant code: PB03979). A.K. and N.Z.F. acknowledge funding from the Marie Skłodowska-Curie Innovative Training Network (ITN) ClickGene (H2020-MSCA-ITN-2014-642023). This project has received funding from the European Union's Horizon 2020 research and innovation programme under the Marie Skłodowska-Curie grant agreement No. 861381 (NATURE-ETN). A.K. also acknowledges funding from Science Foundation Ireland Career Development Award (SFI-CDA;15/CDA/3648) and the Synthesis and Solid-State Pharmaceutical Centre (12/RC/2275_P2).

Institutional Review Board Statement: Not applicable.

Informed Consent Statement: Not applicable.

Data Availability Statement: The data presented in this study are available in this article and supplementary material.

Acknowledgments: This research was performed in Applied Electrochemistry Group Lab, FOCAS Institute, Camden Row, TU Dublin, Dublin 8, Ireland.

Conflicts of Interest: The authors declare no conflict of interest.

Sample Availability: Samples of the compounds are available upon request.

References

1. Brown, G.C. Living too long: The current focus of medical research on increasing the quantity, rather than the quality, of life is damaging our health and harming the economy. *EMBO Rep.* **2015**, *16*, 137–141. [CrossRef]
2. WHO. Latest global cancer data: Cancer burden rises to 18.1 million new cases and 9.6 million cancer deaths in 2018. WHO: Geneva, Switzerland; pp. 1–3. Available online: https://www.iarc.who.int/wp-content/uploads/2018/09/pr263_E.pdf (accessed on 10 June 2018).
3. Hennessy, J.; McGorman, B.; Molphy, Z.; Farrell, N.P.; Singleton, D.; Brown, T.; Kellett, A. A Click Chemistry Approach to Targeted DNA Crosslinking with *cis*-Platinum(II)-Modified Triplex-Forming Oligonucleotides. *Angew. Chem. Int. Ed.* **2021**, *61*, e202110455. [CrossRef]

4. Kellett, A.; O'Connor, M.; McCann, M.; Howe, O.; Casey, A.; McCarron, P.; Kavanagh, K.; McNamara, M.; Kennedy, S.; May, D.D.; et al. Water-soluble bis(1,10-phenanthroline) octanedioate Cu^{2+} and Mn^{2+} complexes with unprecedented nano and picomolar in vitro cytotoxicity: Promising leads for chemotherapeutic drug development. *MedChemComm* **2011**, *2*, 579–584. [CrossRef]
5. McGivern, T.; Afsharpour, S.; Marmion, C. Copper complexes as artificial DNA metallonucleases: From Sigman's reagent to next generation anti-cancer agent? *Inorganica Chim. Acta* **2018**, *472*, 12–39. [CrossRef]
6. Muhammad, N.; Guo, Z. Metal-based anticancer chemotherapeutic agents. *Curr. Opin. Chem. Biol.* **2014**, *19*, 144–153. [CrossRef]
7. Thornton, L.; Dixit, V.; Assad, L.O.; Ribeiro, T.P.; Queiroz, D.; Kellett, A.; Casey, A.; Colleran, J.; Pereira, M.D.; Rochford, G.; et al. Water-soluble and photo-stable silver(I) dicarboxylate complexes containing 1,10-phenanthroline ligands: Antimicrobial and anticancer chemotherapeutic potential, DNA interactions and antioxidant activity. *J. Inorg. Biochem.* **2016**, *159*, 120–132. [CrossRef] [PubMed]
8. Fantoni, N.Z.; Brown, T.; Kellett, A. DNA-Targeted Metallodrugs: An Untapped Source of Artificial Gene Editing Technology. *ChemBioChem* **2021**, *22*, 2184–2205. [CrossRef] [PubMed]
9. Burris, H.A.; Bakewell, S.; Bendell, J.C.; Infante, J.; Jones, S.F.; Spigel, D.R.; Weiss, G.J.; Ramanathan, R.K.; Ogden, A.; Von Hoff, D. Safety and activity of IT-139, a ruthenium-based compound, in patients with advanced solid tumours: A first-in-human, open-label, dose-escalation phase I study with expansion cohort. *ESMO Open* **2016**, *1*, e000154. [CrossRef]
10. Galindo-Murillo, R.; Garcia-Ramos, J.C.; Ruiz-Azuara, L.; Cheatham, T.E.; Cortes-Guzman, F. Intercalation processes of copper complexes in DNA. *Nucleic Acids Res.* **2015**, *43*, 5364–5376. [CrossRef]
11. Leijen, S.; Burgers, S.A.; Baas, P.; Pluim, D.; Tibben, M.; van Werkhoven, E.; Alessio, E.; Sava, G.; Beijnen, J.H.; Schellens, J.H.M. Phase I/II study with ruthenium compound NAMI-A and gemcitabine in patients with non-small cell lung cancer after first line therapy. *Investig. New Drugs* **2014**, *33*, 201–214. [CrossRef] [PubMed]
12. Marzano, C.; Pellei, M.; Tisato, F.; Santini, C. Copper Complexes as Anticancer Agents. *Anti-Cancer Agents Med. Chem.* **2009**, *9*, 185–211. [CrossRef]
13. McStay, N.; Slator, C.; Singh, V.; Gibney, A.; Westerlund, F.; Kellett, A. Click and Cut: A click chemistry approach to developing oxidative DNA damaging agents. *Nucleic Acids Res.* **2021**, *49*, 10289–10308. [CrossRef]
14. Sangeetha Gowda, K.R.; Mathew, B.B.; Sudhamani, C.N.; Naik, H.S.B. Mechanism of DNA Binding and Cleavage. *Biomed.Biotechnol.* **2014**, *2*, 1–9. [CrossRef]
15. Que, B.G.; Downey, K.M.; So, A.G. Degradation of deoxyribonucleic acid by a 1,10-phenanthroline-copper complex: The role of hydroxyl radicals. *Biochemistry* **1980**, *19*, 5987–5991. [CrossRef] [PubMed]
16. Reich, K.A.; Marshall, L.E.; Graham, D.R.; Sigman, D.S. Cleavage of DNA by the 1,10-phenanthroline-copper ion complex. Superoxide mediates the reaction dependent on NADH and hydrogen peroxide. *J. Am. Chem. Soc.* **1981**, *103*, 3582–3584. [CrossRef]
17. Fantoni, N.Z.; Molphy, Z.; Slator, C.; Menounou, G.; Toniolo, G.; Mitrikas, G.; McKee, V.; Chatgilialoglu, C.; Kellett, A. Polypyridyl-Based Copper Phenanthrene Complexes: A New Type of Stabilized Artificial Chemical Nuclease. *Chem.–A Eur. J.* **2018**, *25*, 221–237. [CrossRef]
18. Abreu, F.; Goulart, M.; Brett, A.O. Detection of the damage caused to DNA by niclosamide using an electrochemical DNA-biosensor. *Biosens. Bioelectron.* **2002**, *17*, 913–919. [CrossRef]
19. Kellett, A.; Molphy, Z.; McKee, V.; Slator, C. CHAPTER 4. Recent Advances in Anticancer Copper Compounds. *Met. Based Anticancer Agents* **2019**, *14*, 91–119. [CrossRef]
20. Toniolo, G.; Louka, M.; Menounou, G.; Fantoni, N.Z.; Mitrikas, G.; Efthimiadou, E.K.; Masi, A.; Bortolotti, M.; Polito, L.; Bolognesi, A.; et al. [Cu(TPMA)(Phen)]$(ClO4)_2$: Metallodrug Nanocontainer Delivery and Membrane Lipidomics of a Neuroblastoma Cell Line Coupled with a Liposome Biomimetic Model Focusing on Fatty Acid Reactivity. *ACS Omega* **2018**, *3*, 15952–15965. [CrossRef] [PubMed]
21. Sigman, D.S.; Graham, D.R.; D'Aurora, V.; Stern, A.M. Oxygen-dependent cleavage of DNA by the 1,10-phenanthroline. cuprous complex. Inhibition of Escherichia coli DNA polymerase I. *J. Biol. Chem.* **1979**, *254*, 12269–12272. [CrossRef]
22. Santini, C.; Pellei, M.; Gandin, V.; Porchia, M.; Tisato, F.; Marzano, C. Advances in Copper Complexes as Anticancer Agents. *Chem. Rev.* **2013**, *114*, 815–862. [CrossRef]
23. Fantoni, N.Z.; McGorman, B.; Molphy, Z.; Singleton, D.; Walsh, S.; El-Sagheer, A.H.; McKee, V.; Brown, T.; Kellett, A. Development of Gene-Targeted Polypyridyl Triplex-Forming Oligonucleotide Hybrids. *ChemBioChem* **2020**, *21*, 3563–3574. [CrossRef] [PubMed]
24. Fantoni, N.Z.; Molphy, Z.; O'Carroll, S.; Menounou, G.; Mitrikas, G.; Krokidis, M.G.; Chatgilialoglu, C.; Colleran, J.; Banasiak, A.; Clynes, M.; et al. Polypyridyl-Based Copper Phenanthrene Complexes: Combining Stability with Enhanced DNA Recognition. *Chem.-A Eur. J.* **2020**, *27*, 971–983. [CrossRef]
25. Lauria, T.; Slator, C.; McKee, V.; Müller, M.; Stazzoni, S.; Crisp, A.L.; Carell, T.; Kellett, A. A Click Chemistry Approach to Developing Molecularly Targeted DNA Scissors. *Chem.–A Eur. J.* **2020**, *26*, 16782–16792. [CrossRef]
26. Panattoni, A.; El-Sagheer, A.H.; Brown, T.; Kellett, A.; Hocek, M. Oxidative DNA Cleavage with Clip-Phenanthroline Triplex-Forming Oligonucleotide Hybrids. *ChemBioChem* **2019**, *21*, 991–1000. [CrossRef] [PubMed]
27. Fantoni, N.Z.; El-Sagheer, A.H.; Brown, T. A Hitchhiker's Guide to Click-Chemistry with Nucleic Acids. *Chem. Rev.* **2021**, *121*, 7122–7154. [CrossRef] [PubMed]
28. Larragy, R.; Fitzgerald, J.; Prisecaru, A.; McKee, V.; Leonard, P.; Kellett, A. Protein engineering with artificial chemical nucleases. *Chem. Commun.* **2015**, *51*, 12908–12911. [CrossRef] [PubMed]

29. Fojta, M.; Kubičárová, T.; Paleček, E. Cleavage of Supercoiled DNA by Deoxyribonuclease I in Solution and at the Electrode Surface. *Electroanalysis* **1999**, *11*, 1005–1012. [CrossRef]
30. Ghosh, K.; Kumar, P.; Tyagi, N.; Singh, U.P.; Goel, N. Synthesis, structural characterization and DNA interaction studies on a mononuclear copper complex: Nuclease activity via self-activation. *Inorg. Chem. Commun.* **2011**, *14*, 489–492. [CrossRef]
31. Hirohama, T.; Kuranuki, Y.; Ebina, E.; Sugizaki, T.; Arii, H.; Chikira, M.; Selvi, P.T.; Palaniandavar, M. Copper(II) complexes of 1,10-phenanthroline-derived ligands: Studies on DNA binding properties and nuclease activity. *J. Inorg. Biochem.* **2005**, *99*, 1205–1219. [CrossRef]
32. Molphy, Z.; Prisecaru, A.; Slator, C.; Barron, N.; McCann, M.; Colleran, J.; Chandran, D.; Gathergood, N.; Kellett, A. Copper Phenanthrene Oxidative Chemical Nucleases. *Inorg. Chem.* **2014**, *53*, 5392–5404. [CrossRef]
33. Kellett, A.; Molphy, Z.; Slator, C.; McKee, V.; Farrell, N.P. Molecular methods for assessment of non-covalent metallodrug–DNA interactions. *Chem. Soc. Rev.* **2019**, *48*, 971–988. [CrossRef] [PubMed]
34. Fojta, M.; Kubičárová, T.; Paleček, E. Electrode potential-modulated cleavage of surface-confined DNA by hydroxyl radicals detected by an electrochemical biosensor. *Biosens. Bioelectron.* **2000**, *15*, 107–115. [CrossRef]
35. Labuda, J.; Bučková, M.; Vaníčková, M.; Mattusch, J.; Wennrich, R. Voltammetric Detection of the DNA Interaction with Copper Complex Compounds and Damage to DNA. *Electroanalysis* **1999**, *11*, 101–107. [CrossRef]
36. Paleček, E.; Fojta, M.; Tomschik, M.; Wang, J. Electrochemical biosensors for DNA hybridization and DNA damage. *Biosens. Bioelectron.* **1998**, *13*, 621–628. [CrossRef]
37. Banasiak, A.; Cassidy, J.; Colleran, J. A novel quantitative electrochemical method to monitor DNA double-strand breaks caused by a DNA cleavage agent at a DNA sensor. *Biosens. Bioelectron.* **2018**, *117*, 217–223. [CrossRef] [PubMed]
38. Jopp, M.; Becker, J.; Becker, S.; Miska, A.; Gandin, V.; Marzano, C.; Schindler, S. Anticancer activity of a series of copper(II) complexes with tripodal ligands. *Eur. J. Med. Chem.* **2017**, *132*, 274–281. [CrossRef] [PubMed]
39. Molphy, Z.; Slator, C.; Chatgilialoglu, C.; Kellett, A. DNA oxidation profiles of copper phenanthrene chemical nucleases. *Front. Chem.* **2015**, *3*. [CrossRef] [PubMed]
40. Humphreys, K.J.; Johnson, A.E.; Karlin, K.D.; Rokita, S.E. Oxidative strand scission of nucleic acids by a multinuclear copper(II) complex. *JBIC J. Biol. Inorg. Chem.* **2002**, *7*, 835–842. [CrossRef]
41. Humphreys, K.J.; Karlin, K.D.; Rokita, S.E. Efficient and Specific Strand Scission of DNA by a Dinuclear Copper Complex: Comparative Reactivity of Complexes with Linked Tris(2-pyridylmethyl)amine Moieties. *J. Am. Chem. Soc.* **2002**, *124*, 6009–6019. [CrossRef]
42. Kraft, S.S.N.; Bischof, C.; Loos, A.; Braun, S.; Jafarova, N.; Schatzschneider, U. A [4+2] mixed ligand approach to ruthenium DNA metallointercalators [Ru(tpa)(N–N)](PF6)2 using a tris(2-pyridylmethyl)amine (tpa) capping ligand. *J. Inorg. Biochem.* **2009**, *103*, 1126–1134. [CrossRef] [PubMed]
43. Xu, W.; Louka, F.R.; Doulain, P.E.; Landry, C.A.; Mautner, F.A.; Massoud, S.S. Hydrolytic cleavage of DNA promoted by cobalt(III)–tetraamine complexes: Synthesis and characterization of carbonatobis[2-(2-pyridylethyl)]-(2-pyridylmethyl)aminecobalt(III) perchlorate. *Polyhedron* **2009**, *28*, 1221–1228. [CrossRef]
44. Prisecaru, A.; McKee, V.; Howe, O.; Rochford, G.; McCann, M.; Colleran, J.; Pour, M.; Barron, N.; Gathergood, N.; Kellett, A. Regulating Bioactivity of Cu^{2+} Bis-1,10-phenanthroline Artificial Metallonucleases with Sterically Functionalized Pendant Carboxylates. *J. Med. Chem.* **2013**, *56*, 8599–8615. [CrossRef] [PubMed]
45. Ren, H.; Wu, J.; Xi, C.; Lehnert, N.; Major, T.; Bartlett, H.R.; Meyerhoff, M.E. Electrochemically Modulated Nitric Oxide (NO) Releasing Biomedical Devices via Copper(II)-Tri(2-pyridylmethyl)amine Mediated Reduction of Nitrite. *ACS Appl. Mater. Interfaces* **2014**, *6*, 3779–3783. [CrossRef]
46. Goss, C.A.; Abruna, H.D. Spectral, electrochemical and electrocatalytic properties of 1,10-phenanthroline-5,6-dione complexes of transition metals. *Inorg. Chem.* **1985**, *24*, 4263–4267. [CrossRef]
47. Evans, D.H.; Griffith, D.A. Effect of metal ions on the electrochemical reduction of some heterocyclic quinones. *J. Electroanal. Chem. Interfacial Electrochem.* **1982**, *136*, 149–157. [CrossRef]
48. Kou, Y.-Y.; Xu, G.-J.; Gu, W.; Tian, J.-L.; Yan, S.-P. Synthesis and pH-sensitive redox properties of 1,10-phenanthroline-5,6-dione complexes. *J. Co-ord. Chem.* **2008**, *61*, 3147–3157. [CrossRef]
49. Cory, R.M.; McKnight, D.M. Fluorescence spectroscopy reveals ubiquitous presence of oxidized and reduced quinones in dissolved organic matter. *Environ. Sci. Technol.* **2005**, *39*, 8142–8149. [CrossRef] [PubMed]
50. Alcalde, J.M.; Molero, L.; Cañete, A.; Del Rio, R.; Del Valle, A.M.; Mallavia, R.; Armijo, F. Electrochemical and spectroscopic properties of indolizino[1,2-B] quinole derivates. *J. Chil. Chem. Soc.* **2013**, *58*, 1976–1979. [CrossRef]
51. Mahadevan, S.; Palaniandavar, M. Spectroscopic and Voltammetric Studies on Copper Complexes of 2,9-Dimethyl-1,10-phenanthrolines Bound to Calf Thymus DNA. *Inorg. Chem.* **1998**, *37*, 693–700. [CrossRef]
52. Lucio, A.J.; Shaw, S.K. Pyridine and Pyridinium Electrochemistry on Polycrystalline Gold Electrodes and Implications for CO_2 Reduction. *J. Phys. Chem. C* **2015**, *119*, 12523–12530. [CrossRef]
53. Haynes, W.M. *CRC Handbook of Chemistry and Physics*, 95th ed.; Haynes, W.M., Lide, D.R., Bruno, T.J., Eds.; CRC Press: Boca Raton, FL, USA, 2014.
54. Bard, A.J.; Faulkner, L.R. *Electrochemical Methods. Fundamentals and Applications*, 2nd ed.; John Wiley & Sons: New York, NY, USA, 2000.

55. Carter, M.T.; Rodriguez, M.; Bard, A.J. Voltammetric studies of the interaction of metal chelates with DNA. 2. Tris-chelated complexes of cobalt(III) and iron(II) with 1,10-phenanthroline and 2,2′-bipyridine. *J. Am. Chem. Soc.* **1989**, *111*, 8901–8911. [CrossRef]
56. Chikira, M.; Ng, C.H.; Palaniandavar, M. Interaction of DNA with Simple and Mixed Ligand Copper(II) Complexes of 1,10-Phenanthrolines as Studied by DNA-Fiber EPR Spectroscopy. *Int. J. Mol. Sci.* **2015**, *16*, 22754–22780. [CrossRef] [PubMed]
57. Draksharapu, A.; Boersma, A.J.; Leising, M.; Meetsma, A.; Browne, W.R.; Roelfes, G. Binding of copper(II) polypyridyl complexes to DNA and consequences for DNA-based asymmetric catalysis. *Dalton Trans.* **2014**, *44*, 3647–3655. [CrossRef] [PubMed]
58. Pang, D.-W.; Abruña, H.D. Micromethod for the Investigation of the Interactions between DNA and Redox-Active Molecules. *Anal. Chem.* **1998**, *70*, 3162–3169. [CrossRef]
59. Bolton, J.L.; Dunlap, T. Formation and Biological Targets of Quinones: Cytotoxic versus Cytoprotective Effects. *Chem. Res. Toxicol.* **2016**, *30*, 13–37. [CrossRef]
60. Pinto, A.V.; De Castro, S.L. The Trypanocidal Activity of Naphthoquinones: A Review. *Molecules* **2009**, *14*, 4570–4590. [CrossRef] [PubMed]
61. Steel, A.B.; Herne, T.M.; Tarlov, M.J. Electrochemical Quantitation of DNA Immobilized on Gold. *Anal. Chem.* **1998**, *70*, 4670–4677. [CrossRef] [PubMed]
62. Santra, B.K.; Reddy, P.A.; Neelakanta, G.; Mahadevan, S.; Nethaji, M.; Chakravarty, A.R. Oxidative cleavage of DNA by a dipyridoquinoxaline copper(II) complex in the presence of ascorbic acid. *J. Inorg. Biochem.* **2002**, *89*, 191–196. [CrossRef]
63. Marshall, L.E.; Graham, D.R.; Reich, K.A.; Sigman, D.S. Cleavage of deoxyribonucleic acid by the 1,10-phenanthroline-cuprous complex. Hydrogen peroxide requirement and primary and secondary structure specificity. *Biochemistry* **1981**, *20*, 244–250. [CrossRef]
64. Ausubel, F.M. *Current Protocols in Molecular Biology*; Ausubel, F.M., Brent, R., Kingston, R.E., Moore, D.D., Smith, J.A., Seidman, J.G., Struhl, K., Eds.; John Wiley & Sons: New York, NY, USA, 1989.
65. Pividori, M.I.; Merkoçi, A.; Alegret, S. Electrochemical genosensor design: Immobilisation of oligonucleotides onto transducer surfaces and detection methods. *Biosens. Bioelectron.* **2000**, *15*, 291–303. [CrossRef]

MDPI

Review

Oxidative Crosslinking of Peptides and Proteins: Mechanisms of Formation, Detection, Characterization and Quantification

Eduardo Fuentes-Lemus [1], Per Hägglund [1], Camilo López-Alarcón [2] and Michael J. Davies [1,*]

[1] Department of Biomedical Sciences, Panum Institute, University of Copenhagen, 2200 Copenhagen, Denmark; eduardo.lemus@sund.ku.dk (E.F.-L.); pmh@sund.ku.dk (P.H.)

[2] Departamento de Química Física, Facultad de Química y de Farmacia, Pontificia Universidad Catolica de Chile, Santiago 7820436, Chile; clopezr@uc.cl

* Correspondence: davies@sund.ku.dk; Tel.: +45-2364-9445

Abstract: Covalent crosslinks within or between proteins play a key role in determining the structure and function of proteins. Some of these are formed intentionally by either enzymatic or molecular reactions and are critical to normal physiological function. Others are generated as a consequence of exposure to oxidants (radicals, excited states or two-electron species) and other endogenous or external stimuli, or as a result of the actions of a number of enzymes (e.g., oxidases and peroxidases). Increasing evidence indicates that the accumulation of unwanted crosslinks, as is seen in ageing and multiple pathologies, has adverse effects on biological function. In this article, we review the spectrum of crosslinks, both reducible and non-reducible, currently known to be formed on proteins; the mechanisms of their formation; and experimental approaches to the detection, identification and characterization of these species.

Keywords: crosslink; dimerization; protein oxidation; radicals; di-tyrosine; di-tryptophan; disulfides; thiols; aggregation; proteomics; mass spectrometry

Citation: Fuentes-Lemus, E.; Hägglund, P.; López-Alarcón, C.; Davies, M.J. Oxidative Crosslinking of Peptides and Proteins: Mechanisms of Formation, Detection, Characterization and Quantification. *Molecules* **2022**, *27*, 15. https://doi.org/10.3390/molecules27010015

Academic Editor: Chryssostomos Chatgilialoglu

Received: 29 November 2021
Accepted: 18 December 2021
Published: 21 December 2021

1. Introduction

The formation of covalently linked peptides and proteins plays a key role in many biological processes, both physiologically and pathologically. These can be formed intentionally, such as in the oxidative folding of nascent proteins within mammalian cells in the endoplasmic reticulum or Golgi involving the generation of disulfide bonds from two cysteine (Cys) residues and in the assembly of insect exoskeletons via the crosslinking of two tyrosine (Tyr) residues, or as a result of accidental exposure to oxidizing species (low-molecular mass or enzymes) that chemically link two protein sites. These crosslinks can be formed between different sites within the same molecule (intramolecular or intrachain crosslinks), between two different chains in a single molecule (e.g., the interchain crosslinks in mammalian insulins), or between two separate species (intermolecular crosslinks). Some of these crosslinks play a key role in stabilizing or maintaining proteins structures and can be essential to functional activity [1], whereas others have negative effects of biological function (e.g., altered turnover, lifetime or activity) [2]. Whilst some crosslinks appear to be benign and devoid of adverse effects and end up as targets of catabolic processes (e.g., degradation by proteasomes, lysosomes, other proteases), others are strongly associated with adverse effects and are implicated (in some cases, causally) in the development of pathologies (e.g., [3,4]).

Whilst it is well established that crosslink formation can have major effects on biological systems—either positively or negatively—our knowledge of the full complement of crosslinks formed in biological systems (the 'crosslink-ome', 'X-link-ome') is far from complete [2]. The number of disulfide-containing proteins is large and relatively well defined, with these being particularly abundant in proteins found in biological fluids (e.g., human serum albumin in plasma contains 17 disulfides), in structural proteins such as receptors (e.g., the low-density lipoprotein receptor, LDLR, contains 30 unique intradomain disulfides [5]) and extracellular matrix proteins (e.g., laminin contains ~200 disulfides). Disulfides are also present at more modest levels in intracellular proteins, where they can function as structural elements, allosteric effectors, or be involved in catalytic cycles [1,6]. For proteins containing many disulfides, ensuring correct pairing of Cys residues into disulfides during synthesis and assembly is a complex problem [7].

Unlike the relatively well-characterized complement of disulfide-containing proteins, the number, occurrence and sites of other potential crosslinks is poorly understood, but the subject of active research. Recent developments in our understanding of the chemistry of crosslink formation, and the development of new and more sensitive methods for their detection is driving advances in this field. This article summarizes recent developments, with a focus on crosslinks formed by oxidation (either enzyme-mediated or via the reactions of low-molecular-mass oxidants). Crosslinks formed by deliberately added reagents (e.g., dicarbonyl and related crosslinking agents, such as glutaraldehyde) or those arising from glycation/glycoxidation reactions of sugars are not discussed for reasons of space. These have been discussed elsewhere [8–10].

2. Enzymatic Protein Crosslinking

Multiple enzymes can mediate the crosslinking of proteins, with a few key examples briefly summarized below. Enzyme-generated crosslinks are critical to the formation of many three-dimensional structures as these provide strength and rigidity, if biologically required. Examples include crosslinks formed within the extracellular matrix (ECM) of most, if not all, tissues, such as those formed between matrix proteins, and particularly collagens by the copper-containing lysyl oxidase (LOX) and LOX-like (LOXL) enzymes (reviewed in [11]). LOX oxidizes specific lysine (Lys) and hydroxylysine residues to carbonyls that undergo subsequent reactions to crosslink collagens (e.g., types I and III) and elastin [11–14]. In contrast, the LOXL family of enzymes acts on collagen type IV and drives the assembly of basement membranes [11,15]. Other enzymes also contribute to collagen crosslinking in the ECM with peroxidasin, a member of the heme peroxidase superfamily, mediating the formation of highly specific methionine (Met) to Lys crosslinks within the NC1 domains on collagen via generation of the oxidant hypobromous acid (HOBr). This species reacts rapidly with the Met residue to form an intermediate that then reacts with a suitably positioned Lys residue [16,17] (see also below). This type of crosslinking has been reported across many species [18]. Other members of the peroxidase superfamilies (e.g., horseradish peroxidase, myeloperoxidase, laccase) can also generate crosslinks via enzyme-mediated oxidation of substrates to radicals which then undergo radical–radical coupling. A classic example is oxidative coupling of Tyr and a wide range of other phenols via phenoxyl radical generation [19–21].

Isopeptide crosslinks involving reactions of the Lys side-chain amine with the carboxylic acid of aspartic acid (Asp), or the amides of asparagine (Asn) and glutamine (Gln) residues can be formed spontaneously or enzymatically [22–24]. These can be formed intracellularly, or within the ECM, with their function being associated in the latter case not only with enhanced ECM rigidity, but also in the attachment of bacterial pathogens to host tissue collagens and fibrinogens [25]. Enzymatic isopeptide bonds, formed by factor XIIIa (FXIIIa) between Lys and Gln residues, are critical to the formation of fibrin assemblies in blood clotting. Related crosslinks are generated by tissue transglutaminase enzymes [26]. These crosslinks are used to attach targeting moieties to specific proteins, including ubiquitin to proteins destined for proteosomal degradation [27], and small-ubiquitin-like modifier proteins (SUMOs, via the C-terminal glycine, Gly, residue) in the case of protein transport, targeting and regulation. These crosslinks involve the initial activation of the ubiquitin/SUMO, and subsequent transfer to the target, involving multiple activating (e.g., E1), conjugating (E2) and ligation (E3) enzymes [28]. Multiple glutamic, Glu and Gly residues have also been detected attached to Glu residues in the C-terminal regions of microtubulin, with this reported to be involved in microtubule function [29].

Covalent bonds are also widely and deliberately generated between proteins and co-factors to produce functional enzymes, with examples including the crosslinking of heme, flavin, pyridoxal, biotin, thiamine, molybdopterin and lipoic acid to proteins (see, for example, [30]). Whilst of considerable interest and importance, further discussion of these reactions lies outside the scope of this review, and subsequent sections are focused on oxidant-mediated reactions, though some of these enzyme-mediated crosslinks involve the formation and reactions of oxidants (e.g., by peroxidases and peroxidasin).

An overview of the crosslinks that are discussed further in this review is presented in Figure 1.

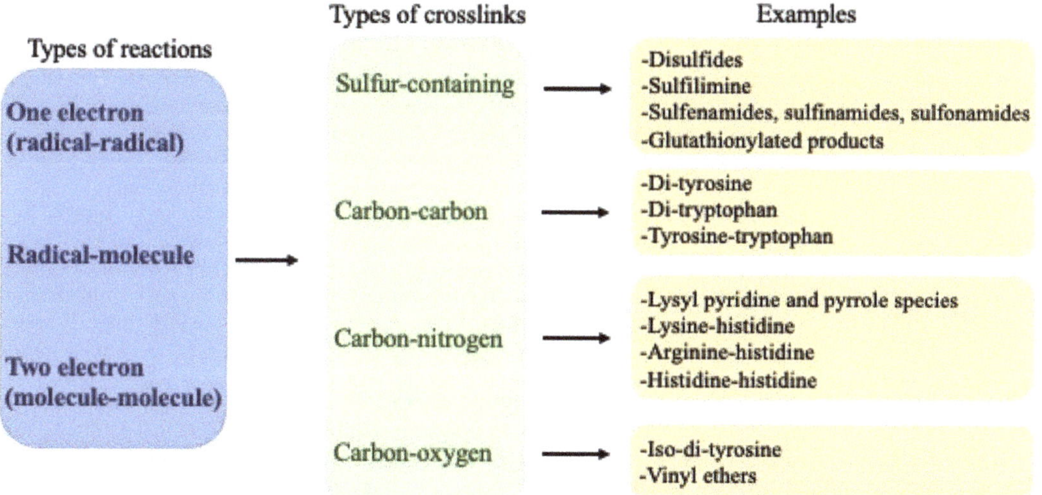

Figure 1. Overview of crosslinks formed on proteins, their nature and mechanisms of formation.

3. One-Electron (Radical–Radical) Reactions

Dimerization of two radicals to form a new covalent bond is typically a very fast process due to the low energy barriers for such reactions. Therefore, they are a major source of crosslinks in peptides and proteins when the radical flux is high and there are limited competing reactions. Most carbon-centered protein radicals ($P^{\bullet}$) formed from aliphatic side-chains by hydrogen–atom abstraction reactions react rapidly with O_2 at diffusion-controlled rates ($k \sim 10^9$ M^{-1} s^{-1}) to give peptide or protein peroxyl radicals ($P\text{-}OO^{\bullet}$) [31]. The rapidity of these reactions limits direct reactions of two $P^{\bullet}$, except in circumstances where the O_2 concentration is low. This is of biological relevance, as hypoxia is a common phenomenon, with endogenous levels of O_2 being typically in the range 3–70 μM [32]. However, lower concentrations are present in situations where demand is great (e.g., high metabolic rates) or perfusion is poor (e.g., in the core of many solid tumors), thereby limiting $P\text{-}OO^{\bullet}$ formation and allowing (P-P) dimer formation [33]. For the limited number of $P^{\bullet}$, where reaction with O_2 is slow or modest, as is the case for Cys-derived thiyl radicals ($RS^{\bullet}$, $k < 10^7$ M^{-1} s^{-1} [34]), tryptophan (Trp) indolyl radicals ($Trp^{\bullet}$, $k < 4 \times 10^6$ M^{-1} s^{-1} [35,36]) and Tyr phenoxyl radicals ($Tyr^{\bullet}$, $k < 10^3$ M^{-1} s^{-1} [37]), formation of disulfides (cystine) from two $RS^{\bullet}$, di-tyrosine from two $Tyr^{\bullet}$, di-tryptophan from two $Trp^{\bullet}$, and crossed dimers between these (e.g., Tyr–Trp) can be generated. The structure and mechanisms of the formation of these species are discussed below.

Light, particularly of wavelengths >~280 nm, which are not absorbed by the ozone layer, can penetrate significantly into biological structures and be absorbed either directly by protein residues, particularly Trp, Tyr and cystine [38], or by other species with high extinction coefficients in the long wavelength UV or visible regions. Energy absorption by non-protein species can give rise to indirect protein oxidation via the formation of excited states (e.g., singlet oxygen, $^1O_2$ and reactive triplets) and/or radicals [38]. Direct UV absorption by proteins can form RS• from homolysis of the –S–S– bond of cystine (with C-S cleavage being an alternative pathway), and Tyr and Trp radicals by photo-ionization of these side-chains. These species can then give rise to crosslinks.

4. Radical–Molecule Reactions

Radical–molecule reactions appear to be a limited pathway for the formation of protein crosslinks, due to the absence of double bonds to which radicals might add in proteins, and limited stability of adducts to aromatic rings. Notable exceptions are the rare amino acids dehydroalanine (DHA; 2-aminoacrylic acid) and dehydroaminobutyric acid (DHB; 2-aminocrotonic acid). These contain a double bond between the α- and β-carbons of the side-chain and are non-proteinogenic species [39], with these being generated via elimination reactions of serine residues (Ser), phospho-Ser and selenocysteine (Sec) residues (in the case of DHA) [39], and from threonine (Thr) and phospho-Thr (in the case of DHB) [40]. DHA can also be formed via cleavage of the carbon–sulfur bonds of the disulfide cystine, via mechanisms involving RS• or nucleophilic elimination reactions [39].

Although radical addition to double bonds is typically rapid and energetically favorable due to low energy barriers, these reactions are rare as the concentrations of both DHA and DHB (with the former more abundant) and the radicals that might undergo addition with them are very low. Nevertheless, some examples are known for radicals that have relatively long lifetimes and modest rates of reaction with O_2 (i.e., Cys thiyl, Tyr phenoxyl, Trp indolyl) [41].

5. Two-Electron (Molecule–Molecule) Reactions

Reactions between two molecules are typically much slower than between two radicals or radical–molecule reactions. However, the concentration of the reactants is often much higher than for reactive intermediates, and consequently, the overall rates of these reactions may be significant—and the yield of products greater—than for the processes outlined above. These reactions are therefore major sources of protein crosslinks. The rate constants for these reactions would be expected to vary enormously—though quantitative data is lacking for most systems—with some reactions involving unstable species (e.g., sulfenic acids (RSOH), S-nitrosothiols (RSNO), unsaturated aldehydes/ketones, quinones) being relatively rapid (i.e., occurring over seconds/minutes).

6. Types of Crosslinks Detected within and between Proteins and Peptides

The following sections and Table 1 summarize various types of crosslinks that have been detected within and between peptides and proteins, the nature of these species, their reversibility, mechanisms of formation and, subsequently, methods available to detect, identify, characterize and quantify these species.

6.1. Sulfur-Containing Crosslinks

6.1.1. Sulfur–Sulfur Crosslinks (Disulfides)

Sulfur–sulfur crosslinks are the most common form of crosslinks present in biological systems, generated by multiple enzymatic and non-enzymatic pathways. Many endogenous (native) disulfides are formed during or shortly after protein synthesis in the endoplasmic reticulum or Golgi, via enzyme-guided reactions. During this process, erroneous (incorrect) disulfides appear to be formed to a significant extent, as judged by the presence of multiple enzymes/repair systems that allow reshuffling of incorrect linkages (e.g., [7,88]). These processes have been widely researched and are not discussed further here.

Multiple other pathways can also generate non-intended disulfides, including thiol–disulfide exchange reactions in which a thiolate anion (RS^-, the ionized and more reactive form of a thiol) reacts with a disulfide, with the formation of a new disulfide and release of a thiol [89] (Figure 2, top section). These reactions are typically slow, as there is little thermodynamic driving force for the reaction, but they can occur at significant rates if the original disulfide is subject to significant strain that is released on reaction with the incoming RS^- [89,90]. The rate of these reactions can also be enhanced, to a very significant extent, within the active sites of enzymes, with di-thiol/disulfide cycles being a common feature of a number of important proteins (e.g., thioredoxins [91]).

Table 1. Examples of major non-disulfide protein crosslinks generated during non-enzymatic oxidative processes and methodologies employed to characterize them.

Crosslinked Residues	Protein(s)	Chemical Nature and/or Mechanism of Formation of the Crosslink	Method(s)	Refs
Tyr-Cys	a) Myoglobin b) Galactose oxidase c) Cysteine dioxygenase	1) Michael addition from thiols (Cys) to oxidized Tyr species (a) 2) Thioether bridge (C-S) (b and c)	Mass spectrometry (a) X-ray crystallography (b, c)	[42–44]
Trp-Cys	Human growth hormone (hGH)	1) Michael addition from N (Trp indole) to DHA (formed from Cys) 2) Thioether bridge (C-S)	Mass spectrometry	[41]
Met-Hydroxy-lysine	Collagen IV	Formation of S=N bridge (sulfilimine bond) induced by peroxidasin/HOBr	Mass spectrometry	[17]
Lys-Cys	Transaldolase	Nitrogen–oxygen–sulfur (NOS) link/redox switch	X-ray crystallography	[45]
Cys-Ser	a) Human growth hormone b) Tyrosine phosphatase 1B	1) Formation of a vinyl ether between Ser and Cys that result in the elimination of the thiol group from Cys (a) 2) Sulfenyl amide (S–N bridge) between Cys-OH and main-chain amide of Ser residue (b)	Mass spectrometry (a) X-ray crystallography (b)	[41,46]
Cys-Phe	hGH	Crosslink between thioaldehyde from Cys and dehydrophenylalanine generated from Phe	Mass spectrometry	[41]
Cys-DHA Cys-DHB	Lens proteins (βB1, βB2, βA3, βA4 and γS crystallins)	Nucleophilic addition from Cys (GSH) to DHA or DHB	Mass spectrometry	[47]
Tyr-Gly	Insulin	Michael addition of primary amines (N-terminal Gly) to oxidized Tyr species	Mass spectrometry	[48]
Trp-Gly	Matrilysin (Matrix metalloproteinase 7)	Crosslink between 3-chloroindolenine (3-Cl-Trp) and the main-chain amide adjacent to a Gly	NMR spectroscopy	[49]
Tyr-His	Insulin	Michael addition from His to oxidized Tyr	Mass spectrometry	[48]

Table 1. *Cont.*

Crosslinked Residues	Protein(s)	Chemical Nature and/or Mechanism of Formation of the Crosslink	Method(s)	Refs
Tyr-Tyr (selected data)	Isolated proteins including: α-lactalbumin, caseins, glucose 6-phosphate dehydrogenase, lysozyme, fibronectin, laminins, tropoelastin, cAMP receptor protein, α-synuclein, calmodulin, insulins, hemoglobin, human Δ25 centrin 2. Human lipoproteins Human plasma proteins, including those from people with chronic renal failure Human atherosclerotic lesions Erythrocytes exposed to H_2O_2 Brain proteins (amyloid-beta and α-synuclein) from Alzheimer's subjects Lipofuscin from aged human brain Urine from people with diabetes Human lens proteins Bacterial spore coat proteins Parasite oocysts	C–C and/or C–O crosslinks via radical–radical reactions	Western blotting UPLC/HPLC with various detection methods Mass spectrometry	[48,50–77]
Trp-Trp	a) α-Lactalbumin b) Superoxide dismutase 1 (hSOD) c) Lysozyme-hSOD d) αB-Crystallin e) Fibronectin	C–C or C–N crosslinks via radical–radical reactions	Mass spectrometry	[50,57,78–80]
Tyr-Trp	a) Cytochrome c peroxidase b) α-Lactalbumin c) Glucose 6-phosphate dehydrogenase d) Lysozyme e) β-Crystallin f) Human cataractous lenses g) Fibronectin	C–C (or C–O and C–N) crosslinks via radical–radical reactions	X-ray crystallography (a) Mass spectrometry (b–g)	[50,53,56,57, 80,81]
His-His	a) Immunoglobulin G1 b) Immunoglobulin G4 c) *N*-Ac-His	Nucleophilic addition of His to oxidized His	Mass spectrometry (a,b) NMR (c)	[82–85]
His-Arg	Ribonuclease A (RNAse)	Nucleophilic addition of Arg to oxidized His	Mass spectrometry	[86]
His-Lys	Immunoglobulin G1	Nucleophilic addition of Lys to oxidized His	Mass spectrometry	[82,84]
His-Cys	Immunoglobulin G1	Nucleophilic addition of Cys to oxidized His	Mass spectrometry	[84]
Tyr-Lys	a) RNAse b) Interferon beta-1a c) Insulin	Michael addition of Lys to oxidized Tyr	Mass spectrometry	[48,86,87]

Standard thiol-disulfide exchange

R-S⁻ + R'-S-S-R" ⟶ R-S-S-R' + HS-R"

or

R-SH + R'-S-S-R" ⟶ R-S-S-R' + HS-R"

Note: Thiolate (R-S⁻) is a better nucleophile than the thiol (R-SH)

Mechanism: S_N2-type nucleophilic substitution via formation of a linear S-S-S like transition state

Non-catalyzed reaction: $k = 0.1 - 10\ M^{-1}\ s^{-1}$ at pH 7

Catalyzed by oxidoreductase enzymes (e.g. thioredoxin, glutaredoxin): $k = 10^4 - 10^6\ M^{-1}\ s^{-1}$

Oxidant-mediated thiol-disulfide exchange

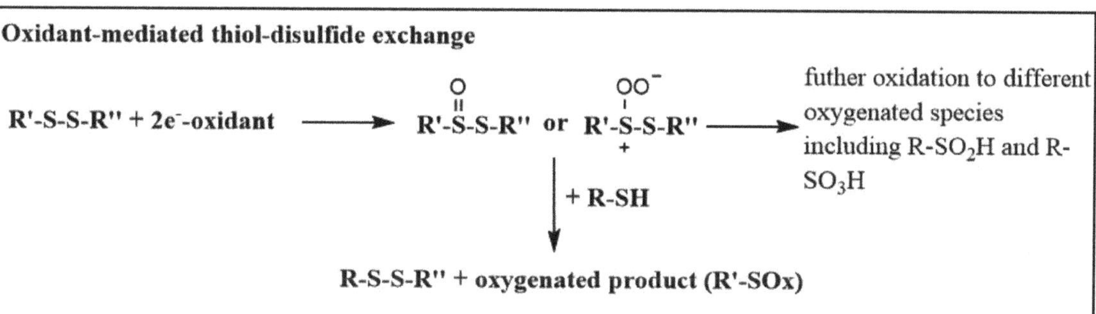

Mechanism: Nucleophilic attack from the thiol (R-SH, or thiolate R-S⁻) to the electron deficient sulfur atom in R'-S(O)-S-R" or R'-S⁺(OO⁻)-S-R" species

Figure 2. Mechanisms of traditional 'thiol–disulfide exchange' (top) and 'oxidant-mediated thiol–disulfide exchange' (bottom) reactions.

Rapid formation of new disulfide bonds has also been reported in so-called 'oxidant-mediated thiol–disulfide exchange reactions', where the original disulfide is initially oxidized, by a range of different species, to a more reactive intermediate (e.g., a thiosulfinate or peroxidic intermediate) that subsequently undergoes rapid reaction with another RS^- [92–94] (Figure 2, lower section). This type of process gives rise to glutathione (and other thiol) adducts to disulfide-containing peptides and proteins (i.e., glutathionylated species [92]), and also new (intermolecular) protein–protein crosslinks [93,94]. In some cases, the residues involved and the exact sites of the new disulfides have been determined by LC-MS peptide mass mapping [93,94] (see also below). The more rapid rate of these reactions relative to 'native' thiol–disulfide exchange arises from the activation of the disulfide via the initial oxidation, which typically involves the formation of an electron-deficient center.

New disulfides can also be formed from thiols via a similar activation (oxidation) process. Thus, reaction of thiols with H_2O_2, HOCl, HOBr, HOSCN, HNO and ONOOH (amongst others) yields short-lived intermediates that undergo subsequent rapid reaction with another RS^-/RSH to give a disulfide (Figure 3). These reactions are facile and rapid in most cases as the displaced group is a stable anion (e.g., HO^-, Cl^-, Br^-, etc.). These are very well established with RS-OH (sulfenic acids) and RS-NO [95], but these are some of the least reactive members of this family, as Cl^- and Br^- are, for example, much better leaving groups than HO^-. Indeed, isolation of RS-Cl, RS-Br and RS-SCN species is difficult due to their high reactivity with nucleophiles, including water. In the latter case, hydrolysis gives RS-OH, which undergoes similar reactions. Some RS-OH and RS-NO species are sufficiently long-lived to be detectable on proteins, though their lifetimes are structure- and environment-dependent [96].

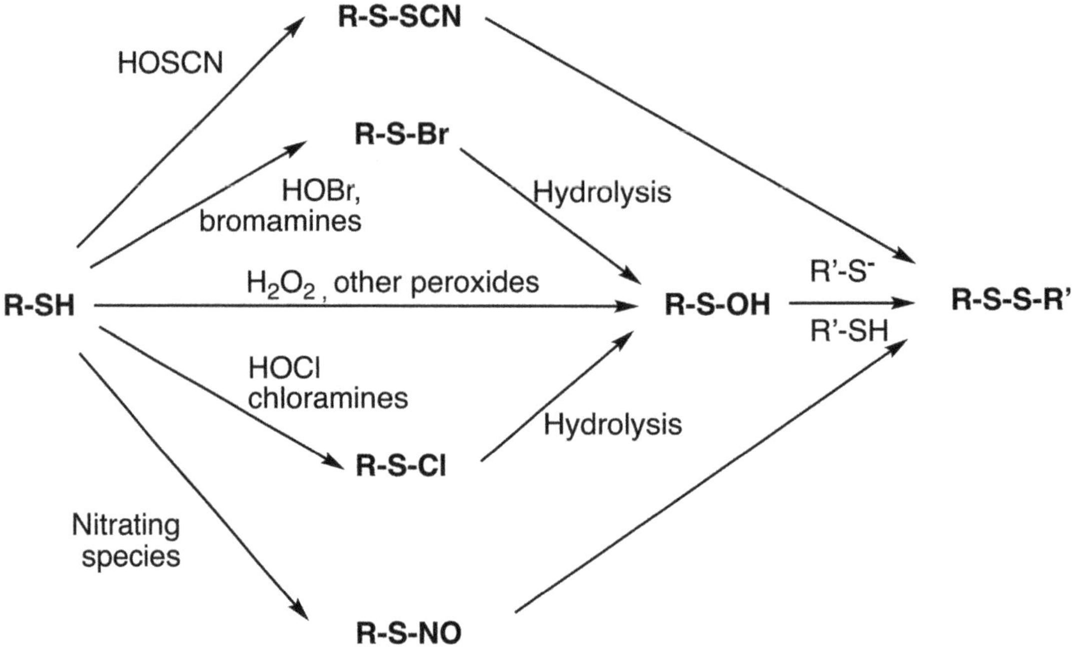

Figure 3. Generation of crosslinks via oxidized thiol residues. Similar reactions of the 'activated' thiols (RS–OH, RS–Cl, RS–Br, RS–SCN, RS–NO) can occur with nitrogen nucleophiles (e.g., RNH_2) to give new S–N bonded species (see text for further details).

Disulfides can also be formed from rapid termination reactions of two $RS^{\bullet}$. However, reaction of $RS^{\bullet}$ with another $R'S^-$ to give the corresponding disulfide radical anion ($RSS'R^{\bullet-}$) is also very rapid, and as the concentration of $R'S^-$ (either on a protein, or a low molecular mass species such as glutathione) is typically much higher than that of $RS^{\bullet}$, the radical anion pathway usually predominates. The resulting $RSS'R^{\bullet-}$ undergoes rapid electron transfer with electron acceptors, including O_2, resulting in the formation of a new disulfide (which may be either a protein–protein species, or a glutathionylated protein) and the superoxide radical anion, $O_2^{\bullet-}$.

6.1.2. Sulfur–Nitrogen (S–N) Crosslinks (Sulfimines, Sulfenamides, Sulfinamides and Sulfonamides)

A wide range of S–N crosslinks are known, with these arising predominantly from molecule reactions of activated sulfur centers with nitrogen nucleophiles, and particularly the side-chain amine of Lys (or hydroxy-Lys) residues (see also Figure 3). Reactions with other nitrogen nucleophiles are known, including the more reactive (when compared to the Lys side-chain) N-terminal amine of peptides or proteins. Reactions at (backbone) amide and guanidine (Arg side-chain) functions have also been reported (see below), but such examples are limited, as these are weaker nucleophiles.

Crosslinking can occur via the thioether group of methionine in the enzymatic (peroxidasin)-induced crosslinking of the NC1 domains of collagens. Reaction of a specific Met residue (Met^{93}) with HOBr formed by peroxidasin generates a transient bromosulfonium ion [–S(Br)$^+$–] that reacts with the amine group of a suitably positioned hydroxy-Lys residue (Hyl[211]) to give an interchain sulfilimine (–S=N–) crosslink. These crosslinks are critical for the generation of functional extracellular matrices in many organisms, with an absence of this enzymatic activity causing tissue dysfunction that is embryonically lethal in some species [16,17]. Reaction with the amine function occurs in competition with the reaction with water (acting as a nucleophile), with consequent methionine sulfoxide formation. This type of crosslink is not unique to peroxidasin-generated HOBr, with other Met-containing peptides undergoing similar sulfilimine crosslink formation with Lys residues, either intra- or intermolecularly [97]. These reactions are also not limited to HOBr, with HOCl, bromamines and chloramines (RNHBr and RNHCl species, respectively) also generating these species in competition with the sulfoxide [97]. In the case of HOCl, these species are usually formed in low yield. For the chloramines, initial oxidation of the sulfur center of the Met to give the chlorosulfonium ion [–S(Cl)$^+$–] appears to be followed by rapid reaction with the nitrogen atom of the original chloramine [97], indicating that these reactions can be highly specific, and that the [–S(Cl)$^+$–] species is short lived.

Related reactions of sulfenyl halides (RS–X) and related species formed from reaction of the –SH group of Cys or GSH with an oxidant (e.g., HOCl, HOBr, chloramines, bromamines, ONOOH) with amine nucleophiles have been reported to give rise to a large family of sulfenamides (RS–NHR′), sulfinamides [RS(O)–NHR′] and sulfonamides [RS(O)$_2$–NHR′]. These have been detected as both intra- and intermolecular crosslinks in both peptides and proteins [98–100]. A well-established example is glutathione sulfonamide (GSA) formation, which has been used as a biomarker of oxidative damage in both cells (e.g., [101,102]) and human tissues and fluids [103,104]. In this case, the linkage is formed intramolecularly with the amide nitrogen of the terminal Gly residue, resulting in an eight-membered ring species [105]. The formation of GSA occurs in competition with reaction of the RS-X with another RSH to give the disulfide (i.e., GSSG in the case of oxidation of GSH) with the ratio of GSA:GSSG dependent on the oxidant [98]. The disulfide is usually the major product, but the GSA yield can be high in some cases [98,105], including in some cells [101,102,106]. The formation of GSA requires three equivalents of oxidant, and the exact sequence of oxidation events is unclear (i.e., whether oxygenation at the sulfur occurs prior to, or after the formation of the S–N bond) [105]. The detection of sulfenamides and sulfinamides in other systems (e.g., S100A8 proteins [99] and peptides [100,107]) suggests that S–N bond formation may be the initial event, with oxidation at the sulfur occurring on reaction with the second and third oxidizing equivalents. These species are poorly reducible when compared to disulfides [99], and only ~50% of the GSH oxidized by HOCl is recovered (presumably from GSSG) on reduction [108]. Evidence has also been reported for model peptides for the involvement of the guanidine group of Arg residues in the formation of S–N crosslinks [100].

There is considerable evidence of the formation of an intramolecular sulfenamide (sulfenyl amide, S–N species) in the enzyme protein tyrosine phosphatase 1B (PTP1B), with the linkage formed between the sulfenic acid (RS–OH) form of the catalytic Cys residue, and the adjacent main chain amide of a Ser residue [46]. This modification appears to protect the Cys residue from irreversible oxidation to sulfinic (RSO$_2$H) or sulfonic (RSO$_3$H) acids and loss of enzyme activity. This species appears to be the stable, inactive 'resting state' form of the enzyme, with the sulfenyl amide being readily reversed by cellular thiols such as GSH, with conversion back to its catalytically active Cys form [46].

6.1.3. Sulfur–Carbon (S–C) Crosslinks

Sulfur–carbon crosslinks can be generated via the addition of RS$^\bullet$ (radical addition) or RS$^-$ (Michael adduction) to the double bonds of DHA and DHB. These are likely to be minor pathways in biological samples due to the low concentrations of DHA/DHB and the occurrence of other alternative reactions of RS$^\bullet$ (see above), though these reactions have been detected with some isolated proteins [109] and also in the nucleus of the lens [63].

Related Michael addition reactions of RS$^-$ with quinones, or $\alpha\beta$-unsaturated aldehydes and related species yield S-C bonds (e.g., [110–112] (Figure 4). The rate constants for these addition reactions are fast for some quinones (k $10^2 - 10^5$ M^{-1} s^{-1} for 1,4-benzoquinone, with lower values for more sterically hindered species [110]), but slower for αβ-unsaturated aldehydes and related species ($k < 1 - 7 \times 10^2$ M^{-1} s^{-1} [111]). As expected, the rate constants vary significantly with the structure of the Michael acceptor (quinone/aldehyde), and in the majority of cases with the thiol pK_a (i.e., higher concentrations of the better nucleophile RS^-) [110,111]. Some of these reactions are reversible, with another RS^- able to displace the added quinone [113]. Whilst these reactions do not directly give protein–protein crosslinks, the added quinone can undergo further reaction with another RS^- at the second double bond, resulting in protein–protein linkages via the original quinone. These reactions are responsible for some of the crosslinking detected in aged and cataractous lenses, with the quinone species being derived from Trp metabolites. These Trp-derived species act as photophysical filters that protect against UV in young lenses but become increasingly oxidized and adducted to lens crystallin proteins with increasing age and light exposure [114].

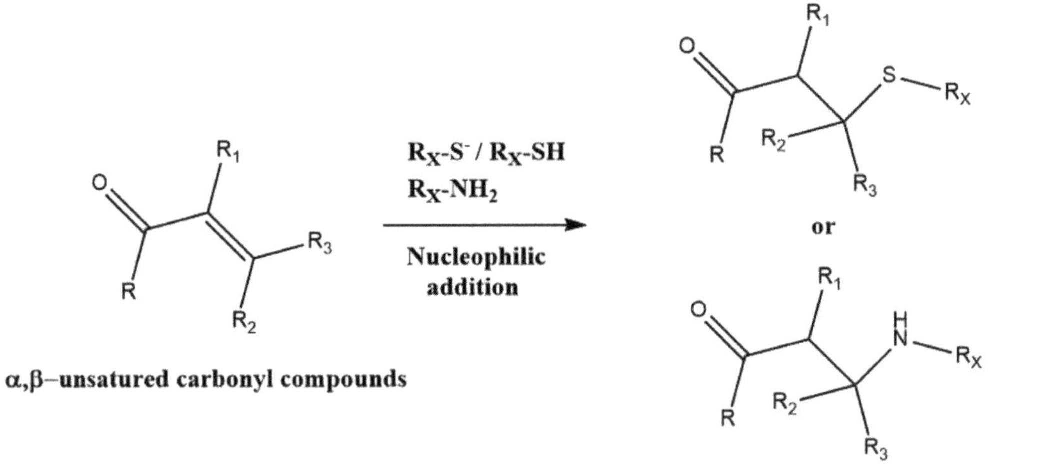

Figure 4. Michael addition reactions of nucleophiles to αβ-unsaturated carbonyl compounds.

Similar reactions can occur with quinones generated on proteins via oxidation of Tyr and Trp, though the latter are less well characterized. Thus, radical and photo-oxidation reactions can convert Tyr to 3,4-hydroxyphenylalanine (DOPA), with this undergoing rapid subsequent oxidation to the quinone (DOPA quinone). This quinone is then susceptible to attack by RS^- on peptides (e.g., GSH) or other proteins, via Michael addition, with this giving glutathionylated proteins or protein–protein crosslinks. These types of reactions have been reported for model peptides and myoglobin subject to oxidative damage [42] and with casein proteins subjected to photo-oxidation (Rossi et al., submitted). Similar reactions can occur within the active site of some enzymes (e.g., copper amine oxidases [115]) where GSH appears to modify the cofactor, which is a redox-active hydroxylated/quinone Tyr or Trp species, possibly via addition [116]. The co-factor in a number of dehydrogenases and oxidase enzymes is cysteine tryptophanylquinone (CTQ), which involves a linkage between Cys residues and an oxygenated Trp [117]—this appears to be formed via initial Trp oxidation in a complex series of enzyme-mediated reactions and subsequent Michael adduction of the Cys to the modified Trp. Related tryptophan tryptophanylquinone cofactors (TTQ), containing Trp-oxygenated Trp crosslinks, are present in selected aliphatic and aromatic amine dehydrogenase enzymes [118].

Evidence has been presented for a radical-mediated crosslink between Tyr^{A19} and Cys^{B20} in photo-oxidized insulin. Irradiation of the protein generates $RS^{\bullet}$ by direct homolysis of one of the three disulfides present in this protein [119]. Subsequent electron/hydrogen transfer from Tyr to one of the RS$^{\bullet}$ would then give the required radical pair for formation of the dimer. Alternatively, photo-induced electron transfer from a photo-excited Tyr to a disulfide could occur, with this yielding Tyr$^{\bullet}$ and a single RS$^{\bullet}$, with the crosslink being formed by subsequent reaction between these two species. A third potential pathway may involve *addition* of RS$^{\bullet}$ formed from Cys^{B20} radical to the intact Tyr^{A19} residue followed by oxidation of the radical adduct [119,120]. The exact structure of the adduct species (i.e., whether this involves an S–O or S–C bond) remains to be resolved. In addition to this Tyr–Cys crosslink, a dithiohemiacetal [–C(SH)–S–C–] crosslink was also detected, involving Cys^{A20} and Cys^{B19}, on photolysis of human insulin in the solid state [121]. The proposed mechanism involves formation of two RS$^{\bullet}$ from the disulfide, followed by disproportionation of the two radicals to give a thiol (RSH) and a thioaldehyde (–C=S), and then molecular addition of the thiol to the thioaldehyde. Similar species have been detected with other photo-oxidized peptides/proteins, including mouse and human growth hormones and some monoclonal antibodies (reviewed in [120]). A variant on the above reaction has been detected when Ser residues are located in the vicinity of the thioaldehyde, with the hydroxyl group of the Ser reacting with the thioaldehyde to give a thiohemiacetal [–C(SH)–O–C–] [41]. Subsequent cleavage of the C–S bond under continued irradiation, and loss of the sulfur center via homolytic or heterolytic reactions, results in a vinylether crosslink (–C=C–O–C) [41]. It is likely that similar reactions occur with Thr residues, and analogous reactions have been detected with a Tyr in place of the Ser (reviewed in [120]).

A complex crosslink involving a nitrogen–oxygen–sulfur (N–O–S) linkage has been reported between a Lys and Cys in a transaldolase reductase from *Neisseria gonorrhoeae*, the pathogen that causes gonorrhea [45]. Interestingly, this crosslink regulates enzyme activity via an allosteric switch, suggesting that its formation may be of biological importance. Whilst the structure of the crosslink has been determined in detail (by X-ray crystallography and others), the mechanism of its formation is unresolved. Multiple mechanisms have been proposed, with each involving initial oxidation at the sulfur to give oxygenated products [45].

6.1.4. Sulfur–Selenium Crosslinks

As sulfur and selenium have considerable similarities in their chemistry, there is evidence for intra- and intermolecular selenium–sulfur and selenium–nitrogen species. Examples of both types are found in two of the major families of enzymes that protect cells against oxidative stress: the glutathione peroxidases (GPx's), which remove H_2O_2 (and lipid hydroperoxides in the case of GPx4), and the thioredoxin reductase (TrxR)/thioredoxin (Trx) redox systems, that maintains thiols in their reduced form. In order to rationalize the resistance of GPx enzymes to deselenation (loss of selenium to give DHA), it has been proposed that internal selenenyl amide (Se–N) linkages with neighboring amides involving five- or eight-membered rings [122] are formed during the catalytic cycle [123]. These selenenyl amides subsequently react with GSH to give a selenenyl GSH (i.e., Se–S) species, which is then repaired at the expense of GSH.

Similar Se–S species are generated during the catalytic cycle of the TrxR/Trx redox system. The resting state of TxrR contains an intramolecular Se–S bond between a Cys and Sec residue, which is reduced by NADPH via an FAD cofactor. The resulting selenate anion (RSe^-) then reacts with the intramolecular disulfide bond of oxidized thioredoxin to give an intermolecular Se–S linkage between the TrxR and Trx. This linkage is then repaired by another (resolving) Cys present in TrxR to regenerate the resting state Se–S species [124].

6.2. Carbon–Carbon

6.2.1. Tyrosine–Tyrosine (Di-Tyrosine, Di-Tyr) Crosslinks

Di-Tyr crosslinks are formed deliberately by some enzymes, post-translationally, to give stability and elasticity to structural proteins (e.g., [125]). Nevertheless, protein exposure to oxidative environments, both biologically [126,127] and in food systems [128,129], can produce additional di-Tyr crosslinks. Thus, exposure of proteins to hydroxyl (HO$^{\bullet}$) and peroxyl (ROO•), peroxynitrous acid/peroxynitrite (ONOOH/ONOO$^-$), nitrogen dioxide (NO_2•), nitrosoperoxycarbonate ($ONOOCO_2^-$), carbonate ($CO_3^{\bullet -}$), lipid hydroperoxides (LOOH) and photoinduced reactions can yield di-Tyr crosslinks [130–132]. These can be formed inter- or intramolecularly between two proteins, from two free Tyr, or between free Tyr and proteins, by radical–radical reactions involving two Tyr• (Figure 5) [130,133]. Electron delocalization over the benzene ring and the phenolic oxygen results in the formation of two regio-isomers, one involving a carbon–oxygen bond (C3–O, iso-dityrosine, *vide infra*) and another with a carbon–carbon (C3–C3, *o,o′*) bond. The latter is thermodynamically preferred (Figure 5) [19,134,135]. The production of di-Tyr crosslinks is limited by alternative reactions, such as that with O_2 (though this is slow, $k < 1 \times 10^3$ M^{-1} s^{-1}, [130]), and also via fast reaction with other radicals such as $O_2^{\bullet -}$ (which gives short-lived peroxides [37,136]) and nitric oxide (NO•) [137], and also with reducing agents such as ascorbate, thiols and polyphenols [138–140]. The extent of dimerization is also likely to be strongly modulated by steric and electronic interactions.

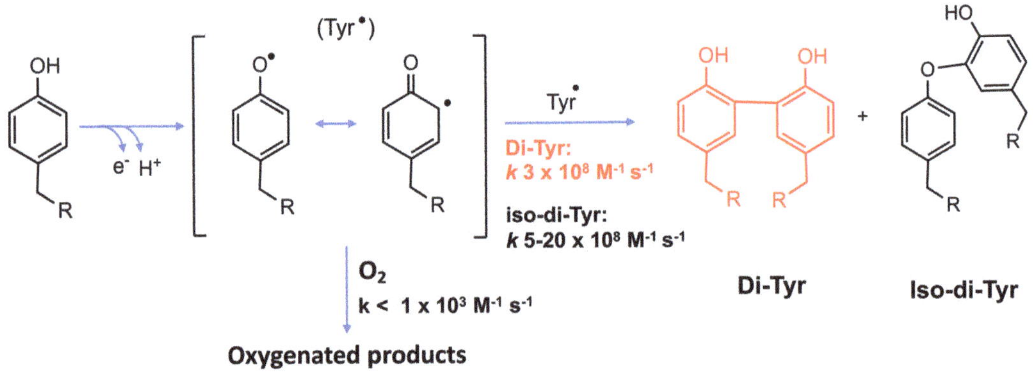

Figure 5. Formation and reactions of Tyr phenoxyl radicals (Tyr•). Tyr• self-react to produce di-Tyr (o,o′-di-Tyr, red; iso-di-Tyr, black) or react with O_2 to generate oxygenated products. Kinetic data from [130].

In pathophysiological contexts, di-Tyr crosslinking has been reported in low-density lipoproteins, arising from oxidative damage within human atherosclerotic lesions [61]; in erythrocytes exposed to a continuous flux of H_2O_2 [62]; in lens proteins, where it appears to originate from photo-oxidation of amino acids [63]; in tissues from a number of neurodegenerative conditions (e.g., amyloid-beta dimers in Alzheimer's disease [64] and in α-synuclein [65,66], which may contribute to Parkinson's disease); in lipofuscin pigments in the aged human brain [67]; in plasma of patients with chronic renal failure; and in urine of people with diabetes [68,69].

Di-Tyr crosslinks have also been detected on synthetic peptides exposed to peroxidase activity (e.g., hemoglobin and H_2O_2 [141]), and a large number of isolated proteins including (amongst others): lipoproteins [61,70,71], glucose 6-phosphate dehydrogenase (G6PDH) [53], ribonuclease A (RNAse A) [86], multiple extracellular matrix proteins [57,58,76], calmodulin [72], lysozyme [54,79], myoglobin [142,143], insulin [48,73,74] and human Δ25 centrin 2 [75].

The large body of evidence for di-Tyr crosslinks in oxidized proteins, fluids and tissues; the stability of this species (which is immune to reduction or repair); the association of this crosslink with the onset and development of human pathologies; and advances in analytical methods for the detection and quantification of di-Tyr (see below) have supported its use as a biomarker of protein oxidation in clinical studies [144,145].

Di-Tyr formation has also been reported in food systems, with both positive and negative effects. Protein crosslinking can give desirable textures and stability to processed foods, while negative effects have been reported with regard to poor protein digestibility and other undesirable properties [128,129]. Di-Tyr crosslinks have been detected in milk powders [146,147], during breadmaking [148], processing of myofibrillar proteins [149], in isolated caseins exposed to riboflavin-mediated photoreactions [51], and lactalbumin exposed to a laccase enzyme system [150] and UV-B light [50]. The presence of di-Tyr in foods, and consequent dietary intake, has resulted in studies on its toxicological properties. Intragastric administration of pure di-Tyr has been associated with metabolic alterations [151], including disrupted glucose metabolism [152], and renal alterations [153]. Interestingly, and of potential biological relevance, processed milks containing di-Tyr (and other Tyr oxidation products) can induce oxidative damage in plasma, liver and brain tissues, as well as spatial learning and memory impairments [154].

Di-Tyr crosslinks have also been detected in a wide variety of other organisms, including in insect cuticles, where di-Tyr crosslinks connect resilin proteins in a three-dimensional network [155]. Peroxidase-catalyzed di-Tyr crosslinks have also been detected in high concentrations in the oocyst walls of the apicomplexan parasite, *Eimeria maxima*, and in some bacterial spore proteins (e.g., those of *Bacillus subtilis*), with these providing a rigid framework that protects the oocyst or spores from harsh environmental conditions [60,77]. On the basis of these data, novel biomaterials have been developed that contain di-Tyr crosslinks that have excellent mechanical properties. Different strategies to induce efficient di-Tyr crosslink formation have been developed, including photoreactions mediated by ruthenium complexes and riboflavin [132].

6.2.2. Tryptophan–Tryptophan (Di-Tryptophan, di-Trp) Crosslinks

Trp is a major target for many oxidants, and a large number of oxidation products have been characterized [156]. Many of these involve the initial formation of (nitrogen-centered) indolyl radicals (Trp$^\bullet$) which, depending on the properties of the protein and environment, can self-react to generate di-Trp crosslinks (Figure 6), though less information is available on these species than di-Tyr. Di-Trp crosslinks have been reported on human superoxide dismutase 1 (hSOD1) upon exposure to H_2O_2 in the presence of carbonate ions [78], with the peroxidase activity of hSOD1 being responsible for the generation of $CO_3^{\bullet-}$ that induce one-electron oxidation of the single, solvent-exposed, Trp32 residue in this protein and consequent intermolecular di-Trp crosslinking. This results in irreversible dimerization and is associated with non-amyloid protein aggregation and the occurrence of amyotrophic lateral sclerosis (ALS) [157]. $CO_3^{\bullet-}$-mediated di-Trp crosslinks have also been reported for both free Trp [158] and lysozyme [79]. In the latter case, oxidation mediated by hSOD1 resulted in the detection of both hetero-dimers of lysozyme with hSOD1 and hSOD1 dimers [79]. These reports indicate a high propensity of $CO_3^{\bullet-}$ to generate di-Trp crosslinks [79]. Di-Trp crosslinks have also been reported on lysozyme exposed to ONOOH/$ONOO^-$ [159] and photo-oxidation mediated by kynurenic acid [160], as well as in solutions of free Trp and *N*-acetyl-Trp exposed to photo-reactions of riboflavin and kynurenic acid, respectively [161,162]. Di-Trp crosslinks have been also detected in lens proteins of patients with nuclear cataracts [80] and in α-crystallin proteins exposed to UV radiation and a photosensitizer [163].

Di-Trp crosslinks have also been detected in aggregates of α-lactalbumin (α-LA) exposed to UV-B light [50], and in FtsZ proteins of *M. jannaschii*, a thermophilic microorganism, exposed to peroxyl radicals [164]. The dimerization of Trp$^\bullet$ is fast (k ~ $2-6 \times 10^8$ M^{-1} s^{-1} [159]), but dependent on the residue environment in peptides and proteins. Due to electron delocalization across the pyrrole and benzene rings, and the presence of chiral carbons, a complex mixture of regio- and stereo-isomers can be formed. This heterogeneity is likely to contribute to the limited number of reports on this type of crosslink, due to the difficulty in identifying and quantifying all of these species. This is exacerbated by a current absence of antibodies against this linkage (unlike di-Tyr, see below). Nonetheless, di-Trp crosslinks appear to be generated in low yields, probably due to the large number of alternative fates of Trp$^\bullet$ (Figure 6) and the low abundance of Trp in proteins (~1–2%). Despite the modest rate constant for addition of O_2 to Trp$^\bullet$ ($k \leq 4 \times 10^6$ M^{-1} s^{-1} [35,36]), oxygenated products and alternative crosslinks (e.g., Trp-Tyr, see below) are commonly detected, and at higher yields than di-Trp. Thus, only a small extent of the aggregation of α-LA induced by UV-B light was associated with di-Trp crosslinks, in contrast to high yields of disulfide bonds [50]. Nonetheless, under anaerobic conditions, or environments with low O_2 levels (e.g., human lens [161]), the formation of di-Trp crosslinks appears to be favored.

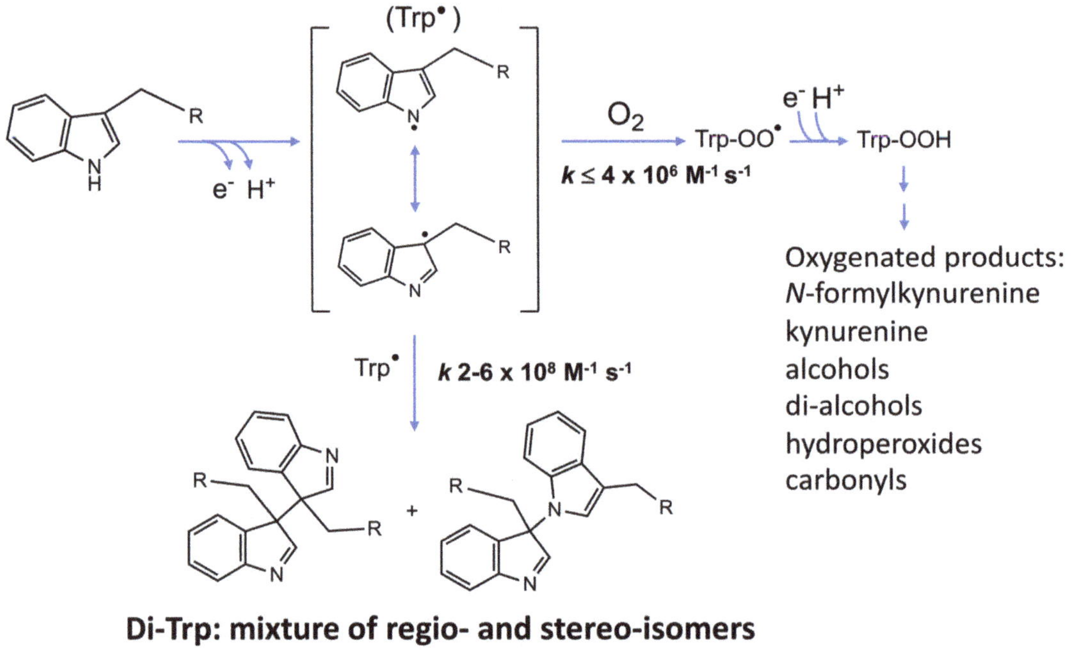

Figure 6. Formation and reactions of Trp indolyl radicals (Trp•). Self-reactions of Trp• produce carbon–carbon (C3–C3) and carbon–nitrogen (C3–N1) di-Trp crosslinks. It should be noted that multiple stereoisomers are potentially formed for both di-Trp dimers. Kinetic constants for self-reactions of Trp• and their reaction with O_2 are from [159] and [36], respectively.

The production of di-Trp crosslinks is also modulated by protein conformations, as reported for the extracellular matrix (ECM) protein fibronectin, when exposed to $ONOO^-$/ONOOH [57]. Fibronectin is polymerized into elastic fibrils (fibrillation) at the cell surface, with this then allowing binding of other ECM components. This process is controlled (amongst others) by alterations to the protein conformation, with the compact structure present in plasma being resistant to fibrillation. However, an extended conformation generated by cell-generated forces (as well as by altered pH and salt concentrations) is more prone to fibrillation. MS analysis of tryptic peptides from fibronectin exposed to $ONOO^-$/ONOOH showed that two of the four crosslinks detected correspond to di-Trp crosslinks [57]. One di-Trp crosslink (Trp445–Trp2264) was detected in both the compact and extended conformations, while another (Trp177–Trp2250) was only detected in the compact state. Interestingly, Trp2250 was oxidized to a high extent in the extended conformation, indicating that compact conformation of fibronectin favors radical–radical reactions (involving Trp177 and Trp2250), while the extended state favors other fates, including formation of 6-nitro-Trp [57].

Trp–Trp bonds have also been detected in the active sites of some enzymes, including amine dehydrogenases (e.g., methylamine dehydrogenase and aromatic amine dehydrogenases), where a tryptophan tryptophanylquinone crosslink is generated between C2 of an unmodified Trp residue, and C5 (on the benzene ring) of a Trp quinone (with the carbonyl groups present at C7 and C8) [165].

6.2.3. Tyrosine–Tryptophan (Tyr–Trp) Crosslinks

Cross-reaction between Tyr• and Trp• can lead to Tyr–Trp crosslinks. These species are less well characterized than those described above, and the precise structure of these species remains to be elucidated (i.e., the nature of the crosslink bond and sites on the two rings that are joined). MS analyses of solutions containing free Tyr and Trp incubated with $CO_3^{\bullet -}$ have provided evidence for three different, isomeric, Tyr–Trp crosslinks, with these likely to involve carbon–carbon bonds, and probably between the ortho (C3) position on the Tyr ring, and C3 on the indole ring of Trp [158].

Tyr–Trp crosslinks have been detected in peptides and isolated proteins exposed to multiple oxidants, including a cytochrome c peroxidase mutant exposed to a heme iron-peroxide reaction [81]; in glucose-6-phosphate dehydrogenase (G6PDH) exposed to ROO• [53]; model peptides exposed to pulsed UV light [166]; in lysozyme treated with ROO• and exposed to photooxidation reactions mediated by riboflavin and rose Bengal [54–56]; in fibronectin exposed to ONOOH/$ONOO^-$ [57]; in protein extracts of the Gram-positive bacterium *Lactococcus lactis* exposed to photooxidation [167]; and proteins extracted from human cataractous lenses [80].

6.2.4. Other Carbon–Carbon Crosslinks

As most protein carbon radicals, P•, react rapidly with O_2 (see above), termination reactions between two P• are limited under most biological conditions. The formation of some P–P crosslinks has, however, been reported in model peptide systems exposed to radiation-induced radicals [33], and similar species may be formed in biological situations where the pO_2 value is low (e.g., in solid tumors, tissues subject to hypoxia). This requires further study.

6.3. *Carbon–Nitrogen (C–N) Crosslinks*

A significant number of carbon–nitrogen crosslinks have been identified, with the majority of these arising from molecule–molecule reactions, with the nitrogen atom acting as a nucleophile, though a small number of radical–radical crosslinks have been characterized.

Delocalization of the unpaired electron over the aromatic rings of Tyr• and Trp•, followed by radical–radical reactions has been proposed, on the basis of MS data, to give di-Trp species with carbon–nitrogen (C3-N1) bonds (Figure 6) [158]. The exact structures of these species are not completely clarified, though it is clear that multiple isobaric species are formed [159,161]. Some of these crosslinks appear to be of pathological relevance, as they have been detected on lens crystallin proteins subjected to photo-oxidation [161] and in human cataractous lenses [80].

The much larger group of molecule–molecule reactions that generate C–N crosslinks primarily involve nucleophilic reactions of the nitrogen centers on Lys and hydroxy-Lys (N^{ε}-amine), Arg (guanidine nitrogen) and His (imidazole nitrogen) with either the carbon atom of carbonyl functions (Schiff base reactions) or other electron-deficient carbon centers (via Michael addition). For the former type of reaction, the carbonyl can be formed enzymatically (e.g., by LOX and LOXL enzymes, see earlier) or as a result of oxidation of alcohols (Ser/Thr) or C–H bonds in the presence of O_2 (e.g., via ROO• and ROOH [31,168]).

In the case of the Cu^{2+}-dependent LOX/LOXL reactions and collagen crosslinking, the initial Schiff base adducts can undergo multiple further condensation reactions that allow several chains to be linked together via a single site [11]. Thus, there is abundant evidence for lysyl pyrrole, hydroxylysyl pyrrole, lysyl pyridinoline and hydroxylysyl pyridinoline species involving three or four collagen chains [169]. The mechanism of formation of these species involves the initial formation of a two-chain crosslink and then further condensation with a Lys/hydroxy-Lys on third and fourth chains. These species are critical to the correct assembly of collagen-containing extracellular matrices in tissues [11].

Similar reactions occur with (Ca^{2+}-dependent) transglutaminases that form isopeptide bonds between the Lys side-chain amine and amide (Gln, Asn) or carboxylate (Asp) residues, with the amide/carboxylate 'activated' via reaction with the Cys residue of a Cys–His–Asp catalytic triad on the enzyme. The reaction of the carbonyl function with the Cys residue generates a thioester that is then susceptible to reaction with the Lys amine [22–24]. Such isopeptide bonds are highly resistant to most proteases and are widely encountered in the formation of insoluble protein polymers/aggregates, such as those found in brain tissue lesions of people with Alzheimer's disease [24,170], in some bacterial spore proteins [60] and in the food industry, where these reactions are used as glues to alter food texture and properties [171,172].

Reactions also occur with α,β-unsaturated aldehyde and ketones and related species via Michael addition reactions (Figure 7). Thus, intra- and interchain crosslinks can be generated by the reaction of Lys/Arg/His residues with an oxidized imidazole group of the amino acid His, which acts as a Michael acceptor. These Lys–His, Arg–His and His–His products have been detected on multiple proteins, and particularly those subjected to photo-oxidation, where His residues are a major initial target of damage (e.g., [82,84–86,173]). They have also been detected in antibodies exposed to (photo)oxidation [82,84]. His-containing crosslinks have also been reported in some collagens, with these probably formed by nucleophilic attack of an imidazole nitrogen on an α,β-unsaturated structure formed via other crosslinking reactions [174].

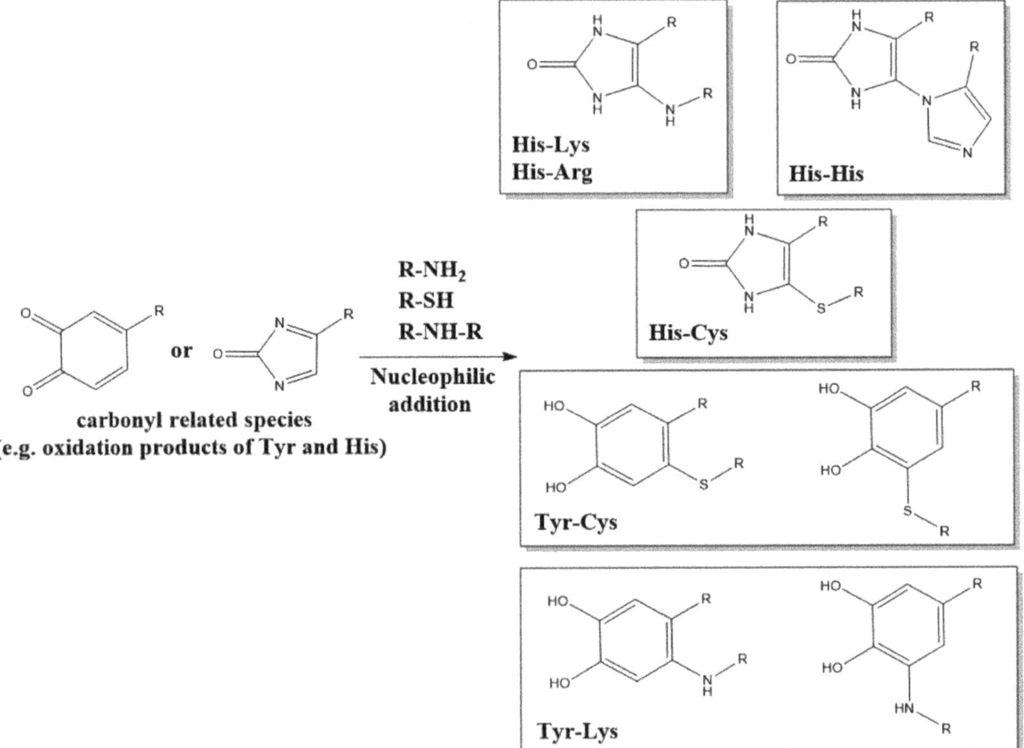

Figure 7. Michael addition reactions of amino acid side-chains to oxidized His and Tyr residues.

Related Michael reactions occur with Tyr-derived quinones formed on oxidation of Tyr residues, with nitrogen nucleophiles, such as the N^{ε}-amine of Lys side-chains undergoing adduction reactions with the oxidized ring (Figure 7). These reactions occur with lower rate constants than the corresponding reaction with the RS^- from Cys [110,175] (i.e., C–S crosslinks predominate over C–N linkages, through this is dependent on the availability and abundance of Cys versus Lys residues, with the latter usually predominating numerically). These reactions are important in the coupling of proteins, including via N-terminal amine groups, by tyrosinase and polyphenol oxidases (e.g., [176]) and in the formation of insect exoskeletons/cuticles of many insects [177], as well as the 'glues' that attach mussels to rocks [178,179].

6.4. Carbon–Oxygen

As alcohols (ROH) are poorer nucleophiles when compared to RS^- and neutral amines (RNH_2), there are limited numbers of known carbon–oxygen crosslinks formed via molecular processes such as Michael reactions. However, such coupling can occur via radical–radical reactions, with the most well-established example being the formation of iso-di-tyrosine (iso-di-Tyr), where two $Tyr^•$ dimerize via the phenolate oxygen of one ring, and C3 on the second ring (Figure 5); this appears to be a minor reaction compared to carbon–carbon coupling (see above) [19]. Iso-di-Tyr crosslinks have been reported in the extensin proteins that contribute to the architecture of wall plant cells [180], and in systems exposed to the myeloperoxidase–H_2O_2 system of human phagocytes, suggesting that these species may be present in tissues subject to acute or chronic inflammation, such as atherosclerosis [19].

7. Secondary Reactions of Crosslinks

In most biological systems, protein crosslinks, and particularly the formation of irreversible covalent crosslinks such as di-Tyr, di-Trp and Tyr–Trp, are considered as 'final' oxidation products [181]. However, over-oxidation of these species is possible, particularly under conditions of extensive oxidative damage, or under environments with long-term protein exposure to oxidants, where secondary one-electron oxidation with formation of radicals such as di-$Tyr^•$, di-$Trp^•$, or Tyr–$Trp^•$ may occur. Such radicals can mediate similar reactions to those described above for $Tyr^•$ and $Trp^•$, including reaction with O_2 to produce oxygenated products (e.g., alcohols and hydroperoxides) and self-reactions to generate trimers and oligomers. Thus, formation of tri-Tyr and pulcherosine crosslinks have been detected in human phagocytes [19], while di-, tri- and tetra-Tyr have been reported in structural proteins of plant parasitic nematodes [182]. In addition, oligomers of Tyr (n = 2–8) have been reported in α-lactalbumin exposed to a horseradish peroxidase–H_2O_2 system [183]. Tri-Trp has been reported in trimers of hSOD1 triggered by $CO_3^{•-}$ [78], while tri-Trp and a di-Trp hydroperoxide (di-Trp-OOH) were reported in solutions of free Trp and riboflavin illuminated with a high-intensity 365 nm light-emitting diode [162].

In contrast, photo-oxidation (at 320 nm) of di-Tyr, in the presence of O_2, has been reported to occur via processes involving $O_2^{•-}$, singlet oxygen ($^1O_2$) and H_2O_2 [184]. These observations were ascribed to the action of di-Tyr crosslinks as photosensitizers that could induce photo-damage to other biomolecules [184]. These findings suggest that di-Tyr crosslinks may be able to extend oxidation processes, opening new pathways of reactions, though these are only likely to be of major impact in systems with very extensive extents of oxidation. The scope and role of these pathways is unexplored, as well as the ability of peroxides such as di-Trp-OOH to extend protein oxidation (in line with the capacity of other hydroperoxides [31]).

8. Detection of Crosslinks, including Advantages and Disadvantages of Different Methods

Despite considerable methodological advancements, analysis of intra- or intermolecular protein crosslinks remains a challenging task, and the mechanisms underlying the formation of some known crosslinks remain to be established. There are also, undoubtedly, more types of crosslinks that remain to be discovered. This lack of knowledge arises partly due to the complexity of some structures, and the relatively poor current 'tool-box' to detect and predict the chemistry of the species and the sites involved. The following section therefore provides an overview of current approaches to the identification and characterization of oxidation-induced crosslinks, with a focus on both mass spectrometry (MS) and techniques (e.g., UPLC, immunological methods) that provide complementary information. It is clearly advisable to use combinations of methods and to investigate both gross modifications (e.g., changes in the molecular mass) and structural/conformational changes in crosslinked vs. native proteins with methods that allow the chemical nature and the residues involved in the formation of these species to be determined (Figure 8). Each of these methods has individual advantages and disadvantages.

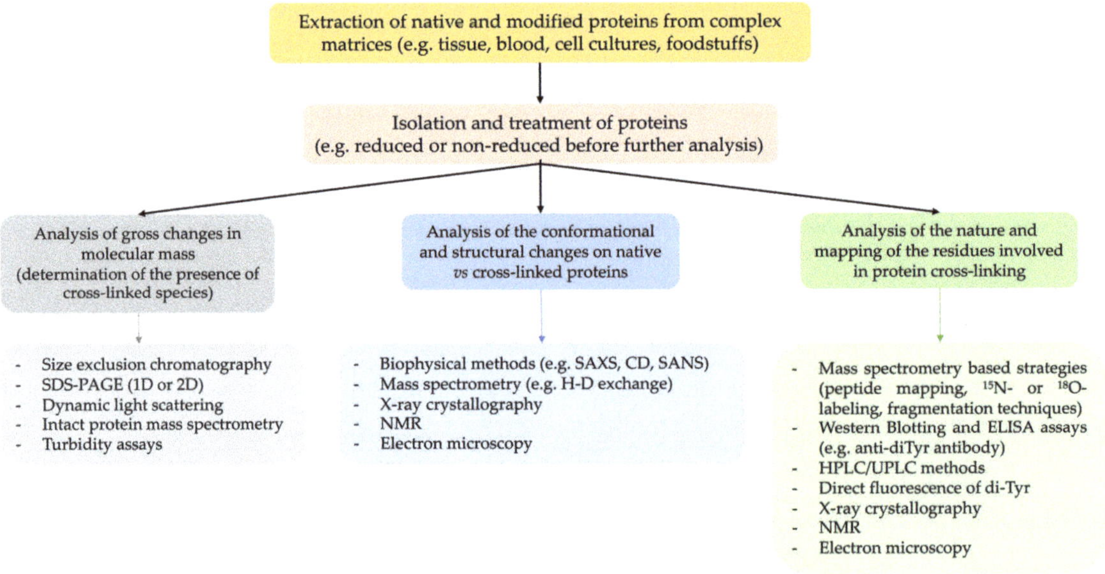

Figure 8. Overview of methods to detect and characterize crosslinked proteins and the sites/types of modifications. Abbreviations used: CD: circular dichroism, SANS: small angle neutron scattering, SAXS: small angle X-ray scattering, H–D: hydrogen–deuterium exchange mass spectrometry.

8.1. Analysis of Changes in Molecular Mass by Electrophoresis and Size Exclusion Chromatography (SEC)

Protein electrophoresis (e.g., SDS-PAGE) and SEC-derived methodologies are excellent tools to assess the presence of crosslinked proteins in samples. Separation by electrophoresis is typically achieved through the use of polyacrylamide gels with different pore sizes. Diverse strategies can be utilized with this approach, such as running gels under native, denaturing and/or reducing conditions. This allows the investigation and differentiation of the contributions of reducible intermolecular bonds (e.g., disulfides between two protein chains) from non-reducible bonds (e.g., carbon–carbon bonds). Similarly, SEC can separate and fractionate soluble proteins based on their hydrodynamic radius and molecular mass, and depending on the mobile phase used, it can also provide information about the nature of the crosslinks (i.e., reducible or non-reducible intermolecular bonds).

Whilst these techniques can provide information on the presence of *inter*molecular bonds, they rarely provide information about the presence of *intra*molecular species. Moreover, data analysis needs to be carried out with care, as multiple proteins may be present in each band/fraction from complex matrices. This complexity can be overcome by using two-dimensional gel electrophoresis, with proteins separated firstly on the basis of their isoelectric point, and subsequently by molecular mass. Both methodologies have been successfully employed in combination with MS-based strategies [54,56,80,84,93,94,141,185], with isolation and enrichment of fractions containing dimers (and species with higher degrees of oligomerization) from non-crosslinked monomers before MS analysis. This approach has been used successfully to characterize crosslinked forms of IgG, α-synuclein and lysozyme [56,84,141]. Two-dimensional gel electrophoresis can also be used for specific detection of proteins connected by intermolecular disulfides [185]. In this approach, the proteins are first separated by non-reducing SDS-PAGE in the first dimension, and then separated by SDS-PAGE under reducing conditions in the second dimension. Proteins lacking intermolecular disulfide bonds line up on the diagonal, whereas proteins connected by intermolecular disulfide bonds dissociate from each other in the second dimension and migrate below the diagonal.

8.2. Analysis of Protein Crosslinks by Western (Immuno-) Blotting and ELISA Assays

The use of antibodies to investigate the formation of oxidation products, including di-Tyr, is a widely used strategy. These are typically examined using immunoblotting or ELISA assays, with the former providing (limited) information on the nature and identity of the proteins on which the crosslinks are present, and whether these are intramolecular (in a monomer) or interchain species. However, there are few well-characterized antibodies against crosslinked species, and these vary significantly in their specificity and selectivity, with some having significant cross-reactivity with other materials. Furthermore, crosslinks buried within highly aggregated species may be poorly, or not, recognized by (large) antibodies. Thus, appropriate control experiments are critical, and both positive and negative data should be validated by alternative methods.

Despite these caveats, immunoblotting has been widely used to detect di-Tyr present in α-synuclein in Lewy Bodies in Parkinson's disease [186], in atherosclerotic plaques [187], and on multiple isolated proteins, including tropoelastin, α-synuclein, caseins, glucose-6-phosphate dehydrogenase, laminins and fibronectin [51,53,58,65,76,188]. Relative yields (i.e., versus the parent) can be achieved by ELISA, and this can serve as a very useful screening tool.

8.3. Direct Detection by Spectrophotometric and Fluorometric Assays

Some crosslinked species, as well as heavily aggregated proteins, can be monitored by spectrophotometry and fluorescence spectroscopy. For example, turbidity changes in solutions can be used (in an approximate manner) to monitor the time-course of protein crosslinking and aggregation [189,190]. However, this approach does not provide information on the type and nature of the crosslinked proteins and is useful only when there is extensive formation of heavily aggregated species, as soluble dimers and oligomers do not contribute significantly to the turbidity of solutions.

In contrast, fluorescence experiments can provide relatively specific data on crosslinks such as di-Tyr (excitation and emission maximum at ~280 and ~410 nm, respectively) [191]. Such data need to be interpreted with care, particularly with intact proteins or complex systems, as other fluorescent or optically absorbing species (e.g., Tyr, Trp, Trp-derived products, co-factors) may be present that distort excitation or emission processes. Thus, depending on the sample complexity, di-Tyr detection by fluorescence can provide useful information, but requires further confirmation using other methodologies. In this context, coupling fluorescence detection methods with HPLC, UPLC or GC separation (see below) has proven to be a valuable strategy to detect and quantify di-Tyr in mixtures of amino acids and oxidation products arising from acid hydrolysis of complex protein and tissue samples [191,192].

8.4. HPLC/UPLC Methodologies

These techniques can provide important quantitative data on both the consumption of the parent amino acid residues, and product formation, including Trp- and Tyr-derived crosslinks [158,191,193,194]. Before analysis, proteins are often isolated by precipitation (e.g., from homogenates or cell lysates) using acids (e.g., trichloroacetic acid) or organic solvents (e.g., ice-cold acetone or ethanol), followed by sample clean-up (e.g., delipidation), and then hydrolysis (with acid, base or enzymes) to give free amino acids and products. The advantages and disadvantages of this approach have been recently reviewed in [192], and therefore, the following is focused solely on the detection and quantification of crosslinked species.

The free amino acids and crosslinked products obtained from hydrolysis are separated by HPLC/UPLC and can be quantified by MS [158,159,193], by fluorescence detection (directly for di-Tyr [51], or by pre-column tagging for parent amino acids using sensitive fluorescent tags such as *o*-phthaldialdehyde [191]), UV absorption, or, for some species, electrochemical oxidation [191,195]. For HPLC/UPLC methodologies coupled to MS detection, heavy atom labeling (usually ^{2}H, ^{13}C, or ^{15}N) can be utilized for accurate quantification [193]. Unfortunately, these approaches do not usually provide information on the *sites* of modification within proteins, or information on which proteins they were located on, if multiple species were present. This drawback, along with the limited number of crosslinked species that can be detected and quantified using these methods (di-Tyr, di-Trp, Trp-Tyr and cystine), makes it sensible to combine this approach with other methods (e.g., peptide mass mapping or immunoblotting).

8.5. Detection and Characterization of Crosslinked Proteins Using Other Biophysical Approaches

Biophysical techniques including circular dichroism (CD), light scattering, small angle neutron scattering (SANS), small angle X-ray scattering (SAXS), X-ray crystallography, NMR spectroscopy, and electron microscopy can provide useful information on protein structure. These methods are sensitive to modified structures, supplying valuable information on changes on morphology (i.e., mass, size and shape), secondary structure and solubility [45,196,197]. Some of these can also yield data on increased electron density between residues, thus supporting the presence of both intra- and intermolecular crosslinked species. This approach has been used in X-ray crystallographic studies, to determine the exact sites of crosslinks in oxidized peroxiredoxin 5, thioredoxin 2 and γS-crystallin, and to elucidate a covalent crosslink between Cys and Lys containing an N–O–S bridge [45,197–199]. However, most of these methods (with the exception of X-ray crystallography, NMR spectroscopy and cryogenic electron microscopy) cannot provide a structure of sufficiently high-resolution to provide definitive identifications and they must therefore be combined with other methodologies. Moreover, these methods are currently limited to homogeneous (single protein) samples that are available in large quantities (mg amounts).

8.6. Mass Spectrometry (MS)-Based Detection and Structural Characterization of Crosslinked Proteins

MS is a highly versatile technique for analysis of protein crosslinks that can be applied to (i) detect crosslinks and quantify their abundance, (ii) localize the specific crosslinking sites within polypeptides and (iii) reveal the identity of the crosslinked proteins. All of these questions cannot, however, be readily answered in a single experiment, and careful consideration must be given to appropriate workflows for specific applications.

8.6.1. MS Analysis of Crosslinked Amino Acids

Crosslinked amino acids can be detected and quantified by GC-MS or LC-MS, with the latter being most commonly used as it limits the need for derivatization and high-temperature conditions. Free crosslinked amino acids (i.e., not protein-bound) such as di-Tyr can be detected in body fluids such as urine, plasma and cerebrospinal fluid [200,201]. As outlined above, crosslinked amino acids can also be released from proteins either through acidic or alkaline hydrolysis, or through the use of unspecific proteases (e.g., pronase that hydrolyses proteins to single amino acids) [61,66,186]. The choice of hydrolysis method depends on the chemical stability of the investigated crosslink. For example, di-Trp crosslinks are unstable to acid hydrolysis (as the indole ring undergoes acid-catalyzed cleavage) and therefore, alkaline hydrolysis or pronase digestion is preferred [159].

8.6.2. MS Analysis of Crosslinked Peptides

Degradation of proteins using specific proteases such as trypsin releases peptides that can be analyzed by MS. This is the principle behind protein identification in standard 'bottom-up' proteomics, where the experimental mass-to-charge (*m/z*) values for the experimentally generated peptides are compared to theoretical values based on the known specificity of the enzyme. This type of workflow can also be applied for the identification of crosslinked peptides, since the mass of the crosslinked peptides can be predicted (for example, di-Tyr yields a mass shift of −2 Da relative to the sum of the two crosslinked peptides). The analysis of crosslinked peptides is, however, often more challenging, since the mass spectra of crosslinked peptides are more complex than those from regular (non-crosslinked) peptides. This is mainly due to the fact that the MS/MS spectra of crosslinked peptides may contain fragments from both of the linked individual peptides. Furthermore, the number of potential combinations of crosslinked peptides to take into consideration may be very large due to the complex chemistry of oxidation-induced crosslinking. This is particularly problematic if complex (cell/tissue) samples are analyzed and if intermolecular crosslinks are considered in database searches against large proteins databases (e.g., the entire human proteome). Due to these issues, there is a significant risk of false positive identifications, and it is therefore important to carefully validate the data, for example, using an ^{18}O-labeling approach. This strategy relies on the ability of trypsin (and other related proteases) to incorporate two ^{18}O atoms from isotope-labeled water molecules into each C-termini of proteolytic peptides [202]. Since crosslinked peptides contain two C-termini, a total of four ^{18}O atoms will be incorporated instead of two ^{18}O for regular (non-crosslinked peptides). Thus, it is possible to distinguish crosslinked peptides from non-crosslinked peptides based on their isotope labeling pattern [167,203]. This strategy has been successfully applied to validate oxidation-induced crosslinks in a wide range of proteins, including glucose-6-phosphate dehydrogenase, lysozyme, RNase A, elastin, fibronectin, cyclic AMP receptor protein, C-reactive protein and β-2-microglobulin (B2M) [53–57,59,86,93,94,203].

Identification of crosslinked peptides can also be facilitated by MS-based fragmentation of the crosslink bond, yielding individual mass spectra of the two formerly linked peptides, which can be analyzed separately. For stable crosslinks such as di-Tyr, this has recently been accomplished by ultraviolet photodissociation (UVPD) fragmentation [141]. For more labile crosslinks, such as disulfides, in-source fragmentation can be utilized [204,205]. Disulfides can also be probed indirectly through MS analysis of alkylated Cys thiol groups released after in vitro disulfide reduction using chemical (e.g., dithiothreitol) or enzymatic (e.g., thioredoxin [206]) treatment.

9. Crosslink Quantification

At present, there are few methods that allow absolute quantification of crosslink concentrations, with a major limitation being the non-availability of pure standards, particularly from commercial sources. Thus, there is a pressing need for further pure crosslink standards for quantitative analyses. Disulfides are a major exception, together with di-Tyr (which is commercially available) and a few species generated via glycation reactions (e.g., pentosidine). Di-Tyr can therefore be quantified in absolute terms using some of the methods outlined above, using the purified material to construct standard curves (e.g., for MS or fluorescence detection, and UPLC/LC separation). Relative concentrations can also be obtained from ELISA assays using an anti-di-Tyr antibody, and approximate concentrations from MS analyses. In nearly all cases, even with isolated proteins where protein turnover is not a confounding factor, the absolute concentrations determined by these approaches give low values, suggesting that there are many other important crosslinks which are either undetected, or which we cannot quantify.

10. Future Perspectives

Irreversible protein crosslinks have been associated with the onset and development of pathological conditions and human diseases. Similarly, reversible crosslinks, such as disulfides, appear to play a key role in cell signaling events, primarily as a result of reversible thiol–disulfide switches or related species (e.g., in the case of protein tyrosine phosphatase-1B, PTP-1B). The growing realization of the importance of these events has increased interest in the detection and quantification of these species together with irreversible crosslinks and other protein oxidation products (such as 3-nitroTyr, 3-chloroTyr, etc.) as biomarkers of biological events. This includes both physiological signaling and the assessment of the extent of oxidative damage for clinical purposes, such as diagnosis, prognosis, prevention or decisions on therapies/treatments. Di-Tyr crosslinks are recognized as reliable biomarkers of several pathologies, particularly neurodegenerative disorders [144]. However, this species only represents a fraction of the total pool of covalent crosslinks generated when proteins are exposed to oxidative conditions. As a consequence, new investigations aimed at understanding the role that this and other crosslinks (e.g., di-Trp or Tyr–Trp) play in controlling biological activity and contributing to human disease, and their use as clinical biomarkers, are necessary. In particular, it will be of interest and importance to determine how individual crosslinks modulate protein lifetime and turnover (e.g., by proteasomes). Current data do not typically allow such an assessment, as the methods used to generate crosslinks also induce multiple other alterations on the proteins, prohibiting the identification of cause and effect relationships. Studies focused on the mechanisms underlying crosslink formation in biological systems as well as the analytical strategies for their detection and quantification should also be considered. Such investigations are also likely to be valuable in other contexts where oxidation is relevant, such as in the processes triggered by physical exercise [207]; in food processing, formulation and degradation; in determining the shelf-life of protein-based medicines and therapeutics; and the development of new biomaterials.

Author Contributions: All the authors contributed to the drafting, editing and final review of the manuscript. All authors have read and agreed to the published version of the manuscript.

Funding: This project received funding from the European Union's Horizon 2020 Research and Innovation Programme under the Marie Skłodowska-Curie grant agreement no. 890681 (to E.F.L.), the Novo Nordisk Foundation (NNF13OC0004294, NNF19OC0058493, and NNF20SA0064214 to M.J.D.), the Carlsberg Foundation (CF19-0451 to P.H.) and Fondecyt (grant no. 1180642 to C.L.A.).

Conflicts of Interest: M.J.D. declares consultancy contracts with Novo Nordisk A/S. This funder had no role in the design of the study; in the collection, analyses or interpretation of data; in the writing of the manuscript, or in the decision to publish the results. The other authors declare no conflict of interest.

Abbreviations

ALS	amyotrophic lateral sclerosis
α-LA	α-lactalbumin
CD	circular dichroism
CTQ	cysteine tryptophanylquinone
DHA	dehydroalanine
DHB	dehydrobutyric acid
Di-Trp	di-tryptophan
Di-Trp-OOH	di-tryptophan hydroperoxides
Di-Tyr	di-tyrosine
DOPA	3,4-hydroxyphenylalanine
ELISA	enzyme-linked immunoassay;
ECM	extracellular matrix
FtsZ	filamenting temperature-sensitive mutant Z
GC-MS	gas chromatography mass spectrometry

G6PDH	glucose 6-phosphate dehydrogenase
GPx's	glutathione peroxidases
GSA	glutathione sulfonamide
GSH	glutathione
GSSG	glutathione disulfide
H_2O_2	hydrogen peroxide
HOBr	the physiological mixture of hypobromous acid and its anion, hypobromite
HOCl	the physiological mixture of hypochlorous acid and its anion, hypochlorite
HPLC	high performance liquid chromatography
hSOD1	human superoxide dismutase 1
Iso-di-Tyr	iso-di-tyrosine
LC-MS	liquid chromatography mass spectrometry
LDLR	low-density-lipoprotein receptor
LOOH	lipid hydroperoxide
LOX	lysyl oxidase
LOXL	lysyl oxidase-like enzyme
MS	mass spectrometry
NMR	nuclear magnetic resonance
$^1O_2$	singlet oxygen
$ONOO^-/ONOOH$	the physiological mixture of peroxynitrite anion and peroxynitrous acid
$P^\bullet$	protein carbon-centered radical
$POO^\bullet$	protein peroxyl radical
P-P	protein dimers
PTP1B	protein tyrosine phosphatase 1B
$RS^\bullet$	thiyl radical RS
RS^-	thiolate anion
R-NHBr	bromamines
R-NHCl	chloramines
ROH	alcohol
ROOH	alkyl hydroperoxide
RNAse A	ribonuclease A
$R\text{-}NH_2$	neutral amine
RS-NHR′	sulfenamide
RSNO	S-nitroso thiol
RSOH	sulfenic acid
RSO_2H	sulfinic acid
RSO_3H	sulfonic acid
RS-(O)-NHR′	sulfinamide
$RS\text{-}(O)_2\text{-}NHR'$	sulfonamide
RS-X	sulfenyl halide
SANS	small angle neutron scattering
SAXS	small angle X-ray scattering
SDS-PAGE	sodium dodecyl sulphate–polyacrylamide gel electrophoresis
SEC	size exclusion chromatography
SUMOs	small-ubiquitin-like modifier proteins
Tri-Trp	tri-tryptophan
$Trp^\bullet$	tryptophan indolyl radical
Trx	thioredoxin
TrxR	thioredoxin reductase
TTQ	tryptophan-tryptophanylquinone
$Tyr^\bullet$	tyrosine phenoxyl radical
UPLC	ultra-performance liquid chromatography
UVPD	ultraviolet photodissociation

References

1. Hogg, P.J. Disulfide bonds as switches for protein function. *Trends Biochem. Sci.* **2003**, *28*, 210–214. [CrossRef]
2. Hägglund, P.; Mariotti, M.; Davies, M.J. Identification and characterization of protein cross-links induced by oxidative reactions. *Expert Rev. Proteom.* **2018**, *18*, 665–681. [CrossRef] [PubMed]

3. Adam, O.; Theobald, K.; Lavall, D.; Grube, M.; Kroemer, H.K.; Ameling, S.; Schafers, H.J.; Bohm, M.; Laufs, U. Increased lysyl oxidase expression and collagen cross-linking during atrial fibrillation. *J. Mol. Cell Cardiol.* **2011**, *50*, 678–685. [CrossRef] [PubMed]
4. Andringa, G.; Lam, K.Y.; Chegary, M.; Wang, X.; Chase, T.N.; Bennett, M.C. Tissue transglutaminase catalyzes the formation of alpha-synuclein crosslinks in Parkinson's disease. *FASEB J.* **2004**, *18*, 932–934. [CrossRef] [PubMed]
5. Martinez-Olivan, J.; Fraga, H.; Arias-Moreno, X.; Ventura, S.; Sancho, J. Intradomain Confinement of Disulfides in the Folding of Two Consecutive Modules of the LDL Receptor. *PLoS ONE* **2015**, *10*, e0132141. [CrossRef] [PubMed]
6. Azimi, I.; Wong, J.W.; Hogg, P.J. Control of mature protein function by allosteric disulfide bonds. *Antioxid. Redox Signal.* **2011**, *14*, 113–126. [CrossRef]
7. Depuydt, M.; Messens, J.; Collet, J.F. How proteins form disulfide bonds. *Antioxid. Redox Signal.* **2011**, *15*, 49–66. [CrossRef]
8. Sinz, A.; Arlt, C.; Chorev, D.; Sharon, M. Chemical cross-linking and native mass spectrometry: A fruitful combination for structural biology. *Protein Sci.* **2015**, *24*, 1193–1209. [CrossRef]
9. Chellan, P.; Nagaraj, R.H. Protein crosslinking by the Maillard reaction: Dicarbonyl-derived imidazolium crosslinks in aging and diabetes. *Arch. Biochem. Biophys.* **1999**, *368*, 98–104. [CrossRef] [PubMed]
10. Nandi, S.K.; Nahomi, R.B.; Rankenberg, J.; Glomb, M.A.; Nagaraj, R.H. Glycation-mediated inter-protein cross-linking is promoted by chaperone-client complexes of alpha-crystallin: Implications for lens aging and presbyopia. *J. Biol. Chem.* **2020**, *295*, 5701–5716. [CrossRef] [PubMed]
11. Pehrsson, M.; Mortensen, J.H.; Manon-Jensen, T.; Bay-Jensen, A.C.; Karsdal, M.A.; Davies, M.J. Enzymatic cross-linking of collagens in organ fibrosis—Resolution and assessment. *Expert Rev. Mol. Diagn.* **2021**, *21*, 1049–1064. [CrossRef] [PubMed]
12. Kagan, H.M. Lysyl oxidase: Mechanism, regulation and relationship to liver fibrosis. *Pathol. Res. Pract.* **1994**, *190*, 910–919. [CrossRef]
13. López, B.; González, A.; Hermida, N.; Valencia, F.; de Teresa, E.; Díez, J. Role of lysyl oxidase in myocardial fibrosis: From basic science to clinical aspects. *Am. J. Physiol. Heart Circ. Physiol.* **2010**, *299*, H1–H9. [CrossRef] [PubMed]
14. Trackman, P.C. Lysyl Oxidase isoforms and potential therapeutic opportunities for fibrosis and cancer. *Expert Opin. Ther. Targets* **2016**, *20*, 935–945. [CrossRef]
15. Bignon, M.; Pichol-Thievend, C.; Hardouin, J.; Malbouyres, M.; Brechot, N.; Nasciutti, L.; Barret, A.; Teillon, J.; Guillon, E.; Etienne, E.; et al. Lysyl oxidase-like protein-2 regulates sprouting angiogenesis and type IV collagen assembly in the endothelial basement membrane. *Blood* **2011**, *118*, 3979–3989. [CrossRef]
16. McCall, A.S.; Cummings, C.F.; Bhave, G.; Vanacore, R.; Page-McCaw, A.; Hudson, B.G. Bromine is an essential trace element for assembly of collagen IV scaffolds in tissue development and architecture. *Cell* **2014**, *157*, 1380–1392. [CrossRef] [PubMed]
17. Bhave, G.; Cummings, C.F.; Vanacore, R.M.; Kumagai-Cresse, C.; Ero-Tolliver, I.A.; Rafi, M.; Kang, J.-S.; Pedchenko, V.; Fessler, L.I.; Fessler, J.H.; et al. Peroxidasin forms sulfilimine chemical bonds using hypohalous acids in tissue genesis. *Nat. Chem. Biol.* **2012**, *8*, 784–790. [CrossRef] [PubMed]
18. Peterfi, Z.; Geiszt, M. Peroxidasins: Novel players in tissue genesis. *Trends Biochem. Sci.* **2014**, *39*, 305–307. [CrossRef] [PubMed]
19. Jacob, J.S.; Cistola, D.P.; Hsu, F.F.; Muzaffar, S.; Mueller, D.M.; Hazen, S.L.; Heinecke, J.W. Human phagocytes employ the myeloperoxidase-hydrogen peroxide system to synthesize dityrosine, trityrosine, pulcherosine, and isodityrosine by a tyrosyl radical-dependent pathway. *J. Biol. Chem.* **1996**, *271*, 19950–19956. [CrossRef] [PubMed]
20. Selinheimo, E.; Autio, K.; Kruus, K.; Buchert, J. Elucidating the mechanism of laccase and tyrosinase in wheat bread making. *J. Agric. Food Chem.* **2007**, *55*, 6357–6365. [CrossRef] [PubMed]
21. Dunford, H.B. Peroxidases. In *Advances in Inorganic Biochemistry*; Elsevier: Amsterdam, The Netherlands, 1982; pp. 41–68.
22. Lorand, L.; Graham, R.M. Transglutaminases: Crosslinking enzymes with pleiotropic functions. *Nat. Rev. Mol. Cell Biol.* **2003**, *4*, 140–156. [CrossRef] [PubMed]
23. Heck, T.; Faccio, G.; Richter, M.; Thony-Meyer, L. Enzyme-catalyzed protein crosslinking. *Appl. MicroBiol. Biotechnol.* **2013**, *97*, 461–475. [CrossRef]
24. Wang, D.-S.; Dickson, D.W.; Malter, J.S. Tissue transglutaminase, protein cross-linking and Alzheimer's disease: Review and views. *Int. J. Clin. Experim. Pathol.* **2008**, *1*, 5–18.
25. Sridharan, U.; Ponnuraj, K. Isopeptide bond in collagen- and fibrinogen-binding MSCRAMMs. *Biophys. Rev.* **2016**, *8*, 75–83. [CrossRef] [PubMed]
26. Siebenlist, K.R.; Mosesson, M.W. Evidence of intramolecular cross-linked A alpha.gamma chain heterodimers in plasma fibrinogen. *Biochemistry* **1996**, *35*, 5817–5821. [CrossRef] [PubMed]
27. Kerscher, O.; Felberbaum, R.; Hochstrasser, M. Modification of proteins by ubiquitin and ubiquitin-like proteins. *Ann. Rev. Cell Dev. Biol.* **2006**, *22*, 159–180. [CrossRef] [PubMed]
28. Scheffner, M.; Nuber, U.; Huibregtse, J.M. Protein ubiquitination involving an E1-E2-E3 enzyme ubiquitin thioester cascade. *Nature* **1995**, *373*, 81–83. [CrossRef]
29. Westermann, S.; Weber, K. Post-translational modifications regulate microtubule function. *Nat. Rev. Mol. Cell Biol.* **2003**, *4*, 938–947. [CrossRef]
30. Cao, Y.; Li, H. Dynamics of protein folding and cofactor binding monitored by single-molecule force spectroscopy. *Biophys. J.* **2011**, *101*, 2009–2017. [CrossRef]
31. Davies, M.J. Protein oxidation and peroxidation. *Biochem. J.* **2016**, *473*, 805–825. [CrossRef] [PubMed]

32. Carreau, A.; El Hafny-Rahbi, B.; Matejuk, A.; Grillon, C.; Kieda, C. Why is the partial oxygen pressure of human tissues a crucial parameter? Small molecules and hypoxia. *J. Cell Mol. Med.* **2011**, *15*, 1239–1253. [CrossRef] [PubMed]
33. Dizdaroglu, M.; Simic, M.G. Isolation and characterization of radiation-induced aliphatic peptide dimers. *Int. J. Radiat. Biol.* **1983**, *44*, 231–239. [CrossRef] [PubMed]
34. Schoneich, C. Thiyl radicals and induction of protein degradation. *Free Radic. Res.* **2016**, *50*, 143–149. [CrossRef]
35. Fang, X.; Jin, F.; Jin, H.; von Sonntag, C. Reaction of the superoxide radical with the N-centred radical derived from N-acetyltryptophan methyl ester. *J. Chem. Soc. Perkin. Trans.* **1998**, 259–263. [CrossRef]
36. Candeias, L.P.; Wardman, P.; Mason, R.P. The reaction of oxygen with radicals from oxidation of tryptophan and indole-3-acetic acid. *Biophys. J.* **1997**, *67*, 229–237. [CrossRef]
37. Hunter, E.P.; Desrosiers, M.F.; Simic, M.G. The effect of oxygen, antioxidants, and superoxide radical on tyrosine phenoxyl radical dimerization. *Free Radic. Biol. Med.* **1989**, *6*, 581–585. [CrossRef]
38. Pattison, D.I.; Rahmanto, A.S.; Davies, M.J. Photo-oxidation of proteins. *PhotoChem. PhotoBiol. Sci.* **2012**, *11*, 38–53. [CrossRef]
39. Siodlak, D. α,β-Dehydroamino acids in naturally occurring peptides. *Amino Acids* **2015**, *47*, 1–17. [CrossRef] [PubMed]
40. Friedman, M. Chemistry, biochemistry, nutrition, and microbiology of lysinoalanine, lanthionine, and histidinoalanine in food and other proteins. *J. Agric. Food Chem.* **1999**, *47*, 1295–1319. [CrossRef] [PubMed]
41. Steinmann, D.; Mozziconacci, O.; Bommana, R.; Stobaugh, J.F.; Wang, Y.J.; Schoneich, C. Photodegradation pathways of protein disulfides: Human growth hormone. *Pharm. Res.* **2017**, *34*, 2756–2778. [CrossRef] [PubMed]
42. Nagy, P.; Lechte, T.P.; Das, A.B.; Winterbourn, C.C. Conjugation of glutathione to oxidized tyrosine residues in peptides and proteins. *J. Biol. Chem.* **2012**, *287*, 26068–26076. [CrossRef] [PubMed]
43. Rokhsana, D.; Howells, A.E.; Dooley, D.M.; Szilagyi, R.K. Role of the Tyr-Cys cross-link to the active site properties of galactose oxidase. *Inorg. Chem.* **2012**, *51*, 3513–3524. [CrossRef]
44. Davies, C.G.; Fellner, M.; Tchesnokov, E.P.; Wilbanks, S.M.; Jameson, G.N. The Cys-Tyr cross-link of cysteine dioxygenase changes the optimal pH of the reaction without a structural change. *Biochemistry* **2014**, *53*, 7961–7968. [CrossRef]
45. Wensien, M.; von Pappenheim, F.R.; Funk, L.M.; Kloskowski, P.; Curth, U.; Diederichsen, U.; Uranga, J.; Ye, J.; Fang, P.; Pan, K.T.; et al. A lysine-cysteine redox switch with an NOS bridge regulates enzyme function. *Nature* **2021**, *593*, 460–464. [CrossRef] [PubMed]
46. Van Montfort, R.L.; Congreve, M.; Tisi, D.; Carr, R.; Jhoti, H. Oxidation state of the active-site cysteine in protein tyrosine phosphatase 1B. *Nature* **2003**, *423*, 773–777. [CrossRef] [PubMed]
47. Wang, Z.; Lyons, B.; Truscott, R.J.; Schey, K.L. Human protein aging: Modification and crosslinking through dehydroalanine and dehydrobutyrine intermediates. *Aging Cell* **2014**, *13*, 226–234. [CrossRef] [PubMed]
48. Torosantucci, R.; Mozziconacci, O.; Sharov, V.; Schöneich, C.; Jiskoot, W. Chemical modifications in aggregates of recombinant human insulin induced by metal-catalyzed oxidation: Covalent cross-linking via michael addition to tyrosine oxidation products. *Pharm. Res.* **2012**, *29*, 2276–2293. [CrossRef]
49. Fu, X.; Kao, J.L.; Bergt, C.; Kassim, S.Y.; Huq, N.P.; d'Avignon, A.; Parks, W.C.; Mecham, R.P.; Heinecke, J.W. Oxidative cross-linking of tryptophan to glycine restrains matrix metalloproteinase activity: Specific structural motifs control protein oxidation. *J. Biol. Chem.* **2004**, *279*, 6209–6212. [CrossRef]
50. Zhao, Z.; Engholm-Keller, K.; Poojary, M.M.; Boelt, S.G.; Rogowska-Wrzesinska, A.; Skibsted, L.H.; Davies, M.J.; Lund, M.N. Generation of aggregates of alpha-lactalbumin by UV-B light exposure. *J. Agric. Food Chem.* **2020**, *68*, 6701–6714. [CrossRef] [PubMed]
51. Fuentes-Lemus, E.; Silva, E.; Leinisch, F.; Dorta, E.; Lorentzen, L.G.; Davies, M.J.; López-Alarcón, C. alpha- and beta-casein aggregation induced by riboflavin-sensitized photo-oxidation occurs via di-tyrosine cross-links and is oxygen concentration dependent. *Food Chem.* **2018**, *256*, 119–128. [CrossRef] [PubMed]
52. Colombo, G.; Clerici, M.; Altomare, A.; Rusconi, F.; Giustarini, D.; Portinaro, N.; Garavaglia, M.L.; Rossi, R.; Dalle-Donne, I.; Milzani, A. Thiol oxidation and di-tyrosine formation in human plasma proteins induced by inflammatory concentrations of hypochlorous acid. *J. Proteom.* **2017**, *152*, 22–32. [CrossRef]
53. Leinisch, F.; Mariotti, M.; Rykaer, M.; López-Alarcón, C.; Hägglund, P.; Davies, M.J. Peroxyl radical- and photo-oxidation of glucose 6-phosphate dehydrogenase generates cross-links and functional changes via oxidation of tyrosine and tryptophan residues. *Free Radic. Biol. Med.* **2017**, *112*, 240–252. [CrossRef] [PubMed]
54. Fuentes-Lemus, E.; Mariotti, M.; Hägglund, P.; Leinisch, F.; Fierro, A.; Silva, E.; Davies, M.J.; López-Alarcón, C. Oxidation of lysozyme induced by peroxyl radicals involves amino acid modifications, loss of activity, and formation of specific crosslinks. *Free Radic. Biol. Med.* **2021**, *167*, 258–270. [CrossRef] [PubMed]
55. Fuentes-Lemus, E.; Mariotti, M.; Hägglund, P.; Leinisch, F.; Fierro, A.; Silva, E.; López-Alarcón, C.; Davies, M.J. Binding of rose bengal to lysozyme modulates photooxidation and cross-linking reactions involving tyrosine and tryptophan. *Free Radic. Biol. Med.* **2019**, *143*, 375–386. [CrossRef] [PubMed]
56. Fuentes-Lemus, E.; Mariotti, M.; Reyes, J.; Leinisch, F.; Hägglund, P.; Silva, E.; Davies, M.J.; López-Alarcón, C. Photo-oxidation of lysozyme triggered by riboflavin is O_2-dependent, occurs via mixed type 1 and type 2 pathways, and results in inactivation, site-specific damage and intra- and inter-molecular crosslinks. *Free Radic. Biol. Med.* **2020**, *152*, 61–73. [CrossRef] [PubMed]

57. Mariotti, M.; Rogowska-Wrzesinska, A.; Hägglund, P.; Davies, M.J. Cross-linking and modification of fibronectin by peroxynitrous acid: Mapping and quantification of damage provides a new model for domain interactions. *J. Biol. Chem.* **2021**, *296*, 100360. [CrossRef]
58. Degendorfer, G.; Chuang, C.Y.; Hammer, A.; Malle, E.; Davies, M.J. Peroxynitrous acid induces structural and functional modifications to basement membranes and its key component, laminin. *Free Radic. Biol. Med.* **2015**, *89*, 721–733. [CrossRef] [PubMed]
59. Leinisch, F.; Mariotti, M.; Andersen, S.H.; Lindemose, S.; Hägglund, P.; Mollegaard, N.E.; Davies, M.J. UV oxidation of cyclic AMP receptor protein, a global bacterial gene regulator, decreases DNA binding and cleaves DNA at specific sites. *Sci. Rep.* **2020**, *10*, 3106. [CrossRef]
60. Ursem, R.; Swarge, B.; Abhyankar, W.R.; Buncherd, H.; de Koning, L.J.; Setlow, P.; Brul, S.; Kramer, G. Identification of native cross-links in Bacillus subtilis spore coat proteins. *J. Proteome Res.* **2021**, *20*, 1809–1816. [CrossRef]
61. Leeuwenburgh, C.; Rasmussen, J.E.; Hsu, F.F.; Mueller, D.M.; Pennathur, S.; Heinecke, J.W. Mass spectrometric quantification of markers for protein oxidation by tyrosyl radical, copper, and hydroxyl radical in low density lipoprotein isolated from human atherosclerotic plaques. *J. Biol. Chem.* **1997**, *272*, 3520–3526. [CrossRef]
62. Giulivi, C.; Davies, K.J. Dityrosine and tyrosine oxidation products are endogenous markers for the selective proteolysis of oxidatively modified red blood cell hemoglobin by (the 19 S) proteasome. *J. Biol. Chem.* **1993**, *268*, 8752–8759. [CrossRef]
63. Truscott, R.J.W.; Friedrich, M.G. The etiology of human age-related cataract. Proteins don't last forever. *Biochim. Biophys. Acta* **2016**, *1860*, 192–198. [CrossRef] [PubMed]
64. Serpell, L.C.; Williams, T.L.; Stewart-Parker, M.; Ford, L.; Skaria, E.; Cole, M.; Bucher, W.G.r.; Morris, K.L.; Sada, A.A.; Thorpe, J.R. A central role for dityrosine crosslinking of Amyloid-β in Alzheimer's disease. *Acta Neuropath Commun.* **2013**, *1*, 83. [CrossRef]
65. Tiwari, M.K.; Leinisch, F.; Sahin, C.; Møller, I.M.; Otzen, D.E.; Davies, M.J.; Bjerrum, M.J. Early events in copper-ion catalyzed oxidation of α-synuclein. *Free Radic. Biol. Med.* **2018**, *121*, 38–50. [CrossRef] [PubMed]
66. Pennathur, S.; Jackson-Lewis, V.; Przedborski, S.; Heinecke, J.W. Mass spectrometric quantification of 3-nitrotyrosine, ortho-tyrosine, and o,o′-dityrosine in brain tissue of 1-methyl-4-phenyl-1,2,3,6- tetrahydropyridine-treated mice, a model of oxidative stress in Parkinson's disease. *J. Biol. Chem.* **1999**, *274*, 34621–34628. [CrossRef] [PubMed]
67. Kato, Y.; Maruyama, W.; Naoi, M.; Hashizume, Y.; Osawa, T. Immunohistochemical detection of dityrosine in lipofuscin pigments in the aged human brain. *FEBS Lett.* **1998**, *439*, 231–234. [CrossRef]
68. Kato, Y.; Dozaki, N.; Nakamura, T.; Kitamoto, N.; Yoshida, A.; Naito, M.; Kitamura, M.; Osawa, T. Quantification of modified tyrosines in healthy and diabetic human urine using liquid chromatography/tandem mass spectrometry. *J. Clin. Biochem. Nutr.* **2009**, *44*, 67–78. [CrossRef]
69. Wu, G.R.; Cheserek, M.; Shi, Y.H.; Shen, L.Y.; Yu, J.; Le, G.W. Elevated plasma dityrosine in patients with hyperlipidemia compared to healthy individuals. *Ann. Nutr. Metab.* **2014**, *66*, 44–50. [CrossRef] [PubMed]
70. Ziouzenkova, O.; Asatryan, L.; Akmal, M.; Tetta, C.; Wratten, M.L.; Loseto-Wich, G.; Jurgens, G.; Heinecke, J.; Sevanian, A. Oxidative cross-linking of ApoB100 and hemoglobin results in low density lipoprotein modification in blood. Relevance to atherogenesis caused by hemodialysis. *J. Biol. Chem.* **1999**, *274*, 18916–18924. [CrossRef]
71. Francis, G.A.; Mendez, A.J.; Bierman, E.L.; Heinecke, J.W. Oxidative tyrosylation of high density lipoprotein by peroxidase enhances cholesterol removal from cultured fibroblasts and macrophage foam cells. *Proc. Natl. Acad. Sci. USA* **1993**, *90*, 6631–6635. [CrossRef]
72. Malencik, D.A.; Anderson, S.R. Dityrosine formation in calmodulin: Cross-linking and polymerization catalyzed by Arthromyces peroxidase. *Biochemistry* **1996**, *35*, 4375–4386. [CrossRef] [PubMed]
73. Aeschbach, R.; Amado, R.; Neukom, H. Formation of dityrosine cross-links in proteins by oxidation of tyrosine residues. *Biochim. Biophys. Acta* **1976**, *439*, 292–301. [CrossRef]
74. Das, A.B.; Nauser, T.; Koppenol, W.H.; Kettle, A.J.; Winterbourn, C.C.; Nagy, P. Rapid reaction of superoxide with insulin-tyrosyl radicals to generate a hydroperoxide with subsequent glutathione addition. *Free Radic. Biol. Med.* **2014**, *70*, 86–95. [CrossRef]
75. Gatin, A.; Billault, I.; Duchambon, P.; Van der Rest, G.; Sicard-Roselli, C. Oxidative radicals ($HO^{\cdot}$ or $N_3^{\cdot}$) induce several di-tyrosine bridge isomers at the protein scale. *Free Radic. Biol. Med.* **2021**, *162*, 461–470. [CrossRef]
76. Degendorfer, G.; Chuang, C.Y.; Kawasaki, H.; Hammer, A.; Malle, E.; Yamakura, F.; Davies, M.J. Peroxynitrite-mediated oxidation of plasma fibronectin. *Free Radic. Biol. Med.* **2016**, *97*, 602–615. [CrossRef]
77. Mai, K.; Smith, N.C.; Feng, Z.P.; Katrib, M.; Šlapeta, J.; Šlapetova, I.; Wallach, M.G.; Luxford, C.; Davies, M.J.; Zhang, X.; et al. Peroxidase catalysed cross-linking of an intrinsically unstructured protein via dityrosine bonds in the oocyst wall of the apicomplexan parasite, Eimeria maxima. *Int. J. Parasitol.* **2011**, *41*, 1157–1164. [CrossRef] [PubMed]
78. Medinas, D.B.; Gozzo, F.C.; Santos, L.F.A.; Iglesias, A.H.; Augusto, O. A ditryptophan cross-link is responsible for the covalent dimerization of human superoxide dismutase 1 during its bicarbonate-dependent peroxidase activity. *Free Radic. Biol. Med.* **2010**, *49*, 1046–1053. [CrossRef]
79. Paviani, V.; Queiroz, R.F.; Marques, E.F.; Di Mascio, P.; Augusto, O. Production of lysozyme and lysozyme-superoxide dismutase dimers bound by a ditryptophan cross-link in carbonate radical-treated lysozyme. *Free Radic. Biol. Med.* **2015**, *89*, 72–82. [CrossRef] [PubMed]

80. Paviani, V.; Junqueira de Melo, P.; Avakin, A.; Di Mascio, P.; Ronsein, G.E.; Augusto, O. Human cataractous lenses contain cross-links produced by crystallin-derived tryptophanyl and tyrosyl radicals. *Free Radic. Biol. Med.* **2020**, *160*, 356–367. [CrossRef] [PubMed]
81. Bhaskar, B.; Immoos, C.E.; Shimizu, H.; Sulc, F.; Farmer, P.J.; Poulos, T.L. A novel heme and peroxide-dependent tryptophan-tyrosine cross-link in a mutant of cytochrome c peroxidase. *J. Mol. Biol.* **2003**, *328*, 157–166. [CrossRef]
82. Liu, M.; Zhang, Z.; Cheetham, J.; Ren, D.; Zhou, Z.S. Discovery and characterization of a photo-oxidative histidine-histidine cross-link in IgG1 antibody utilizing ^{18}O-labeling and mass spectrometry. *Anal. Chem.* **2014**, *86*, 4940–4948. [CrossRef]
83. Powell, T.; Knight, M.J.; O'Hara, J.; Burkitt, W. Discovery of a photoinduced histidine-histidine cross-link in an IgG4 antibody. *J. Am. Soc. Mass Spectrom.* **2020**, *31*, 1233–1240. [CrossRef] [PubMed]
84. Xu, C.F.; Chen, Y.; Yi, L.; Brantley, T.; Stanley, B.; Sosic, Z.; Zang, L. Discovery and characterization of histidine oxidation initiated cross-links in an IgG1 monoclonal antibody. *Anal. Chem.* **2017**, *89*, 7915–7923. [CrossRef]
85. Agon, V.V.; Bubb, W.A.; Wright, A.; Hawkins, C.L.; Davies, M.J. Sensitizer-mediated photooxidation of histidine residues: Evidence for the formation of reactive side-chain peroxides. *Free Radic Biol. Med.* **2006**, *40*, 698–710. [CrossRef] [PubMed]
86. Leinisch, F.; Mariotti, M.; Hägglund, P.; Davies, M.J. Structural and functional changes in RNAse A originating from tyrosine and histidine cross-linking and oxidation. *Free Radic. Biol. Med.* **2018**, *126*, 73–86. [CrossRef] [PubMed]
87. Torosantucci, R.; Sharov, V.S.; Van Beers, M.; Brinks, V.; Schöneich, C.; Jiskoot, W. Identification of oxidation sites and covalent cross-links in metal catalyzed oxidized interferon beta-1a: Potential implications for protein aggregation and immunogenicity. *Mol. Pharm.* **2013**, *10*, 2311–2322. [CrossRef]
88. Oka, O.B.V.; Bulleid, N.J. Forming disulfides in the endoplasmic reticulum. *Biochim Biophys. Acta Mol. Cell Res.* **2013**, *1833*, 2425–2429. [CrossRef]
89. Nagy, P. Kinetics and mechanisms of thiol–disulfide exchange covering direct substitution and thiol oxidation-mediated pathways. *Antioxid. Redox Signal.* **2013**, *18*, 1623–1641. [CrossRef] [PubMed]
90. Singh, R.; Whitesides, G.M. Degenerate intermolecular thiolate-disulfide Interchange involving cyclic five-membered disulfides is faster by ~10^3 than that involving six- or seven-membered disulfides. *J. Am. Chem. Soc.* **1990**, *112*, 6304–6309. [CrossRef]
91. Berndt, C.; Lillig, C.H.; Holmgren, A. Thiol-based mechanisms of the thioredoxin and glutaredoxin systems: Implications for diseases in the cardiovascular system. *Am. J. Physiol. Heart Circ. Physiol.* **2007**, *292*, H1227–H1236. [CrossRef]
92. Carroll, L.; Jiang, S.; Irnstorfer, J.; Beneyto, S.; Ignasiak, M.T.; Rasmussen, L.M.; Rogowska-Wrzesinska, A.; Davies, M.J. Oxidant-induced glutathionylation at protein disulfide bonds. *Free Radic. Biol. Med.* **2020**, *160*, 513–525. [CrossRef]
93. Jiang, S.; Carroll, L.; Mariotti, M.; Hägglund, P.; Davies, M.J. Formation of protein cross-links by singlet oxygen-mediated disulfide oxidation. *Redox Biol.* **2021**, *41*, 101874. [CrossRef]
94. Jiang, S.; Hägglund, P.; Carroll, L.; Rasmussen, L.M.; Davies, M.J. Crosslinking of human plasma C-reactive protein to human serum albumin via disulfide bond oxidation. *Redox Biol.* **2021**, *41*, 101925. [CrossRef]
95. Yang, J.; Carroll, K.S.; Liebler, D.C. The expanding landscape of the thiol redox proteome. *Mol. Cell Proteom.* **2016**, *15*, 1–11. [CrossRef]
96. Gupta, V.; Paritala, H.; Carroll, K.S. Reactivity, selectivity, and stability in sulfenic acid detection: A comparative study of nucleophilic and electrophilic probes. *Bioconjug. Chem.* **2016**, *27*, 1411–1418. [CrossRef]
97. Ronsein, G.E.; Winterbourn, C.C.; Di Mascio, P.; Kettle, A.J. Cross-linking methionine and amine residues with reactive halogen species. *Free Radic. Biol. Med.* **2014**, *70*, 278–287. [CrossRef]
98. Harwood, D.T.; Kettle, A.J.; Winterbourn, C.C. Production of glutathione sulfonamide and dehydroglutathione from GSH by myeloperoxidase-derived oxidants and detection using a novel LC-MS/MS method. *Biochem. J.* **2006**, *399*, 161–168. [CrossRef]
99. Raftery, M.J.; Yang, Z.; Valenzuela, S.M.; Geczy, C.L. Novel intra- and inter-molecular sulfinamide bonds in S100A8 produced by hypochlorite oxidation. *J. Biol. Chem.* **2001**, *276*, 33393–33401. [CrossRef]
100. Fu, X.; Mueller, D.M.; Heinecke, J.W. Generation of intramolecular and intermolecular sulfenamides, sulfinamides, and sulfonamides by hypochlorous acid: A potential pathway for oxidative cross-linking of low-density lipoprotein by myeloperoxidase. *Biochemistry* **2002**, *41*, 1293–1301. [CrossRef]
101. Pullar, J.M.; Vissers, M.C.; Winterbourn, C.C. Glutathione oxidation by hypochlorous acid in endothelial cells produces glutathione sulfonamide as a major product but not glutathione disulfide. *J. Biol. Chem.* **2001**, *276*, 22120–22125. [CrossRef]
102. Carr, A.C.; Winterbourn, C.C. Oxidation of neutrophil glutathione and protein thiols by myeloperoxidase-derived hypochlorous acid. *Biochem. J.* **1997**, *327*, 275–281. [CrossRef]
103. Harwood, D.T.; Kettle, A.J.; Brennan, S.; Winterbourn, C.C. Simultaneous determination of reduced glutathione, glutathione disulphide and glutathione sulphonamide in cells and physiological fluids by isotope dilution liquid chromatography-tandem mass spectrometry. *J. Chromatogr. B Analyt. Technol. Biomed. Life Sci.* **2009**, *877*, 3393–3399. [CrossRef] [PubMed]
104. Kettle, A.J.; Turner, R.; Gangell, C.L.; Harwood, D.T.; Khalilova, I.S.; Chapman, A.L.; Winterbourn, C.C.; Sly, P.D. Oxidation contributes to low glutathione in the airways of children with cystic fibrosis. *Eur. Respir. J.* **2014**, *44*, 122–129. [CrossRef]
105. Winterbourn, C.C.; Brennan, S.O. Characterization of the oxidation products of the reaction between reduced glutathione and hypochlorous acid. *Biochem. J.* **1997**, *326*, 87–92. [CrossRef]
106. Erve, J.C.; Svensson, M.A.; von Euler-Chelpin, H.; Klasson-Wehler, E. Characterization of glutathione conjugates of the remoxipride hydroquinone metabolite NCQ-344 formed in vitro and detection following oxidation by human neutrophils. *Chem. Res. Toxicol.* **2004**, *17*, 564–571. [CrossRef] [PubMed]

107. Shetty, V.; Spellman, D.S.; Neubert, T.A. Characterization by tandem mass spectrometry of stable cysteine sulfenic acid in a cysteine switch peptide of matrix metalloproteinases. *J. Am. Soc. Mass Spectrom* **2007**, *18*, 1544–1551. [CrossRef]
108. Winterbourn, C.C. Comparative reactivities of various biological compounds with myeloperoxidase-hydrogen peroxide-chloride, and similarity of the oxidant to hypochlorite. *Biochim. Biophys. Acta* **1985**, *840*, 204–210. [CrossRef]
109. Mozziconacci, O.; Kerwin, B.A.; Schoneich, C. Reversible hydrogen transfer reactions of cysteine thiyl radicals in peptides: The conversion of cysteine into dehydroalanine and alanine, and of alanine into dehydroalanine. *J. Phys. Chem. B* **2011**, *115*, 12287–12305. [CrossRef]
110. Shu, N.; Lorentzen, L.G.; Davies, M.J. Reaction of quinones with proteins: Kinetics of adduct formation, effects on enzymatic activity and protein structure, and potential reversibility of modifications. *Free Radic. Biol. Med.* **2019**, *137*, 169–180. [CrossRef] [PubMed]
111. Sauerland, M.; Mertes, R.; Morozzi, C.; Eggler, A.L.; Gamon, L.F.; Davies, M.J. Kinetic assessment of Michael addition reactions of alpha, beta-unsaturated carbonyl compounds to amino acid and protein thiols. *Free Radic. Biol. Med.* **2021**, *169*, 1–11. [CrossRef] [PubMed]
112. Li, W.W.; Heinze, J.; Haehnel, W. Site-specific binding of quinones to proteins through thiol addition and addition-elimination reactions. *J. Am. Chem. Soc.* **2005**, *127*, 6140–6141. [CrossRef]
113. Miura, T.; Kakehashi, H.; Shinkai, Y.; Egara, Y.; Hirose, R.; Cho, A.K.; Kumagai, Y. GSH-mediated S-transarylation of a quinone glyceraldehyde-3-phosphate dehydrogenase conjugate. *Chem. Res. Toxicol.* **2011**, *24*, 1836–1844. [CrossRef]
114. Truscott, R.J. Age-related nuclear cataract-oxidation is the key. *Exp. Eye Res.* **2005**, *80*, 709–725. [CrossRef] [PubMed]
115. McIntire, W.S. Quinoproteins. *FASEB J.* **1994**, *8*, 513–521. [CrossRef]
116. Nappi, A.J.; Vass, E. The effects of glutathione and ascorbic acid on the oxidations of 6-hydroxydopa and 6-hydroxydopamine. *Biochim. Biophys. Acta* **1994**, *1201*, 498–504. [CrossRef]
117. Datta, S.; Mori, Y.; Takagi, K.; Kawaguchi, K.; Chen, Z.W.; Okajima, T.; Kuroda, S.; Ikeda, T.; Kano, K.; Tanizawa, K.; et al. Structure of a quinohemoprotein amine dehydrogenase with an uncommon redox cofactor and highly unusual crosslinking. *Proc. Natl. Acad. Sci. USA* **2001**, *98*, 14268–14273. [CrossRef]
118. Davidson, V.L.; Wilmot, C.M. Posttranslational biosynthesis of the protein-derived cofactor tryptophan tryptophylquinone. *Ann. Rev. Biochem.* **2013**, *82*, 531–550. [CrossRef]
119. Mozziconacci, O.; Williams, T.D.; Kerwin, B.A.; Schoneich, C. Reversible intramolecular hydrogen transfer between protein cysteine thiyl radicals and alpha C-H bonds in insulin: Control of selectivity by secondary structure. *J. Phys. Chem. B* **2008**, *112*, 15921–15932. [CrossRef] [PubMed]
120. Schoneich, C. Thiyl radical reactions in the chemical degradation of pharmaceutical proteins. *Molecules* **2019**, *24*, 4357. [CrossRef]
121. Mozziconacci, O.; Haywood, J.; Gorman, E.M.; Munson, E.; Schoneich, C. Photolysis of recombinant human insulin in the solid state: Formation of a dithiohemiacetal product at the C-terminal disulfide bond. *Pharm. Res.* **2012**, *29*, 121–133. [CrossRef] [PubMed]
122. Orian, L.; Mauri, P.; Roveri, A.; Toppo, S.; Benazzi, L.; Bosello-Travain, V.; De Palma, A.; Maiorino, M.; Miotto, G.; Zaccarin, M.; et al. Selenocysteine oxidation in glutathione peroxidase catalysis: An MS-supported quantum mechanics study. *Free Radic. Biol. Med.* **2015**, *87*, 1–14. [CrossRef]
123. Mauri, P.; Benazzi, L.; Flohe, L.; Maiorino, M.; Pietta, P.G.; Pilawa, S.; Roveri, A.; Ursini, F. Versatility of selenium catalysis in PHGPx unraveled by LC/ESI-MS/MS. *Biol. Chem.* **2003**, *384*, 575–588. [CrossRef] [PubMed]
124. Zhong, L.; Arner, E.S.; Holmgren, A. Structure and mechanism of mammalian thioredoxin reductase: The active site is a redox-active selenolthiol/selenenylsulfide formed from the conserved cysteine-selenocysteine sequence. *Proc. Natl. Acad. Sci. USA* **2000**, *97*, 5854–5859. [CrossRef] [PubMed]
125. Heinecke, J.W.; Shapiro, B.M. Respiratory burst oxidase of fertilization. *Proc. Natl. Acad. Sci. USA* **1989**, *86*, 1259–1263. [CrossRef]
126. Sies, H.; Berndt, C.; Jones, D.P. Oxidative Stress. *Ann. Rev. Biochem.* **2017**, *86*, 715–748. [CrossRef]
127. Sies, H. Oxidative stress: Concept and some practical aspects. *Antioxidants* **2020**, *9*, 852. [CrossRef]
128. Hellwig, M. The chemistry of protein oxidation in food. *Angew. Chem. Int. Ed. Engl.* **2019**, *58*, 16742–16763. [CrossRef]
129. Xiong, Y.L.; Guo, A. Animal and plant protein oxidation: Chemical and functional property significance. *Foods* **2020**, *10*, 40. [CrossRef] [PubMed]
130. Houée-Lévin, C.; Bobrowski, K.; Horakova, L.; Karademir, B.; Schöneich, C.; Davies, M.J.; Spickett, C.M. Exploring oxidative modifications of tyrosine: An update on mechanisms of formation, advances in analysis and biological consequences. *Free Radic. Res.* **2015**, *49*, 347–373. [CrossRef] [PubMed]
131. Gross, A.J.; Sizer, I.W. The oxidation of tyramine, tyrosine, and related compounds by peroxidase. *J. Biol. Chem.* **1959**, *234*, 1611–1614. [CrossRef]
132. Liu, C.; Hua, J.; Ng, P.F.; Fei, B. Photochemistry of bioinspired dityrosine crosslinking. *J. Mater. Sci. Technol.* **2021**, *63*, 182–191. [CrossRef]
133. Heinecke, J.W.; Li, W.; Francis, G.A.; Goldstein, J.A. Tyrosyl radical generated by myeloperoxidase catalyzes the oxidative cross-linking of proteins. *J. Clin. Investig.* **1993**, *91*, 2866–2872. [CrossRef]
134. Giulivi, C.; Traaseth, N.J.; Davies, K.J. Tyrosine oxidation products: Analysis and biological relevance. *Amino Acids* **2003**, *25*, 227–232. [CrossRef] [PubMed]

135. Karmakar, S.; Datta, A. Understanding the reactivity of $CO_3^{\cdot-}$ and $NO_2^{\cdot}$ radicals toward S-containing and aromatic amino acids. *J. Phys. Chem. B* **2017**, *121*, 7621–7632. [CrossRef] [PubMed]
136. Winterbourn, C.C.; Parsons-Mair, H.N.; Gebicki, S.; Gebicki, J.M.; Davies, M.J. Requirements for superoxide-dependent tyrosine hydroperoxide formation in peptides. *Biochem. J.* **2004**, *381*, 241–248. [CrossRef]
137. Folkes, L.K.; Bartesaghi, S.; Trujillo, M.; Wardman, P.; Radi, R. The effects of nitric oxide or oxygen on the stable products formed from the tyrosine phenoxyl radical. *Free Radic. Res.* **2021**, *55*, 141–153. [CrossRef]
138. Gebicki, J.M.; Nauser, T.; Domazou, A.; Steinmann, D.; Bounds, P.L.; Koppenol, W.H. Reduction of protein radicals by GSH and ascorbate: Potential biological significance. *Amino Acids* **2010**, *39*, 1131–1137. [CrossRef] [PubMed]
139. Nauser, T.; Koppenol, W.H.; Gebicki, J.M. The kinetics of oxidation of GSH by protein radicals. *Biochem. J.* **2005**, *392*, 693–701. [CrossRef] [PubMed]
140. Nauser, T.; Gebicki, J.M. Antioxidants and radical damage in a hydrophilic environment: Chemical reactions and concepts. *Essays Biochem.* **2020**, *64*, 67–74. [CrossRef]
141. Mukherjee, S.; Kapp, E.A.; Lothian, A.; Roberts, A.M.; Vasil'Ev, Y.V.; Boughton, B.A.; Barnham, K.J.; Kok, W.M.; Hutton, C.A.; Masters, C.L.; et al. Characterization and identification of dityrosine cross-linked peptides using tandem mass spectrometry. *Anal. Chem.* **2017**, *89*, 6136–6145. [CrossRef] [PubMed]
142. Tew, D.; Ortiz de Montellano, P.R. The myoglobin protein radical. Coupling of Tyr-103 to Tyr-151 in the H_2O_2-mediated cross-linking of sperm whale myoglobin. *J. Biol. Chem.* **1988**, *263*, 17880–17886. [CrossRef]
143. Das, A.B.; Nagy, P.; Abbott, H.F.; Winterbourn, C.C.; Kettle, A.J. Reactions of superoxide with the myoglobin tyrosyl radical. *Free. Radic. Biol. Med.* **2010**, *48*, 1540–1547. [CrossRef]
144. Kehm, R.; Baldensperger, T.; Raupbach, J.; Hohn, A. Protein oxidation—Formation mechanisms, detection and relevance as biomarkers in human diseases. *Redox Biol.* **2021**, *42*, 101901. [CrossRef]
145. DiMarco, T.; Giulivi, C. Current analytical methods for the detection of dityrosine, a biomarker of oxidative stress, in biological samples. *Mass Spectrom. Rev.* **2007**, *26*, 108–120. [CrossRef] [PubMed]
146. Chen, Z.; Leinisch, F.; Greco, I.; Zhang, W.; Shu, N.; Chuang, C.Y.; Lund, M.N.; Davies, M.J. Characterisation and quantification of protein oxidative modifications and amino acid racemisation in powdered infant milk formula. *Free Radic. Res.* **2019**, *53*, 68–81. [CrossRef] [PubMed]
147. Fenaille, F.; Parisod, V.; Vuichoud, J.; Tabet, J.C.; Guy, P.A. Quantitative determination of dityrosine in milk powders by liquid chromatography coupled to tandem mass spectrometry using isotope dilution. *J. Chromatogr. A* **2004**, *1052*, 77–84. [CrossRef] [PubMed]
148. Rodriguez-Mateos, A.; Millar, S.J.; Bhandari, D.G.; Frazier, R.A. Formation of dityrosine cross-links during breadmaking. *J. Agric. Food Chem.* **2006**, *54*, 2761–2766. [CrossRef] [PubMed]
149. Ma, J.B.; Wang, X.Y.; Li, Q.; Zhang, L.; Wang, Z.; Han, L.; Yu, Q.L. Oxidation of myofibrillar protein and crosslinking behavior during processing of traditional air-dried yak (Bos grunniens) meat in relation to digestibility. *LWT Food Sci. Technol.* **2021**, *142*, 110984. [CrossRef]
150. Ma, L.; Li, A.L.; Li, T.Q.; Li, M.; Wang, X.D.; Hussain, M.A.; Qayum, A.; Jiang, Z.M.; Hou, J.C. Structure and characterization of laccase-crosslinked alpha-lactalbumin: Impacts of high pressure homogenization pretreatment. *LWT Food Sci. Technol.* **2020**, *118*, 108843. [CrossRef]
151. Yang, Y.; Zhang, H.; Yan, B.; Zhang, T.; Gao, Y.; Shi, Y.; Le, G. Health effects of dietary oxidized tyrosine and dityrosine administration in mice with nutrimetabolomic strategies. *J. Agric. Food Chem.* **2017**, *65*, 6957–6971. [CrossRef] [PubMed]
152. Ding, Y.Y.; Tang, X.; Cheng, X.R.; Wang, F.F.; Li, Z.Q.; Wu, S.J.; Kou, X.R.; Shi, Y.H.; Le, G.W. Effects of dietary oxidized tyrosine products on insulin secretion via the thyroid hormone T3-regulated TR beta 1-Akt-mTOR pathway in the pancreas. *RSC Adv.* **2017**, *7*, 54610–54625. [CrossRef]
153. Li, Z.L.; Shi, Y.; Ding, Y.; Ran, Y.; Le, G. Dietary oxidized tyrosine (O-Tyr) stimulates TGF-beta1-induced extracellular matrix production via the JNK/p38 signaling pathway in rat kidneys. *Amino Acids* **2017**, *49*, 241–260. [CrossRef] [PubMed]
154. Li, B.W.; Mo, L.; Yang, Y.H.; Zhang, S.; Xu, J.B.; Ge, Y.T.; Xu, Y.C.; Shi, Y.H.; Le, G.W. Processing milk causes the formation of protein oxidation products which impair spatial learning and memory in rats. *RSC Adv.* **2019**, *9*, 22161–22175. [CrossRef]
155. Schaefer, J.; Kramer, K.J.; Garbow, J.R.; Jacob, G.S.; Stejskal, E.O.; Hopkins, T.L.; Speirs, R.D. Aromatic cross-links in insect cuticle: Detection by solid-state ^{13}C and ^{15}N NMR. *Science* **1987**, *235*, 1200–1204. [CrossRef]
156. Bellmaine, S.; Schnellbaecher, A.; Zimmer, A. Reactivity and degradation products of tryptophan in solution and proteins. *Free Radic. Biol. Med.* **2020**, *160*, 696–718. [CrossRef]
157. Coelho, F.R.; Iqbal, A.; Linares, E.; Silva, D.F.; Lima, F.S.; Cuccovia, I.M.; Augusto, O. Oxidation of the tryptophan 32 residue of human superoxide dismutase 1 caused by its bicarbonate-dependent peroxidase activity triggers the non-amyloid aggregation of the enzyme. *J. Biol. Chem.* **2014**, *289*, 30690–30701. [CrossRef]
158. Figueroa, J.D.; Zarate, A.M.; Fuentes-Lemus, E.; Davies, M.J.; López-Alarcón, C. Formation and characterization of crosslinks, including Tyr-Trp species, on one electron oxidation of free Tyr and Trp residues by carbonate radical anion. *RSC Adv.* **2020**, *10*, 25786–25800. [CrossRef]
159. Carroll, L.; Pattison, D.I.; Davies, J.B.; Anderson, R.F.; López-Alarcón, C.; Davies, M.J. Formation and detection of oxidant-generated tryptophan dimers in peptides and proteins. *Free Radic. Biol. Med.* **2017**, *113*, 132–142. [CrossRef]

160. Zhuravleva, Y.S.; Sherin, P.S. Influence of pH on radical reactions between kynurenic acid and amino acids tryptophan and tyrosine. Part II. Amino acids within the protein globule of lysozyme. *Free Radic. Biol. Med.* **2021**, *174*, 211–224. [CrossRef]
161. Sormacheva, E.D.; Sherin, P.S.; Tsentalovich, Y.P. Dimerization and oxidation of tryptophan in UV-A photolysis sensitized by kynurenic acid. *Free Radic. Biol. Med.* **2017**, *113*, 372–384. [CrossRef]
162. Silva, E.; Barrias, P.; Fuentes-Lemus, E.; Tirapegui, C.; Aspee, A.; Carroll, L.; Davies, M.J.; López-Alarcón, C. Riboflavin-induced Type 1 photo-oxidation of tryptophan using a high intensity 365nm light emitting diode. *Free Radic. Biol. Med.* **2019**, *131*, 133–143. [CrossRef]
163. Sherin, P.S.; Zelentsova, E.A.; Sormacheva, E.D.; Yanshole, V.V.; Duzhak, T.G.; Tsentalovich, Y.P. Aggregation of α-crystallins in kynurenic acid-sensitized UVA photolysis under anaerobic conditions. *Phys. Chem. Chem. Phys.* **2016**, *18*, 8827–8839. [CrossRef] [PubMed]
164. Reyes, J.S.; Fuentes-Lemus, E.; Aspee, A.; Davies, M.J.; Monasterio, O.; López-Alarcón, C.M. *Jannaschii* FtsZ, a key protein in bacterial cell division, is inactivated by peroxyl radical-mediated methionine oxidation. *Free Radic. Biol. Med.* **2021**, *166*, 53–66. [CrossRef]
165. Barik, S. The uniqueness of tryptophan in biology: Properties, metabolism, interactions and localization in proteins. *Int. J. Mol. Sci.* **2020**, *21*, 8776. [CrossRef] [PubMed]
166. Leo, G.; Altucci, C.; Bourgoin-Voillard, S.; Gravagnuolo, A.M.; Esposito, R.; Marino, G.; Costello, C.E.; Velotta, R.; Birolo, L. Ultraviolet laser-induced cross-linking in peptides. *Rapid Commun. Mass Spectrom.* **2013**, *27*, 1660–1668. [CrossRef]
167. Mariotti, M.; Leinisch, F.; Leeming, D.J.; Svensson, B.; Davies, M.J.; Hägglund, P. Mass-spectrometry-based identification of cross-links in proteins exposed to photo-oxidation and peroxyl radicals using ^{18}O labeling and optimized tandem mass spectrometry fragmentation. *J. Proteome Res.* **2018**, *17*, 2017–2027. [CrossRef]
168. Gebicki, J.M. Protein hydroperoxides as new reactive oxygen species. *Redox Rep.* **1997**, *3*, 99–110. [CrossRef] [PubMed]
169. Hanson, D.A.; Eyre, D.R. Molecular site specificity of pyridinoline and pyrrole cross-links in type I collagen of human bone. *J. Biol. Chem.* **1996**, *271*, 26508–26516. [CrossRef]
170. Wilhelmus, M.M.M.; Grunberg, S.C.S.; Bol, J.G.J.M.; Van Dam, A.M.; Hoozemans, J.J.M.; Rozemuller, A.J.M.; Drukarch, B. Transglutaminases and transglutaminase-catalyzed cross-links colocalize with the pathological lesions in Alzheimer's disease brain. *Brain Pathol.* **2009**, *19*, 612–622. [CrossRef]
171. De Jong, G.A.H.; Koppelman, S.J. Transglutaminase catalyzed reactions: Impact on food applications. *J. Food Sci.* **2002**, *67*, 2798–2806. [CrossRef]
172. Griffin, M.; Casadio, R.; Bergamini, C.M. Transglutaminases: Nature's biological glues. *Biochem. J.* **2002**, *368*, 377–396. [CrossRef] [PubMed]
173. Shen, H.-R.; Spikes, J.D.; Smith, C.J.; Kopeček, J. Photodynamic cross-linking of proteins: IV. Nature of the His-His bond(s) formed in the rose bengal-photosensitized cross-linking of N-benzoyl-L-histidine. *J. PhotoChem. PhotoBiol. A Chem.* **2000**, *130*, 1–6. [CrossRef]
174. Tanzer, M.L.; Housley, T.; Berube, L.; Fairweather, R.; Franzblau, C.; Gallop, P.M. Structure of two histidine-containing crosslinks from collagen. *J. Biol. Chem.* **1973**, *248*, 393–402. [CrossRef]
175. Li, Y.; Jongberg, S.; Andersen, M.L.; Davies, M.J.; Lund, M.N. Quinone-induced protein modifications: Kinetic preference for reaction of 1,2-benzoquinones with thiol groups in proteins. *Free Radic. Biol. Med.* **2016**, *97*, 148–157. [CrossRef] [PubMed]
176. Marmelstein, A.M.; Lobba, M.J.; Mogilevsky, C.S.; Maza, J.C.; Brauer, D.D.; Francis, M.B. Tyrosinase-mediated oxidative coupling of tyrosine tags on peptides and proteins. *J. Am. Chem. Soc.* **2020**, *142*, 5078–5086. [CrossRef]
177. Rzepecki, L.M.; Nagafuchi, T.; Waite, J.H. α,β-Dehydro-3,4-dihydroxyphenylalanine derivatives: Potential schlerotization intermediates in natural composite materials. *Arch. Biochem. Biophys.* **1991**, *285*, 17–26. [CrossRef]
178. Burzio, L.A.; Waite, J.H. Cross-linking in adhesive proteins: Studies with model decapeptides. *Biochemistry* **2000**, *39*, 11147–11153. [CrossRef]
179. Rzepecki, L.M.; Waite, J.H. Wresting the muscle from mussel beards: Research and applications. *Mol. Mar. Biol. Biotechnol.* **1995**, *4*, 313–322.
180. Mishler-Elmore, J.W.; Zhou, Y.; Sukul, A.; Oblak, M.; Tan, L.; Faik, A.; Held, M.A. Extensins: Self-assembly, crosslinking, and the role of peroxidases. *Front. Plant. Sci.* **2021**, *12*, 664738. [CrossRef]
181. Breusing, N.; Grune, T. Biomarkers of protein oxidation from a chemical, biological and medical point of view. *Exp. Gerontol.* **2010**, *45*, 733–737. [CrossRef] [PubMed]
182. Lopezllorca, L.V.; Fry, S.C. Dityrosine, trityrosine and tetratyrosine, potential cross-links in structural proteins of plant-parasitic nematodes. *Nematologica* **1989**, *35*, 165–179. [CrossRef]
183. Dhayal, S.K.; Sforza, S.; Wierenga, P.A.; Gruppen, H. Peroxidase induced oligo-tyrosine cross-links during polymerization of alpha-lactalbumin. *Biochim Biophys. Acta Proteins Proteom.* **2015**, *1854*, 1898–1905. [CrossRef] [PubMed]
184. Reid, L.O.; Vignoni, M.; Martins-Froment, N.; Thomas, A.H.; Dantola, M.L. Photochemistry of tyrosine dimer: When an oxidative lesion of proteins is able to photoinduce further damage. *PhotoChem. PhotoBiol. Sci.* **2019**, *18*, 1732–1741. [CrossRef]
185. Sommer, A.; Traut, R.R. Diagonal polyacrylamide-dodecyl sulfate gel electrophoresis for the identification of ribosomal proteins crosslinked with methyl-4-mercaptobutyrimidate. *Proc. Natl. Acad. Sci. USA* **1974**, *71*, 3946–3950. [CrossRef]

186. Al-Hilaly, Y.K.; Biasetti, L.; Blakeman, B.J.F.; Pollack, S.J.; Zibaee, S.; Abdul-Sada, A.; Thorpe, J.R.; Xue, W.-F.; Serpell, L.C.; Goedert, M.; et al. The involvement of dityrosine crosslinking in α-synuclein assembly and deposition in Lewy bodies in Parkinson's disease. *Sci. Rep.* **2016**, *6*, 39171. [CrossRef]
187. Kato, Y.; Wu, X.; Naito, M.; Nomura, H.; Kitamoto, N.; Osawa, T. Immunochemical detection of protein dityrosine in atherosclerotic lesion of apo-E-deficient mice using a novel monoclonal antibody. *Biochem. Biophys. Res. Commun.* **2000**, *275*, 11–15. [CrossRef]
188. Degendorfer, G.; Chuang, C.Y.; Mariotti, M.; Hammer, A.; Hoefler, G.; Hägglund, P.; Malle, E.; Wise, S.G.; Davies, M.J. Exposure of tropoelastin to peroxynitrous acid gives high yields of nitrated tyrosine residues, di-tyrosine cross-links and altered protein structure and function. *Free Radic. Biol. Med.* **2017**, *115*, 219–231. [CrossRef] [PubMed]
189. Huang, Y.R.; Hua, Y.F.; Qiu, A.Y. Soybean protein aggregation induced by lipoxygenase catalyzed linoleic acid oxidation. *Food Res. Int.* **2006**, *39*, 240–249. [CrossRef]
190. Cui, X.H.; Xiong, Y.L.L.; Kong, B.H.; Zhao, X.H.; Liu, N. Hydroxyl radical-stressed whey protein isolate: Chemical and structural properties. *Food Bioprocess. Technol.* **2012**, *5*, 2454–2461. [CrossRef]
191. Hawkins, C.L.; Morgan, P.E.; Davies, M.J. Quantification of protein modification by oxidants. *Free Radic. Biol. Med.* **2009**, *46*, 965–988. [CrossRef] [PubMed]
192. Hawkins, C.L.; Davies, M.J. Detection, identification and quantification of oxidative protein modifications. *J. Biol. Chem.* **2019**, *294*, 19683–19708. [CrossRef] [PubMed]
193. Gamon, L.F.; Guo, C.; He, J.; Hägglund, P.; Hawkins, C.L.; Davies, M.J. Absolute quantitative analysis of intact and oxidized amino acids by LC-MS without prior derivatization. *Redox Biol.* **2020**, *36*, 101586. [CrossRef]
194. Desmons, A.; Thioulouse, E.; Hautem, J.Y.; Saintier, A.; Baudin, B.; Lamaziere, A.; Netter, C.; Moussa, F. Direct liquid chromatography tandem mass spectrometry analysis of amino acids in human plasma. *J. Chromatogr. A* **2020**, *1622*, 461135. [CrossRef] [PubMed]
195. Hensley, K.; Maidt, M.L.; Yu, Z.; Sang, H.; Markesbery, W.R.; Floyd, R.A. Electrochemical analysis of protein nitrotyrosine and dityrosine in the Alzheimer brain indicates region-specific accumulation. *J. NeuroSci.* **1998**, *18*, 8126–8132. [CrossRef]
196. Verzini, S.; Shah, M.; Theillet, F.X.; Belsom, A.; Bieschke, J.; Wanker, E.E.; Rappsilber, J.; Binolfi, A.; Selenko, P. Megadalton-sized dityrosine aggregates of alpha-synuclein retain high degrees of structural disorder and internal dynamics. *J. Mol. Biol.* **2020**, *432*, 166689. [CrossRef]
197. Thorn, D.C.; Grosas, A.B.; Mabbitt, P.D.; Ray, N.J.; Jackson, C.J.; Carver, J.A. The structure and stability of the disulfide-linked gammaS-crystallin dimer provide insight into oxidation products associated with lens cataract formation. *J. Mol. Biol.* **2019**, *431*, 483–497. [CrossRef]
198. Evrard, C.; Capron, A.; Marchand, C.; Clippe, A.; Wattiez, R.; Soumillion, P.; Knoops, B.; Declercq, J.P. Crystal structure of a dimeric oxidized form of human peroxiredoxin 5. *J. Mol. Biol.* **2004**, *337*, 1079–1090. [CrossRef] [PubMed]
199. Smeets, A.; Evrard, C.; Landtmeters, M.; Marchand, C.; Knoops, B.; Declercq, J.P. Crystal structures of oxidized and reduced forms of human mitochondrial thioredoxin 2. *Protein Sci.* **2005**, *14*, 2610–2621. [CrossRef]
200. Marvin, L.F.; Delatour, T.; Tavazzi, I.; Fay, L.B.; Cupp, C.; Guy, P.A. Quantification of o,o'-dityrosine, o-nitrotyrosine, and o-tyrosine in cat urine samples by LC/ electrospray ionization-MS/MS using isotope dilution. *Anal. Chem.* **2003**, *75*, 261–267. [CrossRef]
201. Abdelrahim, M.; Morris, E.; Carver, J.; Facchina, S.; White, A.; Verma, A. Liquid chromatographic assay of dityrosine in human cerebrospinal fluid. *J. Chromatogr. B Biomed. Sci.* **1997**, *696*, 175–182. [CrossRef]
202. Rose, K.; Savoy, L.A.; Simona, M.G.; Offord, R.E.; Wingfield, P. C-terminal peptide identification by fast atom bombardment mass spectrometry. *Biochem. J.* **1988**, *250*, 253–259. [CrossRef] [PubMed]
203. Liu, M.; Zhang, Z.; Zang, T.; Spahr, C.; Cheetham, J.; Ren, D.; Zhou, Z.S. Discovery of undefined protein cross-linking chemistry: A comprehensive methodology utilizing ^{18}O-labeling and mass spectrometry. *Anal. Chem.* **2013**, *85*, 5900–5908. [CrossRef] [PubMed]
204. Cramer, C.N.; Kelstrup, C.D.; Olsen, J.V.; Haselmann, K.F.; Nielsen, P.K. Generic workflow for mapping of complex disulfide bonds using in-source reduction and extracted ion chromatograms from data-dependent mass spectrometry. *Anal. Chem.* **2018**, *90*, 8202–8210. [CrossRef]
205. Massonnet, P.; Haler, J.R.N.; Upert, G.; Smargiasso, N.; Mourier, G.; Gilles, N.; Quinton, L.; De Pauw, E. Disulfide connectivity analysis of peptides bearing two intramolecular disulfide bonds using MALDI in-source decay. *J. Am. Soc. Mass Spectrom.* **2018**, *29*, 1995–2002. [CrossRef]
206. Hägglund, P.; Bunkenborg, J.; Maeda, K.; Svensson, B. Identification of thioredoxin disulfide targets using a quantitative proteomics approach based on isotope-coded affinity tags. *J. Proteome Res.* **2008**, *7*, 5270–5276. [CrossRef]
207. Gomez-Cabrera, M.C.; Carretero, A.; Millan-Domingo, F.; Garcia-Dominguez, E.; Correas, A.G.; Olaso-Gonzalez, G.; Vina, J. Redox-related biomarkers in physical exercise. *Redox. Biol.* **2021**, *42*, 101956. [CrossRef] [PubMed]

 molecules

MDPI

Review

The Reductive Dehydroxylation Catalyzed by IspH, a Source of Inspiration for the Development of Novel Anti-Infectives

Hannah Jobelius [1,†], Gabriella Ines Bianchino [1,†], Franck Borel [2], Philippe Chaignon [1] and Myriam Seemann [1,*]

1 Equipe Chimie Biologique et Applications Thérapeutiques, Institut de Chimie de Strasbourg UMR 7177, Université de Strasbourg/CNRS, 4, rue Blaise Pascal, 67070 Strasbourg, France; hjobelius@unistra.fr (H.J.); gibianchino@unistra.fr (G.I.B.); p.chaignon@unistra.fr (P.C.)

2 Institut de Biologie Structurale, Université Grenoble Alpes/CEA/CNRS, 38000 Grenoble, France; franck.borel@ibs.fr

* Correspondence: mseemann@unistra.fr

† These authors contributed equally to this work.

Abstract: The non-mevalonate or also called MEP pathway is an essential route for the biosynthesis of isoprenoid precursors in most bacteria and in microorganisms belonging to the Apicomplexa phylum, such as the parasite responsible for malaria. The absence of this pathway in mammalians makes it an interesting target for the discovery of novel anti-infectives. As last enzyme of this pathway, IspH is an oxygen sensitive [4Fe-4S] metalloenzyme that catalyzes $2H^+/2e^-$ reductions and a water elimination by involving non-conventional bioinorganic and bioorganometallic intermediates. After a detailed description of the discovery of the [4Fe-4S] cluster of IspH, this review focuses on the IspH mechanism discussing the results that have been obtained in the last decades using an approach combining chemistry, enzymology, crystallography, spectroscopies, and docking calculations. Considering the interesting druggability of this enzyme, a section about the inhibitors of IspH discovered up to now is reported as well. The presented results constitute a useful and rational help to inaugurate the design and development of new potential chemotherapeutics against pathogenic organisms.

Keywords: MEP pathway; antibiotics; IspH; LytB; [4Fe-4S] cluster; reductive dehydroxylation; bioorganometallic intermediate; inhibitors

Citation: Jobelius, H.; Bianchino, G.I.; Borel, F.; Chaignon, P.; Seemann, M. The Reductive Dehydroxylation Catalyzed by IspH, a Source of Inspiration for the Development of Novel Anti-Infectives. *Molecules* **2022**, 27, 708. https://doi.org/10.3390/molecules27030708

Academic Editor: René Csuk

Received: 5 December 2021
Accepted: 18 January 2022
Published: 21 January 2022

1. IspH, an Enzyme Involved in the Biosynthesis of Isoprenoids

Isoprenoids, also known as terpenoids, constitute one of the largest classes of natural products with over 50,000 molecules. All terpenoids are derived from a common five-carbon precursor unit and are biosynthesized by the addition of one or more molecules of isopentenyl diphosphate (IPP, **1**) to its isomer dimethylallyl diphosphate (DMAPP, **2**) (Figure 1) [1]. The great structural diversity of isoprenoids is caused by different numbers of isoprene units, cyclization and oxidation reactions [2].

Isoprenoids are present in all living organisms and are involved in numerous important biological processes. For instance, in most eukaryotes, sterols as for example cholesterol **3** play the role of membrane stabilizers, and in vertebrates, they play the role of precursors of steroid hormones and bile acids. In phototrophic organisms, carotenoids (β-carotene, **4**) and chlorophylls (chlorophyll a, **5**) are essential for the conversion of light into chemical energy. Menaquinone **6** or ubiquinone **7** are used for electron transport in cellular respiration. Isopentenyladenosine **8** is a modification found in some tRNAs in bacteria and eukaryotes. Bactoprenol **9** is a crucial isoprenoid for bacteria as it is responsible for the transport of the biosynthetic sugar precursors of peptidoglycan, an essential constituent of the bacterial envelop.

From early on, humans took advantage of isoprenoids as odor compounds in fragrance or as essential oils with one famous representative being menthol **10**. More recently, taxol **11** was shown to be an efficient drug to treat some cancers.

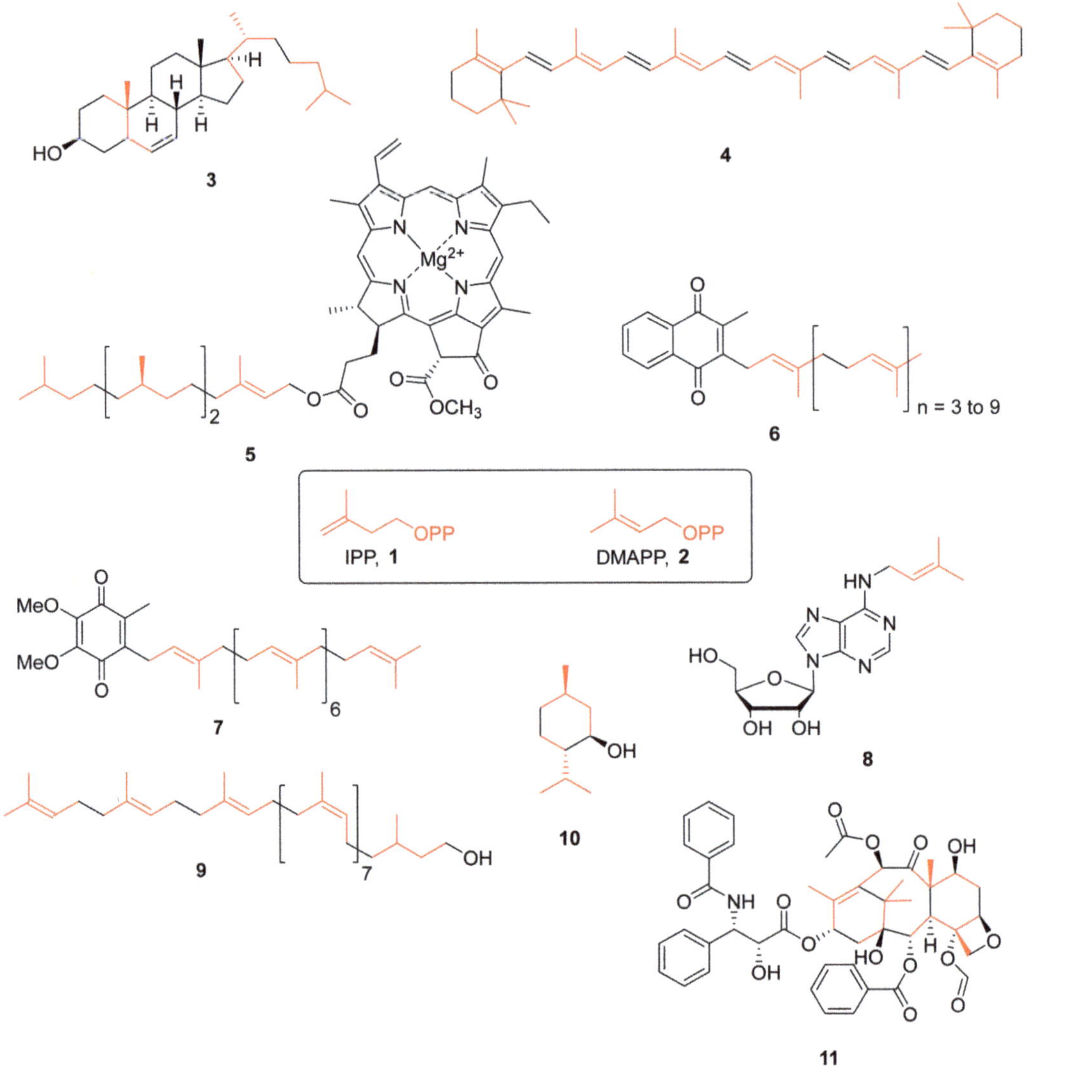

Figure 1. Some examples of isoprenoids and their biological precursors. The common five-carbon precursors are highlighted in red.

Since the late 1950s, the mevalonate pathway was the only known metabolic pathway for the synthesis of IPP **1** and DMAPP **2** [3]. Importantly, this pathway is present in most eukaryotes including humans. One of its enzymes (HMGCoA reductase) is the target of drugs called statins that are given to patients suffering from hypercholesterolemia.

In the mid-1990s, an alternative pathway called the methylerythritol 4-phosphate (MEP) pathway was identified in bacteria, apicomplexan parasites and plant plastids for the biosynthesis of IPP and DMAPP [4–6].

The MEP pathway (Scheme 1) starts with the condensation of pyruvate **12** and glyceraldehyde 3-phosphate **13** to form 1-deoxy-D-xylulose-5-phosphate (DXP, **14**) catalyzed by the enzyme DXS. DXP is then converted into 2-*C*-methyl-D-erythritol-4-phosphate (MEP, **15**) by IspC (also called DXR). A sequence of three enzyme-catalyzed steps (IspD, IspE, and IspF) further converts MEP **15** into 2-*C*-methyl-D-erythritol-2,4-cyclodiphosphate (MEcPP, **18**) via cytidine diphosphate intermediates (4-diphosphocytidyl-2*C*-methyl-D-erythritol-2-phosphate **16** and 4-diphosphocytidyl-2*C*-methyl-D-erythritol-2-phosphate **17**). IspG/GcpE catalyzes the penultimate reaction by reducing and opening the cyclic diphosphate intermediate **18** to form (*E*)-4-hydroxy-3-methylbut-2-en-1-yl diphosphate (HMBPP, **19**).

In the last step of this biosynthetic route, IspH, also called LytB (EC.1.17.7.4), catalyzes the conversion of HMBPP **19** into a mixture of IPP **1** and DMAPP **2**. IspH, the subject of this review, catalyzes a two-electron reduction and OH-group elimination by using an oxygen sensitive $[4Fe\text{-}4S]^{2+}$ center.

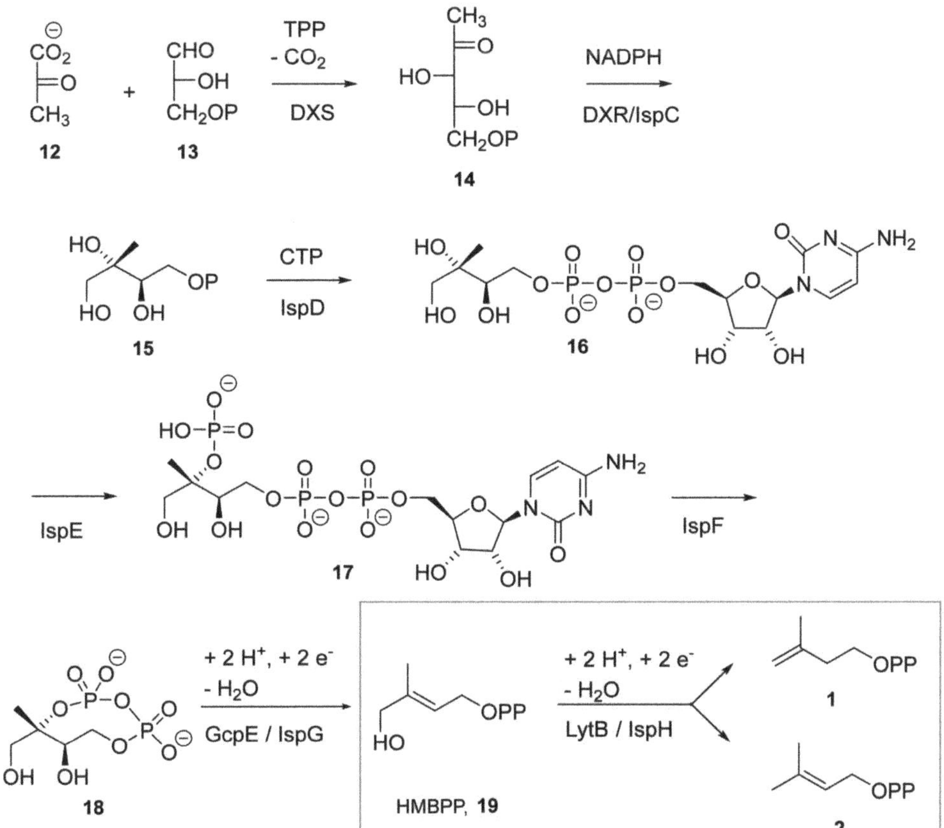

Scheme 1. Methylerythritol phosphate pathway. The reaction catalyzed by IspH is framed.

2. Occurrence of IspH

IspH as an enzyme of the MEP pathway occurs in most bacteria, plant chloroplasts, green algae, and apicomplexan but is absent in humans [7].

Interestingly, IspH is present in numerous disease-causing microbes including those posing a major challenge for new drug development. Indeed, antibiotic resistance is a growing concern as some infections have already become impossible to treat due to resistance. The worldwide emergence of carbapenemase-producing bacteria represents a health menace as carbapenems are often the last option for the treatment of patients infected by these bacteria [8]. Therefore, in 2017, the WHO published for the first time a list of antibiotic-resistant priority pathogens to guide research and development of new antibiotics as strains that cannot be fought by any antibiotic are emerging worldwide [8,9]. Importantly, 9 out of 12 of the classified bacteria contain an IspH enzyme (Table 1) highlighting its potential as a new target for antibiotic development.

Mycobacterium tuberculosis, the bacteria responsible for tuberculosis is another health threat. The WHO estimated that 1.8 million people worldwide died from tuberculosis each year with 250,000 cases from antibiotic resistant *M. tuberculosis* [8]. Interestingly, *M. tuberculosis* synthesizes its isoprenoids according to the MEP pathway and therefore relies on a functional IspH for survival. Actually, it was reported that *M. tuberculosis* contained two IspH homologs called LytB1 and LytB2 with both showing activity when expressed in *E. coli* [10]. However, it was shown that LytB2 was essential for *M. tuberculosis'* viability as LytB1 was unable to complement for the loss of LytB2 [11].

Malaria is transmitted to humans through bites from female mosquitoes infected by the apicomplexan *Plasmodium falciparum* parasite and is causing nearly half a million deaths worldwide each year, mostly children less than 5 years of age [12]. In the context of fighting this life-threatening disease, fosmidomycin, an inhibitor of the *P. falciparum* enzyme DXR, the second enzyme of the MEP pathway, is under clinical trial as an antimalarial agent in combination with piperaquine [13], making the MEP pathway an attractive target for drug development. As an enzyme of the MEP pathway, IspH is not only present in *P. falciparum* but also in *Toxoplasma gondii*, the parasite responsible for toxoplasmosis.

Table 1. Occurrence of IspH in the bacteria classified as priority by the WHO.

Priority	Bacteria	IspH
critical	*Acinetobacter baumannii*, carbapenem-resistant	+
	Pseudomonas aeruginosa, carbapenem-resistant	+
	Enterobacteriaceae, carbapenem-resistant, ESBL-producing	+
high	*Enterococcus faecium*, vancomycin-resistant	– [a]
	Staphylococcus aureus, methicillin-resistant, vancomycin-intermediate and resistant	–
	Helicobacter pylori, clarithromycin-resistant	+
	Campylobacter spp., fluoroquinolone-resistant	+
	Salmonellae, fluoroquinolone-resistant	+
	Neisseria gonorrhoeae, cephalosporin-resistant, fluoroquinolone-resistant	+
medium	*Streptococcus pneumoniae*, penicillin-non-susceptible	–
	Haemophilus influenzae, ampicillin-resistant	+
	Shigella spp., fluoroquinolone-resistant	+

[a] partial IspH protein with incomplete catalytic site; rely on MVA pathway [14].

Since the discovery of IspH two decades ago, an intense research activity has been devoted to the understanding of its mechanism with the aim of developing unprecedented therapeutics with new modes of action that are urgently needed to combat infectious diseases.

3. The Discovery of the IspH Metalloenzyme

3.1. The Lytb Gene

The *lytB* gene was first described in *Escherichia coli* by the group of Ishiguro in 1993 as a gene involved in penicillin resistance [15] and was further reported to be present in other bacteria [16]. The group of Gantt revealed that *lytB* was an essential gene for a *Synechocystis* strain (cyanobacterium) as its disruption was lethal and that the growth of this mutant could only be restored by the addition of the alcohol analogs of IPP and DMAPP (isopentenol and dimethylallyl alcohol) in the media. Furthermore, the overexpression of this *Synechocystis* gene in an *E. coli* mutant engineered to produce carotenoids led to increased accumulation of these isoprenoids in *E. coli* highlighting the role of *lytB* in isoprenoid biosynthesis and most probably in the MEP pathway [17]. *LytB* was further shown to be an essential gene for *E. coli* and to be involved in the MEP pathway [18,19]. The group of Jomaa reported that an *E. coli* mutant harboring a deletion of the *lytB* gene and engineered for the utilization of exogenous mevalonate (to rescue the cells and allow the synthesis of crucial isoprenoids when the endogenous MEP pathway is disrupted), accumulates an immunogenic compound stimulating the proliferation of $V\gamma9/V\delta2$ T cells. This compound was identified as HMBPP **19** after its isolation followed by a complete characterization using NMR spectroscopy and mass spectrometry, affording the first evidence that HMBPP was the substrate of the *lytB* gene product [20].

3.2. The IspH Catalyzed Reaction

A very elegant in vivo approach was used by Rohdich, Eisenreich, and collaborators to identify the reaction catalyzed by the *lytB* gene product. An *E. coli* strain overexpressing the *xylB* gene coding for D-xylulose kinase, known to phosphorylate 1-deoxy-D-xylulose (DX) in DXP, and all the genes of the MEP pathway including *lytB* was constructed. After feeding of [U-$^{13}C_6$]DX to this strain followed by ^{13}C-NMR analysis of the bacterial crude extract, it was shown that this *E. coli* mutant accumulated [U-$^{13}C_6$]IPP and [U-$^{13}C_6$]DMAPP in a 5:1 ratio [21]. When *lytB* was absent in the construction, the mutant accumulated [U-$^{13}C_6$]HMBPP from [U-$^{13}C_6$]DX [22]. Together, these results indicate that the *lytB* gene product, further called LytB or IspH, catalyzes the conversion of HMBPP into IPP and DMAPP in a mixture of 5:1.

E. coli IspH has a mass of about 36 kDa [21]. First activity assays performed with purified *E. coli* IspH displayed no activity whereas the cell-free extract of *E. coli* expressing the recombinant IspH was active suggesting the involvement of auxiliary enzymes [23]. Purified *E. coli* IspH was further shown to convert HMBPP into IPP and DMAPP with an activity of 3 nmol min^{-1} mg^{-1} when assayed under anaerobic conditions in the presence of a reduction system such as photoreduced deazaflavin or NADPH/flavodoxin/flavodoxin reductase [24]. An approximatively 3000 times higher activity was reported by Jomaa and collaborators for recombinant IspH from the thermophilic eubacteria *Aquifex aeolicus* when purified in a glove-box under anaerobic conditions and using reduced methyl viologen as reduction system [25]. The authors pointed out the fact that the enzyme's activity could not be maintained after contact with oxygen indicative of the enzyme's oxygen sensitivity.

3.3. IspH, a [4Fe-4S] Metalloenzyme

The nature of the iron-sulfur cluster of the IspH enzyme has been the subject of debate for about a decade. The cysteine residue is a common ligand for iron centers of which many examples are known in literature [26]. Rohdich et al. assumed the presence of such a cluster due to three conserved cysteines which they found in ten different species containing the IspH protein [23]. The iron-sulfur cluster of IspH was first experimentally evidenced in the UV visible spectrum as a broad absorption maximum at a wavelength of 420 nm [25]. By comparison with spectroscopic data of other holoenzymes such as IspG [27], which is the enzyme prior to IspH in the MEP pathway, a $[4Fe\text{-}4S]^{2+}$ cluster was proposed [24].

The early propositions concerning the $[4Fe\text{-}4S]^{2+}$ cluster were backed up by Rohmer and collaborators conducting EPR spectroscopic measurements after reconstitution of the Fe/S cluster by incubation of the apoenzyme with $FeCl_3$, Na_2S, and dithiothreitol under anaerobic conditions [28]. $[4Fe\text{-}4S]^{2+}$ itself does not show any signals on the EPR spectrum due to its diamagnetic nature. Upon reduction of the chemically reconstituted enzyme with dithionite, however, a paramagnetic species appeared having g values of 1.92(1) and 2.03(7). By analogy to earlier studies of iron-sulfur clusters measured by EPR spectroscopy, the microwave power half-saturation $P_{1/2}$ was used as means to distinguish between a $[2Fe\text{-}2S]^+$ and a $[4Fe\text{-}4S]^+$ species [29,30]. With a rather high $P_{1/2}$ value of 153 mW, a $[4Fe\text{-}4S]^+$ cluster was proposed [28]. Iron species that could be detected on the enzyme purified under aerobic conditions were attributed to $[3Fe\text{-}4S]^+$ (g = 2.02) and to ferric ions (g = 4.29) and exhibited signals already known from literature [31]. The discussion about the nature of the iron-sulfur cluster continued owing to new results from Rohdich et al. [32]. In their work, recombinant *E. coli* IspH protein was expressed in the presence of an *isc* operon, enabling the assembly of the full iron-sulfur cluster in vivo, and purified under anaerobic conditions. They found the same absorption maximum in the UV visible spectrum at about 410 nm as had been found before [28], but with a lower absorption coefficient (here: ε = 11,800 M^{-1} cm^{-1}, compared to ε = 18,750 M^{-1} cm^{-1} in the previous study) suggesting a lower concentration of the iron-sulfur cluster. The corresponding EPR spectrum showed an anisotropic signal with g-values at 2.032(3) and 2.003(3). Its symmetry led them to the conclusion that a $[3Fe\text{-}4S]^+$ was the active species as they found a high catalytic rate of 550 nmol mg^{-1} min^{-1} when using NADPH/flavodoxin reductase/flavodoxin as reduction system.

In order to solve this controversy, two groups independently investigated *E. coli* IspH purified under anaerobic conditions using Mössbauer spectroscopy [33,34]. Irrespective of its magnetic state, Mössbauer spectroscopy offers the possibility to gain structural information about the iron-sulfur cluster [35]. The analyses of the Mössbauer spectra indicated concordantly the structure of a $[4Fe\text{-}4S]^{2+}$ cluster with a unique fourth iron site. The spectra could be simulated by the sum of three components with a ratio of 2:1:1 as shown in Table 2 and indicated an unusual $[4Fe\text{-}4S]^{2+}$ center [33]. Two iron centers share one electron with an oxidation state of +2.5 each as derived from their isomer shift of δ = 0.42 mm s^{-1} and exhibit a tetrahedral coordination sphere with four sulfur ligands; besides the three sulfide ions, a cysteine residue is coordinated to each site as confirmed by X-ray structure analysis [36,37]. The two remaining iron sites consist of a high-spin Fe^{3+} ion and a high-spin Fe^{2+} ion, respectively [33]. The latter was assumed to have a coordination sphere with five or six ligands, most likely three sulfur and three nitrogen or oxygen ligands. Schünemann and collaborators further used Mössbauer spectroscopy to unequivocally access the nature of the Fe/S cluster of IspH existing in vivo in *E. coli* [33]. The authors recorded Mössbauer spectra of *E. coli* cells overexpressing IspH and of a WT strain, both strains grown under the same conditions. The difference of the spectra obtained for each type of cells could be simulated using the Mössbauer parameters obtained for pure ^{57}Fe IspH. This proves that IspH assembles a $[4Fe\text{-}4S]^{2+}$ cluster in vivo that is maintained when the protein is purified under anaerobic conditions. More details about the coordination sphere on this unique fourth iron atom were revealed by nuclear resonance vibrational spectroscopy (NRVS) on substrate-free IspH by the groups of Schünemann and Seemann [38]. NRVS is an advanced Mössbauer technique using synchrotron based radiation that is sensitive to vibrational modes involving iron [39]. By comparison of the experimental spectrum with DFT calculations obtained on an IspH structure harboring an incomplete Fe/S cluster (PDB: 3F7T) in which they inserted the missing iron, the authors concluded that the unique iron site was in a hexa-coordinated form with three water molecules. Together, these results highlight that IspH harbors an unusual [4Fe-4S] cluster linked to the three conserved cysteines with a unique iron site of oxidation state +2 that is linked to three sulfides of the cluster and to three water molecules.

The fact that the [4Fe-4S] cluster was the active species in the IspH enzyme was further confirmed by Duin et al. [40]. They could show that the enzyme's activity was linearly dependent on the content of its cofactor.

It is now firmly established that IspH is an oxygen sensitive $[4Fe\text{-}4S]^{2+}$ enzyme that catalyzes a reductive dehydroxylation and that the electrons needed for this reduction are provided in vivo by a biological system such as NADPH/flavodoxin reductase/flavodoxin in the case of *E. coli* or NADPH/ferredoxin reductase/ferredoxin in the case of *P. falciparum* [41].

Table 2. Results of Mössbauer and NRVS spectroscopy on the $[4Fe\text{-}4S]^{2+}$ cluster [33,38]. The Mössbauer spectra were recorded at T = 77 K and B = 0 T. The ΔE_Q and δ values of the HMBPP-bound IspH are found in parentheses.

	ΔE_Q [mm s^{-1}]	δ [mm s^{-1}]	Oxidation State and Coordination
Component 1 (50%)	1.21 (1.33)	0.42 (0.42)	Tetrahedrally coordinated $Fe^{2.5+}$, mixed valence iron pairs, 3 S^{2-} and 1 Cys ligand
Component 2 (25%)	0.89 (0.92)	0.37 (0.38)	Tetrahedrally coordinated high-spin Fe^{3+}, 3 S^{2-} and 1 Cys ligand
Component 3 (25%)	1.97 (1.00)	0.89 (0.64)	Hexa-coordinated high-spin Fe^{2+}, 3 S^{2-} and 3 H_2O (tetrahedrally coordinated, 3 S^{2-} and OH_{HMBPP})

4. The Mechanism of the Reductive Dehydroxylation

4.1. Protein Conformation in the Crystalline State

Crystal structures can give hints for the elucidation of the mechanism, but the propositions are usually confirmed by studies in solution. During the last two decades, 30 IspH crystallographic structures were deposited into the PDB database (Figure 2, Table 3). IspH structures from three different organisms are available. The first structure solved at the end of 2008 by Ermler and collaborators was the one from *Aquifex aeolicus*, a Gram-negative hyperthermophilic bacteria [36]. In 2009, Gräwert et al. released the first IspH structure of *E. coli* [42]. The last organism for which the crystallographic structure of IspH has been described is the one from the apicomplexan parasite *P. falciparum* [43]. For *A. aeolicus* and *E. coli* the full-length protein was crystallized whereas for *P. falciparum* the 217 first amino acids were not included in the construct used for the production and the crystallization of the enzyme.

The unprecedented fold of all IspH enzymes looks like a three-leaf clover and is composed of three structurally similar domains that do not display detectable similarity [42]. Each domain comprises a central parallel beta sheet containing four beta strands surrounded by three or four alpha helices. The active site is buried in a hydrophobic cavity embedded between the three domains. The [4Fe-4S] cluster, required for enzyme activity [28], is bound to the protein via three highly conserved cysteines [42]. Each domain contributes to the cluster coordination since these cysteines are located at the N-terminus of the first alpha helix of each of the three domains. *E. coli* and *P. falciparum* display an additional extended C-terminal loop containing a short beta sheet positioned above the active site pocket to protect the cluster from the surrounding solvent. In *A. aeolicus*, the dimerization of the protein is proposed to achieve the same role [44]. However, it should be noted that the crystal structure of the thermophilic enzyme does not reveal evidence of dimerization [36].

Among the 30 IspH structures solved only the one of *A. aeolicus* was obtained without a ligand into the binding pocket. This structure contained an incomplete [3Fe-4S] cluster. It should be mentioned that no crystal structure of substrate-free IspH in its [4Fe-4S] form has been reported to date. The inability to obtain such a structure is most probably due to the instability of the apical iron linked to three water molecules [38], which dissociates during the crystallization process. Comparison of this empty structure with the others reveals a conformational change. Indeed, upon substrate binding, a motion of the third domain of IspH results in the closing of the active site [42], and from the *P. falciparum* structure [43] it appears that a single sulfate ion bound in the substrate cavity is sufficient to induce the closing of the active site.

Table 3. Crystallization conditions and structural data of the reported IspH.

PDB ID	Release Date	Source Organism	Crystallization Method (Vapor difusion)	Crystal Growth Procedure	pH	Space Group	Res. (Å)	Cluster	Ligand ID	Ref
3DNF	30 December 2008	*A. aeolicus*	Hanging drop	0.1 M Tris-Hcl, 10% PEG 8000	8.0	P 1 21 1	1.65	*Fe3 S4*	-	[36]
4N7B	20 November 2013	*P. falciparum*	Sitting drop	0.1 M Na citrate, 0.1 M $(NH_4)_2SO_4$, 30% PEG 4000	5.6	P 31 2 1	2.20	*Fe3 S4*	SO4	[43]
3F7T	7 July 2009	*E. coli*	Hanging drop	1.6 M Potassium Phosphate	8.0	P 32 2 1	1.80	*Fe3 S4*	POP	[42]
3KEF	12 January 2010	*E. coli*	Sitting drop	0.1 M Bis-Tris, 0.2 M Li_2SO_4, 25% PEG 3350	6.5	P 21 21 21	1.70	*Fe3 S4*	DMA	[37]
3KEL							1.80	*Fe3 S4*	POP	
3KE8							1.70	**Fe4 S4**	EIP	
3KE9							1.90	**Fe4 S4**	IPE	
3KEM					7.0	C 1 2 1	2.00	*Fe3 S4*	IPE	
3T0G	30 November 2011	*E. coli*	Hanging drop	0.1 M Bis-Tris, 0.2 M Li_2SO_4, 25% PEG3350	6.5	P 21 21 21	2.10	*Fe3 S4*	H6P	[45]
3T0F							1.90	*Fe3 S4*	H6P	
3SZU							1.40	*Fe3 S4*	H6P	
3SZL							1.60	**Fe4 S4**	H6P	
3SZO							1.60	**Fe4 S4**	H6P	
3URK	5 September 2012	*E. coli*	Sitting drop	0.1 M Bis-Tris, 0.2 M $(NH_4)_2SO_4$, 25% PEG3350	6.5	P 21 21 21	1.50	**Fe4 S4**	0CG	[46]
3UTC							1.90	**Fe4 S4**	0JX	
3UTD							1.70	*Fe3 S4*	OCJ	
3UV3							1.60	**Fe4 S4**	0CM	
3UV6							1.70	**Fe4 S4**	0CH	
3UV7							1.60	**Fe4 S4**	0CN	
3UWM							1.80	**Fe4 S4**	0K2/0JX	
3ZGL	9 January 2013	*E. coli*	Hanging drop	0.1M Bis-Tris, 0.2 M Li_2SO_4, 24% PEG 3350	6.5	C 1 2 1	1.68	**Fe4 S4**	10E	[47]
3ZGN							1.95	**Fe4 S4**	10G	
4H4C	23 January 2013	*E. coli*	Sitting drop	0.1 M Bis-Tris, 0.2 M $(NH_4)_2SO_4$, 25% PEG3350	6.5	P 21 21 21	1.80	**Fe4 S4**	10D	[48]
4H4D							1.35	**Fe4 S4**	10E	
4H4E							1.70	**Fe4 S4**	10G	
4EB3	6 February 2013	*E. coli*	Sitting drop	0.1 M Bis-Tris, 0.2 M $(NH_4)_2SO_4$, 25% PEG3350	6.5	P 21 21 21	1.90	**Fe4 S4**	0O3	[49]
4MUY	11 June 2014	*E. coli*	Sitting drop	0.1 M Bis-Tris, 0.2 M $(NH_4)_2SO_4$, 25% PEG3350	6.5	P 21 21 21	1.80	*Fe3 S4*	2E5	[50]
4MV0							1.90	*Fe3 S4*	2E6	
4MV5	18 June 2014						1.90	*Fe3 S4*	2E7	
4MUX							1.70	**Fe4 S4**	2E4	

4.2. First Step: Binding of the Hydroxyl Group of HMBPP to the Iron-Sulfur Cluster

The proposed mechanistic course of the IspH enzyme is depicted in Scheme 2. Formally, the mechanism of IspH involves the elimination of water and the transfer of two electrons as well as two protons.

The active site of this reaction includes the [4Fe-4S] cluster that has already been described in detail in Section 3.3. Synthetic models for such 3:1 site-differentiated clusters were recently provided by the Suess group and it was shown that the unique fourth iron site could easily undergo ligand exchange [51] while the other ligands of the cluster were cysteine thiolates that are soft bases making stronger coordination bonds. It was proposed that the coordination geometry and the exchange at the unique iron site was controlled by steric effects provided by the ligands of the adjacent iron sites. Therefore, the opening of the coordination sphere of the apical iron required for its hexacoordination in IspH might be controlled by steric effects provided by the position of the cysteine ligands. Furthermore, broken-symmetry density functional theory (BS-DFT) studies performed on IspH indicated that the Fe-S bonds of the unique fourth iron site could be distorted which led to greater flexibility in ligand binding [52]. The three water molecules are good leaving groups contrary to the deprotonated cysteines exhibiting strong coordination bonds, suggesting a role of the apical iron of IspH in catalysis.

Docking studies of HMBPP using the *A. aeolicus* IspH structure in its $[3Fe\text{-}4S]^+$ form in which the missing iron was computationally added revealed that the OH of HMBPP could bind to the fourth iron [36]. This observation was first experimentally confirmed using

Mössbauer spectroscopy by adding the substrate HMBPP to ^{57}Fe-LytB in solution [33]. A change in the isomer shift from δ_3 = 0.89 mm s^{-1} to $\delta_{3'}$ = 0.64 mm s^{-1} of the component related to the unique fourth iron site of the cluster is observed in the presence of HMBPP, confirming the binding of the substrate to this iron site. The quadrupole splitting of this iron is also lowered from $\Delta E_{Q,3}$ = 1.97 mm s^{-1} to $\Delta E_{Q,3'}$ = 1.00 mm s^{-1} which indicates a change in the geometry of this iron ion. Upon substrate binding, the three water molecules of **I** are released and instead, the hydroxyl group of HMBPP is bound to the unique fourth iron site which is now tetrahedrally coordinated (intermediate **II**).

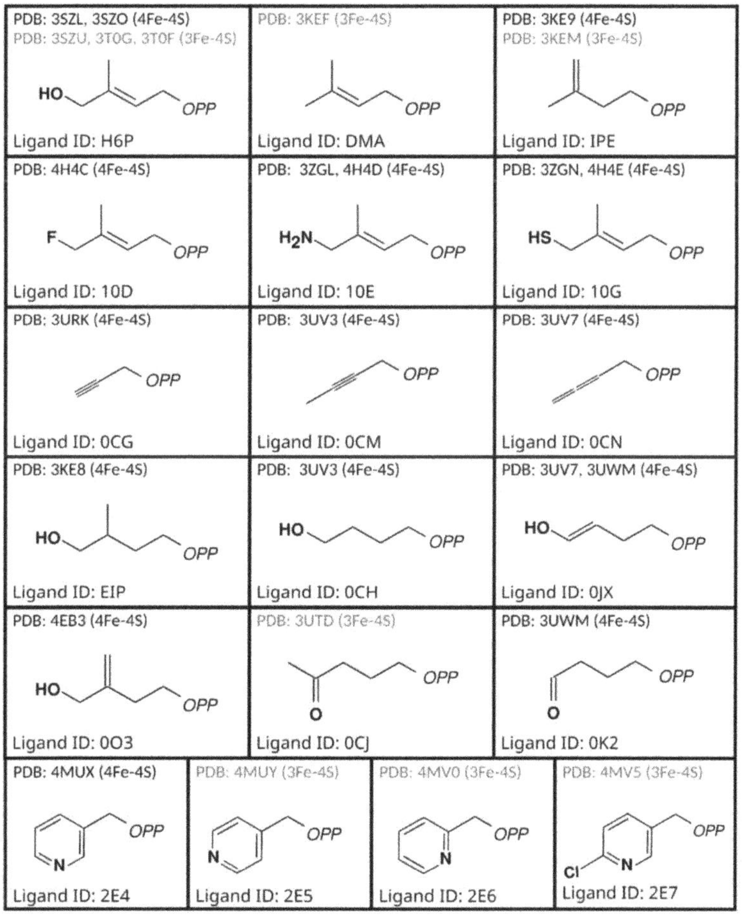

PDB: 3SZL, 3SZO (4Fe-4S) PDB: 3SZU, 3T0G, 3T0F (3Fe-4S) HO, OPP Ligand ID: H6P	PDB: 3KEF (3Fe-4S) OPP Ligand ID: DMA	PDB: 3KE9 (4Fe-4S) PDB: 3KEM (3Fe-4S) OPP Ligand ID: IPE
PDB: 4H4C (4Fe-4S) F, OPP Ligand ID: 10D	PDB: 3ZGL, 4H4D (4Fe-4S) H2N, OPP Ligand ID: 10E	PDB: 3ZGN, 4H4E (4Fe-4S) HS, OPP Ligand ID: 10G
PDB: 3URK (4Fe-4S) OPP Ligand ID: 0CG	PDB: 3UV3 (4Fe-4S) OPP Ligand ID: 0CM	PDB: 3UV7 (4Fe-4S) OPP Ligand ID: 0CN
PDB: 3KE8 (4Fe-4S) HO, OPP Ligand ID: EIP	PDB: 3UV3 (4Fe-4S) HO, OPP Ligand ID: 0CH	PDB: 3UV7, 3UWM (4Fe-4S) HO, OPP Ligand ID: 0JX
PDB: 4EB3 (4Fe-4S) HO, OPP Ligand ID: 0O3	PDB: 3UTD (3Fe-4S) O, OPP Ligand ID: 0CJ	PDB: 3UWM (4Fe-4S) O, OPP Ligand ID: 0K2

PDB: 4MUX (4Fe-4S) N, OPP Ligand ID: 2E4	PDB: 4MUY (3Fe-4S) N, OPP Ligand ID: 2E5	PDB: 4MV0 (3Fe-4S) N, OPP Ligand ID: 2E6	PDB: 4MV5 (3Fe-4S) Cl, N, OPP Ligand ID: 2E7

Figure 2. Structures of IspH deposited in the PDB database with their corresponding ligand.

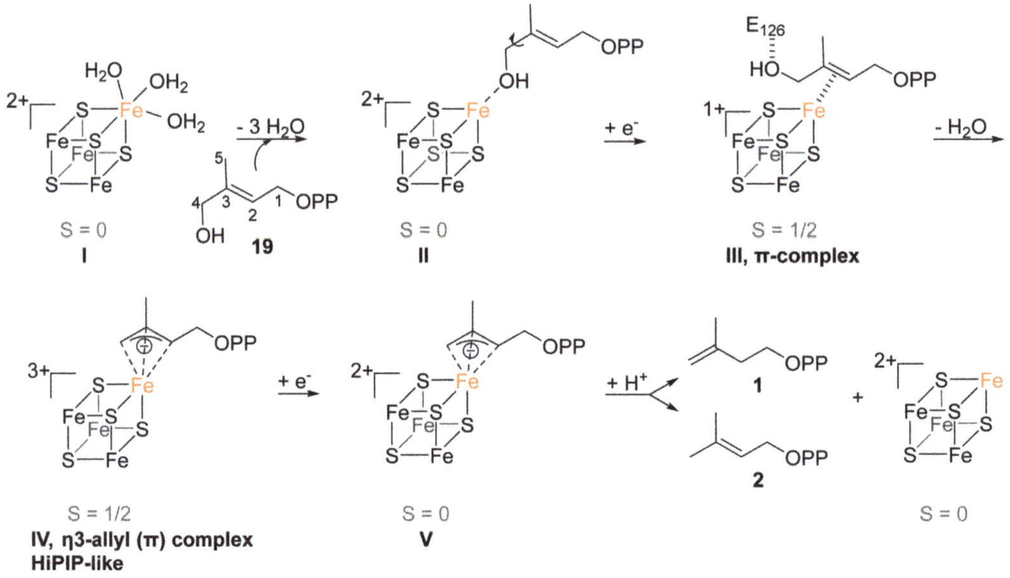

Scheme 2. Proposition for the mechanism of the IspH catalyzed reductive dehydroxylation.

E. coli wild type or IspH mutants have been further crystallized with various ligands such as substrate, substrate analogs, products, or inhibitors (Figure 2, Table 3) [37,45–50]. In 2010, Groll, Bacher, and collaborators reported the first *E. coli* IspH structure with an intact [4Fe-4S] cluster as a complex with a substrate molecule (HMBPP; PDB ligand ID: H6P) [37]. (The ligand reported in the PDB file (3KE8) is not the substrate but an analog with the PDB ligand ID EIP, Figure 2, Table 3). The binding of HMBPP does not induce any significant backbone differences compared to the previously described structures. The diphosphate moiety inserts into a polar pocket composed of the side chains H41, 74, and 124, S225 and 269, T168, N227, and Q166 and is engaged in an extensive hydrogen bonding network. The substrate adopts a hairpin-like conformation to allow its C4 oxygen to form a coordination bond with the apical iron of the [4Fe-4S] cluster. The structures of the complex with each of the two reaction products (DMAPP and IPP) are also available [37]. Here again, the two molecules retain the same binding mode and their hairpin conformations but they are no longer bound to the apical iron.

The X-ray analysis of IspH in complex with its substrate confirmed that the hydroxyl moiety of HMBPP or its deprotonated form (the alkoxide) is bound to the unique fourth iron atom. With regards to the question of protonation, Blachly, Noodleman, and co-workers gave a plausible answer. By applying BS-DFT calculations, the protonated OH-form of HMBPP yields Mössbauer values comparable to the simulated values from experimental spectra [53]. Indeed, when HMBPP is bound to the apical iron as an alcohol, the HO-Fe distance is elongated with shortening of the distances of this iron and the S^{2-} giving to the apical iron more ferrous character as revealed by Mössbauer spectroscopy.

Together these results show that the first step in the mechanism of IspH is the binding of the OH of the substrate to the unique iron site accompanied by the release of three water molecules. It is not clear what triggers the departure of the three water molecules but conformational change of the protein upon binding of the HMBPP substrate might generate movement of the cysteine ligands leading to reduction of the space around the apical iron and therefore favoring tetrahedral coordination geometry. Similar steric effects were reported in models [51]. Furthermore, the hydrogen bonding network revealed in the

IspH structure in complex with the substrate will also increase the stability of this complex compared to the substrate-free enzyme harboring loosely bound water molecules.

4.3. Second Step: Reduction of the [4Fe-4S]$^{2+}$ Cluster and Rotation of the CH_2OH Group

Systematic studies by Liu et al. with different organic redox mediators along with the reducing agent dithionite revealed that the enzyme's activity was dependent on the redox potential of these redox mediators [34]. The enzyme was most efficient when using a system of dithionite and a mediator with a redox potential of −450 mV vs. NHE while no detectable activity was found for redox mediators bearing redox potentials above −250 mV vs. NHE or below −720 mV vs. NHE. By using this chemical system, the authors reported an activity of 31.6 µmol min $^{-1}$ mg $^{-1}$ which is much higher than the activity of about 800 nmol min $^{-1}$ mg $^{-1}$ that is now frequently described for *E. coli* IspH using the natural NADPH/flavodoxin reductase/flavodoxin as reduction system [32,33].

After binding of the substrate to the apical iron atom, a rotation of the CH_2OH group has been proposed upon reduction of the [4Fe-4S]$^{2+}$ cluster to form intermediate **III**. The first who observed this rotation to be a part of the catalytic course were Oldfield et al. who conducted docking calculations of the substrate with reduced IspH [54]. After molecular mechanics optimization in silico, the hydroxyl group did not bind to the iron-sulfur cluster anymore but pointed towards the opposite direction as is depicted in Figure 3.

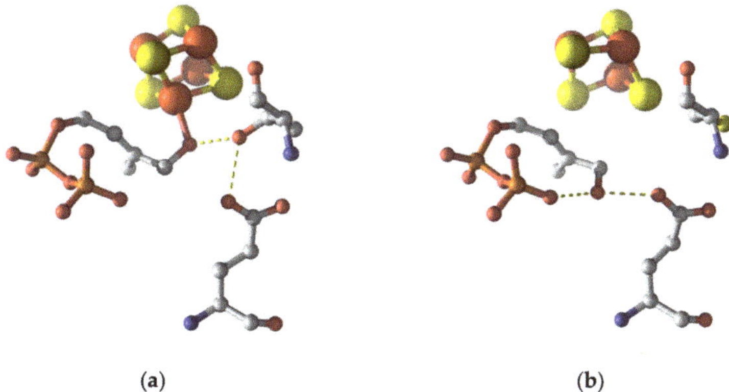

Figure 3. Crystallographic structures depicting the rotation of the hydroxymethyl group [45].(**a**) The IspH WT was incubated with the substrate HMBPP. The full iron-sulfur cluster is present and the CH_2OH group is coordinated to the apical iron, stabilized by T167. (**b**) The IspH T167C mutant was incubated with the substrate HMBPP. The CH_2OH group has rotated about almost 180° and is now stabilized by the diphosphate group as well as the glutamate E126. The crystal structure only shows a [3Fe-4S] cluster.

Intermediate **III** was then observed experimentally by Groll and collaborators who provided crystal structures of *E. coli* IspH enzyme with bound HMBPP [45]. The authors compared the structures before and after irradiation with X-rays as synchrotron irradiation is known to reduce some metalloenzymes [55,56]. Based on the measured electron density of the crystal, two HMBPP conformers were suggested: the early intermediate in which the OH was bound to the apical iron and a new conformer with the CH_2OH group partially rotated away from the apical iron. The hydroxyl group was now stabilized intramolecularly by the pyrophosphate group as well as by the amino acid residue E126 of IspH, which in turn formed a hydrogen bond with T167. However, the interpretation was still ambiguous as the initial binding of the hydroxyl group to the iron-sulfur cluster could also be detected.

Independently, Dickschat and collaborators conducted feeding experiments with deuterated isotopologues of the precursor of the DOX pathway, 1-deoxy-D-xylulose [57]. These in vivo experiments were conducted on the bacteria *Streptomyces avermitilis* that are known to produce the isoprenoid pentalenene for which the biosynthetic route was deciphered. After conversion, the configuration of the IPP and DMAPP products was determined by tracing back the course of the deuterium atoms in pentalenene. Given that only one of the two possible deuterated IPP analogs was formed, a rotation of the CH_2OH group seemed likely to be involved in the formation of the required intermediate.

More evidence for this rotation was given by HYSCORE experiments on the *E. coli* IspH E126Q mutant incubated with ^{17}O-labeled HMBPP ([4-^{17}O]HMBPP) [49]. In previous experiments, this mutant exhibited only 0.3% activity compared to the wild type [45]. It was proposed that E126 played a role in the proton transfer step, so mutation to glutamine might hinder the subsequent protonation needed for water elimination. After reduction of the $[4Fe\text{-}4S]^{2+}$ cluster of IspH, an intermediate species was detected with a very low ^{17}O hyperfine coupling constant (A_{iso} = 1 MHz) compared to those of direct Fe-O bonds [49]. This led the authors to the conclusion that the substrate's oxygen atom must have had a greater distance to the apical iron atom, most likely as a result of a rotation of the CH_2OH group. The same authors had previously suggested that based on ^{13}C-ENDOR experiments using the inactive E126A IspH mutant and [U-$^{13}C_6$]HMBPP a π- or π/σ-complex between the reduced cluster and the C2-C3 double bond of the substrate was formed. A similar complex had already been confirmed for a nitrogenase FeMo cofactor mutant in complex with allyl alcohol or ethylene [58,59]. The *g*-values of the IspH E126Q as well as the IspH E126A mutant were comparable with those of the Fe alkenyl complexes [49,54].

Similarly to the in vivo experiments, deuterated substrates were later prepared for more straightforward in vitro experiments with IspH [60]. HMBPP was deuterated on the C4-position to yield the two stereoisomers (*S*)-[4-$^2H_1$]HMBPP and (*R*)-[4-$^2H_1$]HMBPP, respectively. After incubation of each substrate with IspH in the presence of its naturally occurring reducing system, the configuration of the deuterium atoms on the products was determined by NMR spectroscopy. The *S*- and the *R*-configured substrates both yield [4-$^2H_1$]DMAPP without any stereocenters. However, the (*S*)-configured substrate only yields (*E*)-[4-$^2H_1$]IPP whereas the (*R*)-configured substrate only yields (*Z*)-[4-$^2H_1$]IPP. With regard to these results, Seemann and coworkers concluded that after reduction of the iron-sulfur cluster the CH_2OH group rotated by almost 180° compared to the initial binding of the hydroxyl moiety to the apical iron atom as suggested from the previous structural data [45].

The driving force of this rotation may be explained by the fact that the apical ferrous ion becomes even softer upon reduction of the $[4Fe\text{-}4S]^{2+}$ cluster [60]. According to the HSAB concept [61], this iron is more prone to form a complex with the soft alkene than with the harder hydroxyl group. The flexibility of the [4Fe-4S] cluster has been described in great detail by Holm et al. [62] and the reduced form of the cluster exhibits a greater core volume and distortions facilitating the ligand exchange. An η^2-ring conformation of HMBPP was suggested by Blachly, Noodleman and coworkers [52]. The rotated CH_2OH-group may be stabilized by forming intramolecular hydrogen bonds with the diphosphate moiety which was found to be the energetically most favorable configuration according to BS-DFT calculations when the [4Fe-4S] is reduced by one electron. Another possibility could be the formation of an η^2-trans state where the CH_2OH group is stabilized by the nearby E126. When the [4Fe-4S] is reduced by one electron both η^2 configurations were showed to be lower in energy than the configuration having the hydroxyl of the substrate bound to the apical iron.

An alternative pathway where rotation of the CH_2OH-group occurs before the reduction has also been proposed based on computed reduction potentials obtained from BS-DFT calculations [52].

4.4. Third Step: Protonation of the Hydroxyl Group and Elimination of Water

Mechanistic studies with substrate analogs were conducted by Liu et al. in order to get insights into the presumed C-O bond cleavage of the substrate's hydroxyl group [63]. Replacing the diphosphate group at the C1 atom with a pyrophosphonate group (see

Scheme 3, compound **20**) still leads to the IPP analog after incubation with IspH and the naturally occurring reducing system as confirmed by ^{1}H NMR spectroscopy. Thus, C-O bond cleavage at the C1 site of HMBPP can be excluded despite the fact that diphosphate is a better leaving group than the hydroxyl group. When incubating the fluorine analog **21** with the system mentioned above both IPP and DMAPP are obtained. Hence, the mechanism is thought to be analogous to that for HMBPP. As the fluorine atom exhibits a high electronegativity, only a heterolytic cleavage of the C-F bond comes into consideration which is assumed for the C4-O bond as well. Furthermore, isotopic labeling of the C1 hydrogen atoms with deuterium (compound **22**) or substitution of hydrogen with fluorine (compound **23**) still allows some turnover of the substrate, and at least one of the two products (the IPP analog in case of compound **23**) are synthesized [64]. Therefore, it is unlikely that the C1 atom is deprotonated or otherwise directly involved in the mechanistic pathway.

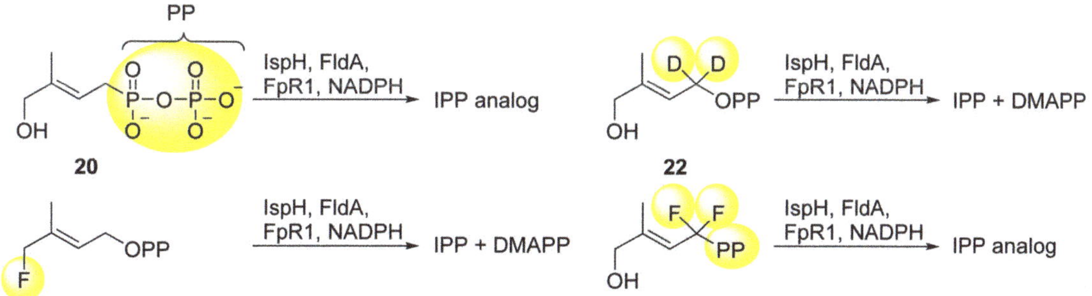

Scheme 3. Substrate analogs of HMBPP used for the study of C-O bond cleavage [63,64]. Discrepancies from the natural substrate HMBPP are highlighted in yellow.

As mentioned earlier, a possible proton source for the formation of water after hydroxyl elimination may be the glutamate residue E126. It lies in proximity to the hydroxyl group [45] and upon mutation to glutamine, aspartate, or alanine, the respective mutants E126Q, E126D, and E126A exhibit almost no activity [37,42,54]. The proton may be transferred to this glutamate residue by the adjacent histidine residue H124. It was shown that the respective mutants alanine (H124A) or phenylalanine (H124F) had a very low activity also [40,54]. Quantum mechanistics/molecular mechanistics (QM/MM) investigations support evidence that the dehydroxylation is triggered by a proton transfer from protonated E126 to the hydroxyl group of HMBPP mediated by a conserved water molecule [65].

4.5. Fourth Step: Formation of a HiPIP-like Cluster

The next intermediate (**IV**) was first observed by Duin et al. by *cw* EPR spectroscopy [40]. It was trapped by incubation of one electron-reduced IspH with its substrate without further electron source for complete turnover. The resulting paramagnetic species had g-values of 2.173, 2.013, and 1.997, comparable to those of ferredoxin:thioredoxin reductase that contains a $[4Fe\text{-}4S]^{3+}$ cluster [66]. The authors proposed that the intermediate in the IspH catalyzed reaction was linked to $[4Fe\text{-}4S]^{3+}$, but further interpretation of the data remained obscure.

Oldfield et al. clarified the nature of this intermediate. By freeze-quenching a reaction mixture containing IspH, dithionite as the only reducing agent, and the substrate, the reaction was slowed down and the trapped intermediate **IV** could be analyzed by EPR spectroscopy. Eventually, they measured the same intermediate as did Duin et al. [49]. The g-tensor was found to be g = [2.171, 2.010, 1.994] and was also similar to that found in other high-potential iron-sulfur proteins (HiPIP) [67]. The authors confirmed the presence of a $[4Fe\text{-}4S]^{3+}$ cluster, which implicates that formally a two-electron transfer from the cluster to the substrate and the elimination of the hydroxyl group would occur.

Support for this $[4Fe\text{-}4S]^{3+}$-intermediate was provided by the same group after HYSCORE experiments on one-electron reduced IspH incubated with HMBPP [68]. In these experiments, the electron source for the second electron transfer was again removed before incubation with the substrate, thus a possible reaction intermediate could be trapped. For the HYSCORE measurements, the oxygen of the substrate's hydroxyl group had been isotopically labeled with ^{17}O. As no ^{17}O hyperfine coupling was detected, it was suggested that in the intermediate under discussion the hydroxyl group had already been eliminated. Furthermore, in another experiment one of the respective C2, C3, and C4 atoms of HMBPP had been labeled with ^{13}C. In this way, possible hyperfine couplings to the unique fourth Fe species of the cluster could be measured. The ^{13}C labeling led to hyperfine coupling constants of A_{iso} = 1.8 MHz (C2), A_{iso} = 3.1 MHz (C3) and A_{iso} = 3.0 MHz (C4) (Figure 4). By DFT calculations, Fe-C distances of 2.17, 2.10, and 2.14 Å for the respective C2, C3, and C4 atoms were estimated in silico. As simulated A_{iso} values were in the range of the experimental values, the proposed distances were taken as a good approximation and suggest an allyl anion that forms an η^3-complex with the $[4Fe\text{-}4S]^{3+}$ cluster. The Fe-C distances of 2.6–2.7 Å measured previously in the crystal structure of IspH in complex with the "converted substrate" reported by Gräwert et al. presumably represent this same intermediate [37]. The HiPIP-like cluster in complex with the delocalized π-system is considered as one of the rare examples of a bioorganometallic intermediate found in nature.

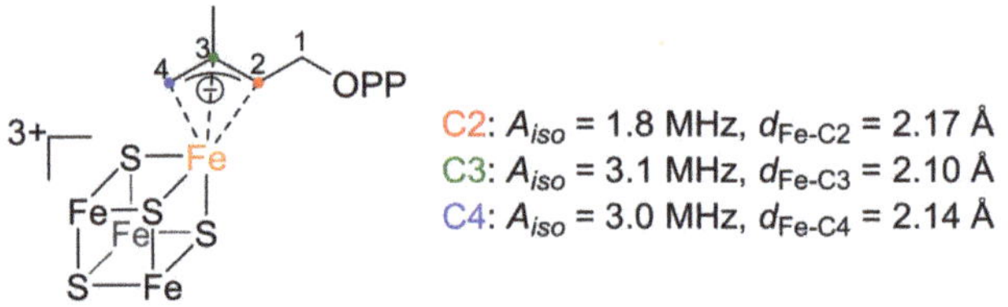

Figure 4. Schematic representation of the HiPIP-like η^3-allyl (π/σ) complex. The hyperfine coupling constants A_{iso} obtained after HYSCORE measurements of ^{13}C-labelled HMBPP as well as the iron-carbon distances $d_{Fe\text{-}C}$ estimated by DFT calculations are given for the C2, C3, and C4 atom [68].

The formation of the η^3-complex occurs after two-electron transfer from the cluster to the substrate and the elimination of the hydroxyl group. It is still unclear if the OH leaves the substrate as OH or as water after protonation by E126 and if the two-electron transfers are concomitant with the water formation.

4.6. Fifth Step: Protonation and Formation of the Two Products IPP and DMAPP

IspH catalyzes the reductive dehydroxylation of HMBPP into IPP and DMAPP with a ratio of about 4–6:1 [23]. This ratio is different from the equilibrium provided by isopentenyl diphosphate isomerase that favors the thermodynamically more stable DMAPP with a ratio of 3:7 (IPP:DMAPP) [24]. Obviously, protonation on the C2 atom of the allyl anion complex yields IPP while on the C4 atom, it yields DMAPP. The source of proton is still a matter of debate and several hypotheses for the proton mediator were done.

The proton source was identified early by Rohmer and co-workers without knowledge of the IspH-catalyzed mechanism [69,70] using the bacterium *Zymomonas mobilis* that due to its peculiar metabolic pathways (metabolization of glucose according to the Entner-Doudoroff pathway and no conversion of pyruvate into glyceraldehyde phosphate) is able to synthesize the two possible $NAD(P)^2H$ isotopomers from [1-^{2}H]glucose. After feeding *Z. mobilis* with [1-^{2}H]glucose, two pentacyclic triterpenes of the hopanoid series were analyzed by ^{1}H- and ^{13}C-NMR spectroscopy. These hopanoids exhibited deuterium atoms on definite positions of the carbon skeleton that could be traced back to the C2 and C4 positions of IPP and DMAPP. While the deuterium atom on the C4 position could be explained by an earlier reduction step in the MEP pathway catalyzed by DXR, the deuterium atom on the C2 position revealed the signature of a further reduction step in the formation of IPP and DMAPP that could not be explained at that time. The authors also observed that the deuterium was introduced in the pro-*S* position of IPP. Later on, the authors concluded that the deuterium was introduced by IspH and that NADPH was the proton donor for this catalytic step at least for this bacterium [28,71]. Feeding experiments of different methylerythritol isotopomers in *E. coli* [69] or of labelled deoxyxylulose in *E. coli* and in several plant cells [71] by Arigoni and co-workers and by the groups of Rohmer and Bach shed light on the introduction of a deuterium on C2 of IPP in the pro-*S* position [71–74].

The stereochemical course of the IPP formation was further investigated by Rohdich, Eisenreich et al. using enzymology on IspH [75]. The authors conducted ^{13}C-NMR spectroscopy on isotopically labeled HMBPP (**24**) after incubation with IspH in D_2O followed by incubation with isopentenyl diphosphate isomerase in H_2O (Scheme 4a). The *E. coli* IPP isomerase is known to eliminate the pro-*R*-proton of IPP. The resulting DMAPP exhibits a deuterium atom on the C2 atom which thus corresponds to the pro-*S*-proton of IPP. Hence, it is proposed that protonation occurs only from the *si* face of the allyl anion. Based on the crystallographic structure, the diphosphate moiety of the substrate may serve as proton mediator on this site as it is in close vicinity to the C2 and C4 atoms [37,42].

In line with these assumptions, the product formation of the ^{13}C isotopically labeled substrate analog **25** was investigated and it was found that only the IPP analog was formed [76] showing that the protonation was sterically restricted to the C4 atom. The authors explained this observation by the higher distance between the C5 of **25** and its diphosphate once **25** was in the active site of IspH that would prevent the diphosphate to protonate C5.

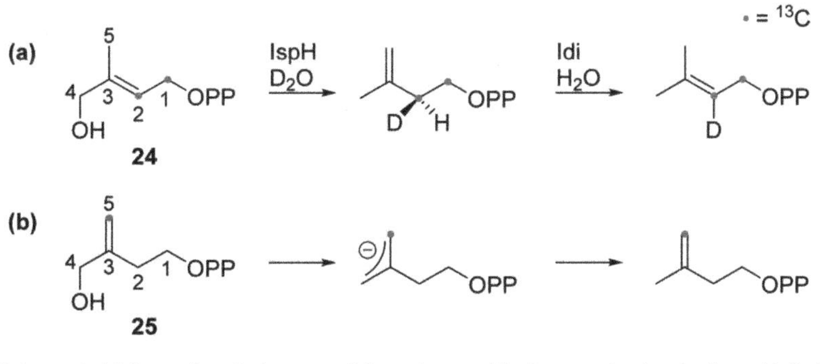

• = ^{13}C

(a) 24 —IspH, D_2O→ ... —Idi, H_2O→ ...

(b) 25 → ... → ...

Scheme 4. (**a**) Stereochemical course of deuterium and hydrogen after incubation with IspH in D_2O and isopentenyl diphosphate isomerase in H_2O [75]. (**b**) Substrate analog **25** used by Liu et al. and the product formed in the IspH catalyzed reaction [76].

5. IspH Inhibitors

Since the first breakthrough into the IspH mechanism, it has been readily exploited for the generation of potent inhibitors to develop unprecedented and safe therapeutics to fight against multiresistant bacteria and the malaria parasite *Plasmodium falciparum*.

Most of the compounds which have been classified as IspH inhibitors interact with the apical iron of the [4Fe-4S] cluster through different mechanisms. The group of Hirsch demonstrated that the active site and the neighboring pockets of IspH are druggable and, moreover, molecules larger than the natural substrate might be able to access this spacious site [77]. All the inhibitors studied so far can be divided into four classes: substrate analogs, pyridine diphosphates, alkyne derivatives, and non-diphosphate compounds.

5.1. Analogs of the HMBPP Substrate as Inhibitors

The first step in the IspH mechanism is the binding of the hydroxyl group of the substrate to the apical iron (Fe^{2+}) site of the $[4Fe\text{-}4S]^{2+}$ cluster. Replacement of this OH by a softer ligand such as a thiol was envisioned. Furthermore, substrate analogs not able to undergo the OH elimination step such as amino analogs were considered. In this context, Poulter, Seemann, and collaborators conceived and tested two HMBPP analogs where a thiol or an amino group replaces the hydroxyl group of the substrate (Figure 5) [78,79]. A complete kinetic investigation under anaerobic conditions revealed that the thiol analog **26** and the amino analog **27** were extremely potent inhibitors of *E. coli* IspH displaying competitive mode of inhibition with K_i values in the nanomolar range (K_i = 24 nM for **26** and K_i = 54 nM for **27**). Furthermore, **27** presented a slow binding inhibition mode with the slow step being the formation of the enzyme-inhibitor complex. This might be explained by the fact that the amino group undergoes a deprotonation at physiological pH to bind to the apical iron, therefore, the authors assumed that the slow step was the deprotonation of the amine [78]. Previous Mössbauer spectroscopy studies showed a transition from octahedral to tetrahedral geometry for the apical iron when adding **26** or **27** to the enzyme, as it was already observed when HMBPP is in complex with IspH [79]. This finding validates that these inhibitors are bound via their amino and thiol group to the unique iron of the [4Fe-4S] cluster. The same binding mode was observed in the crystal structures of *E. coli* IspH in complex with these two potent inhibitors (AMBPP **27**; PDB ligand ID: 10E and TMBPP **26**; PDB ligand ID: 10G), confirming that **27** is bound to the apical iron of the $[4Fe\text{-}4S]^{2+}$ cluster by its amino group and **26** is bound to this iron by its thiol group [47,48].

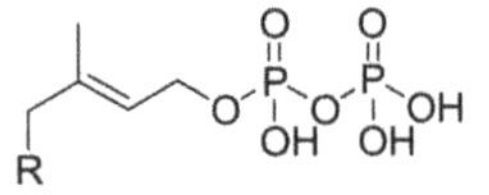

26: R = -SH IC_{50} = 210 nM (*E. coli*)
27: R = -NH_2 IC_{50} = 150 nM (*E. coli*)

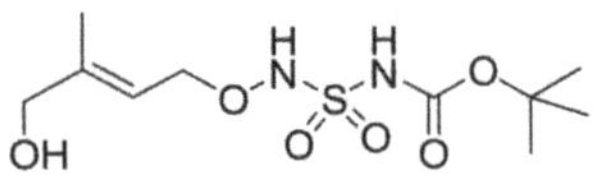

28: 61 % residual IspH activity at 1 mM (*A. aeolicus*)

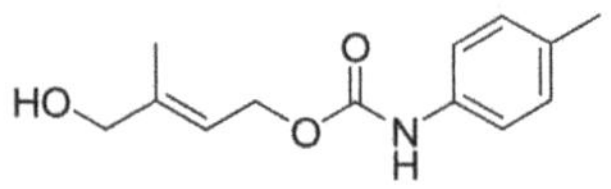

29: 62 % residual IspH activity at 1 mM (*A. aeolicus*)

Figure 5. Substrate analogs as inhibitors of IspH.

Nuclear resonance vibrational spectroscopy (NRVS) data were further recorded by Schünemann and collaborators for IspH bound to either of these inhibitors [38]. The measured NRVS data and those calculated based on the X-ray structure reported by Span et al. [48] showed reasonable agreement and the deviation was attributed to the presence of different IspH conformations in protein crystals and in solution. These differences were better explained once Borel and collaborators reported new structures of *E. coli* IspH in complex with these inhibitors in which the authors ascertained that the apical iron was present with an occupancy of 100% [47] as 35–45% of IspH molecules in the previously reported [48] *E. coli* IspH structure in complex with the amino analog **27** seemed to have an incomplete cluster. Furthermore, a detailed analysis of the *E. coli* IspH:**26** complex revealed that **26** bound in each monomer adopted two different conformations [47]: a classical one, superimposed on HMBPP already described, and one in which the C4 atom of **26** was pulled away from the diphosphate group and E126. The fact that IspH can accommodate more than one substrate conformer was proposed to be related to the selectivity of the protonation and consequently to the genesis of the two products. The presence of two conformers was further exploited to generate in silico putative IspH inhibitors with enhanced affinity by mimicking the space occupied by both conformers by a cyclopentyl group linked to a thiol and a diphosphate moiety [47]. Interestingly, the best virtual inhibitor candidate (**30**) that arose from these in silico evolution experiments is displayed in Figure 6 and revealed binding of the thiol moiety to the apical iron of the $[4Fe\text{-}4S]^{2+}$ cluster and formation of a salt bridge between an amino group and E126. However, no data of the potency of inhibition of **30** on IspH was reported.

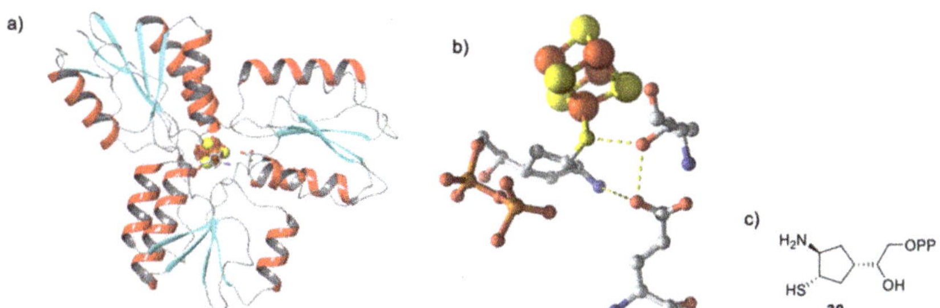

Figure 6. In silico docking result of the IspH inhibitor candidate **30** designed based on the structure of *E. coli* IspH:**26** (PDB ID: 3ZGN) and showing the best docking score. (**a**) Docking result of **30** binding to IspH; (**b**) active site; (**c**) chemical structure of **30**.

Two more substrate analogs have been designed by the group of Van Calenbergh and the group of Jomaa by replacing the diphosphate of HMBPP with a *N*-acyl-*N′*-oxy sulfamate function leading to **28** and a carbamate moiety leading to **29** (Figure 5). These compounds, tested at 1 mM concentration, showed a very low inhibition of IspH from *A. aeolicus* [80]. The poor inhibition potency of both molecules might be related to the fact that they lack the diphosphate group which is involved in important interactions with different histidine residues, such as His42 and His124 that are highly conserved in IspH of several species [36].

5.2. Pyridine Diphosphate as Inhibitors

Wang et al. studied a set of pyridine derivatives substituted in ortho, meta, and para (**31–44**) (Figure 7), as potential inhibitors of IspH from *A. aeolicus* [81]. In some compounds, the diphosphate has been replaced by the bioisostere bisphosphonate, which is more stable to phosphatase hydrolysis in vivo [82]. ^{14}N and ^{15}N HYSCORE investigations by Oldfield and co-workers demonstrate that inhibitor **38** binds to the apical iron probably via a η^1 bonding between the nitrogen and the [4Fe-4S] cluster, whereas no ^{14}N hyperfine interactions were observed with ortho and para pyridinyl analogs [83]. Furthermore, crystallographic studies reported by Span, Groll, Oldfield, and collaborators confirmed that the pyridine is close to the apical iron (2.3–2.4 Å), and the authors observed a continuous electron density between the nitrogen of the pyridine derivative and the apical iron. According to these observations and quantum chemical calculations, Span et al. provided evidence that the binding occurs through a η^2 coordination [50]. Moreover, the authors explored the orientation that the pyridine ring adopts in relation to the [4Fe-4S] cluster. X-ray crystallographic studies on *E. coli* IspH demonstrated that the nitrogen of the pyridine points in a direction likely represented in Figure 8. This orientation is presumably the result of the interaction of **38** with two pockets of the catalytic site, a lipophilic one characterized by four Val15, Val43, Val73, Val99 residues, and a hydrophilic site composed of Glu126, Thr167, Thr168, Thr200, and Asn227 residues [50].

Considering the insights acquired about the pyridine binding mode, it seems clearer why **38** and **39** showed the best inhibitory potency in this class. For example, **41** is not an inhibitor probably due to electron-withdrawing effect of the chlorine in *ortho*-position, which reduces the capacity of the nitrogen as electron donor Lewis-base. For similar reasons, the cation pyridinium analogs **42** and **43** did not show any biological activity.

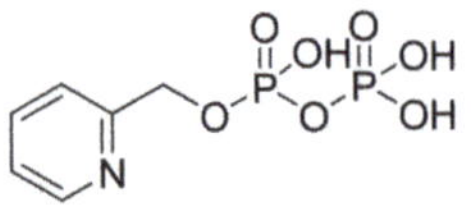

31: IC_{50} = 1.2 mM (*A. aeolicus*) 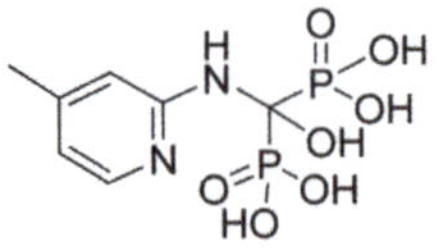
32: IC_{50} = 67 μM (*A. aeolicus*) 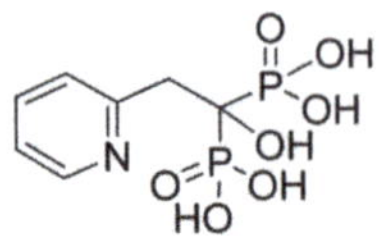
33: IC_{50} = 485 μM (*A. aeolicus*) 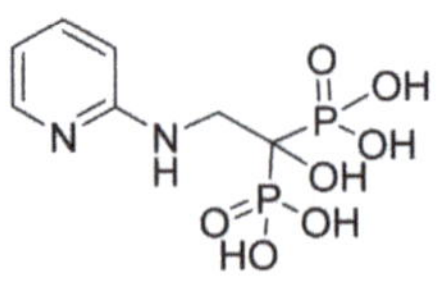
34: IC_{50} > 1 mM (*A. aeolicus*)

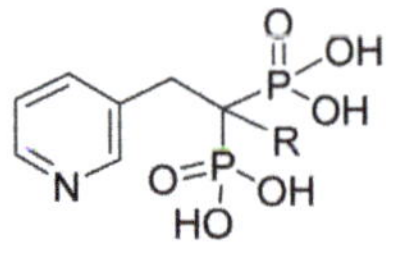

35: R = -H IC_{50} > 1 mM (*A. aeolicus*)
36: R = -OH IC_{50} > 1 mM (*A. aeolicus*)
37: R = -CH_3 IC_{50} > 1 mM (*A. aeolicus*)

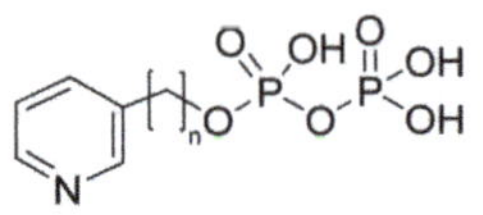

38: n = 1 IC_{50} = 38 μM (*A. aeolicus*)
39: n = 2 IC_{50} = 9.1 μM (*A. aeolicus*)
40: n = 3 IC_{50} = 463 μM (*A. aeolicus*)

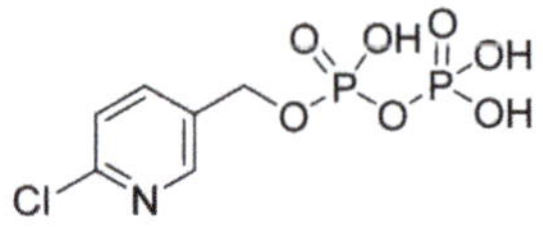

41: IC_{50} = 3.7 mM (*A. aeolicus*)

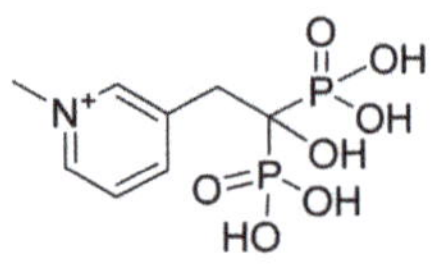

42: IC_{50} > 1mM (*A. aeolicus*)
43: IC_{50} = 540 μM (*A. aeolicus*) 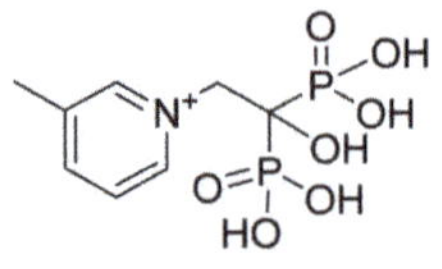
44: IC_{50} = 149 μM (*A. aeolicus*) 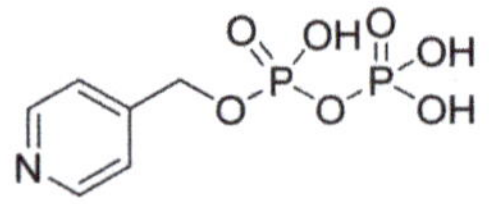

Figure 7. Compounds **31–44**, tested as inhibitors of IspH from *A. aeolicus*.

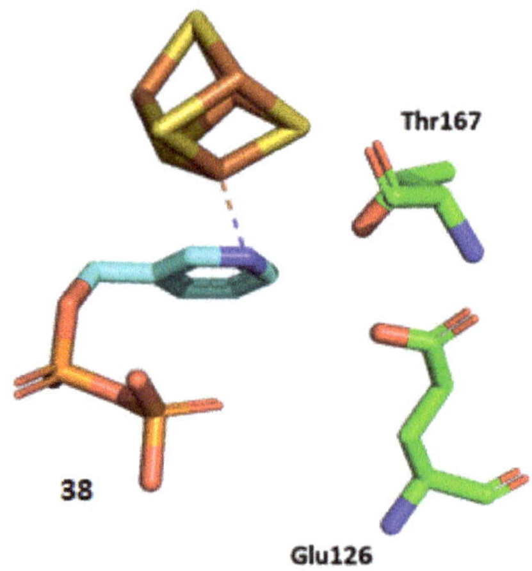

Figure 8. Structural representation of *Ec*-IspH in complex with compound **38** (PDB code 4MUX) [50]. The amino acids are shown as sticks, as well as the [4Fe-4S] cluster and the ligands.

5.3. Alkyne Derivatives as Inhibitors

Wang et al. designed a series of alkyne diphosphates **45–50** and a phosphonate analog **51** (Figure 9), which were first screened by EPR spectroscopy. Furthermore, EPR investigations and specifically ^{13}C-ENDOR results of reduced *A. aeolicus* IspH with [$^{13}C_3$]-**48** revealed strong ^{13}C hyperfine coupling interactions (6 MHz) [54], allowing to conclude that the acetylenic diphosphates bind through the alkyne fragment to the [4Fe-4S]$^+$ cluster by π or π-σ interactions. The *A. aeolicus* IspH inhibition assays demonstrated that **49** is the best inhibitor of this category (IC_{50} = 0.45 μM). The substitution of the diphosphate with a phosphonate for **51** led to a dramatic drop in inhibition potency, which might be due to the fact that the diphosphate fits better in the conserved cationic pocket seen in previous crystallographic investigations [37]. Moreover, Span et al. examined the compounds **48–50** bound to the oxidized IspH from *E. coli* by Mössbauer spectroscopy [46]. Previous investigations [33] helped to easily interpret the results obtained by the authors, then stating that the apical iron undergoes a conversion from octahedral to tetrahedral geometry with a 3S1O coordination, likewise what happens with the natural substrate HMBPP. Co-crystallographic structures of **48** with oxidized IspH were in accordance with the unexpected Mössbauer spectroscopy results and demonstrated that the binding between the alkyne residue and the fourth iron is not direct, indeed a water molecule is in between [46]. This result was further validated using NRVS measurements of the IspH-**48** complex [84]. NRVS studies of the free-substrate IspH had previously shown that the apical iron was bound to three sulfide ions of the cluster and to three molecules of water [38]. This finding was analog for the NRVS results of IspH in its $[4Fe\text{-}4S]^{2+}$ state in presence of **48**, where the fourth iron was in a tetrahedral geometry not linked to the acetylene residue but to a single molecule of water.

On the other hand, due to the impossibility of studying the oxidized IspH by EPR, the inhibitor **48** was examined by EPR in complex with the reduced IspH. The results obtained by O'Dowd et al. [84] revealed plausible π-interactions between the inhibitor **48** and the reduced [4Fe-4S] cluster.

Surprisingly, X-ray structures of the best inhibitor **49** showed that IspH converts this compound to an aldehyde (Scheme 5), revealing two ligand conformations. One is an η^1-O-enolate in complex with the apical iron, and the other one is a cyclic conformation of the aldehyde which binds to the Glu126 residue. Moreover, **50** was shown to react with the oxidized IspH, yielding in this case a ketone (Scheme 5). These unexpected products were also confirmed by ^{1}H-NMR spectroscopy and electrospray ionization mass spectrometry (ESI-MS) [46].

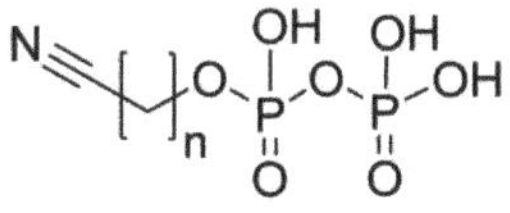

45: n = 1 IC_{50} = 245 μM (*A. aeolicus*)
46: n = 2 IC_{50} = 53 μM (*A. aeolicus*)

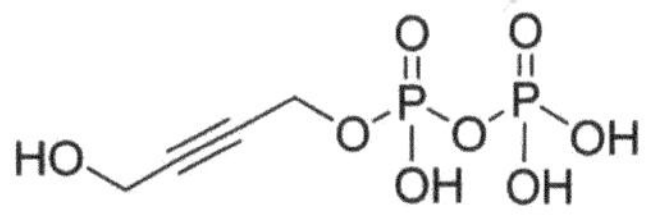

47: IC_{50} = 26 μM (*A. aeolicus*)

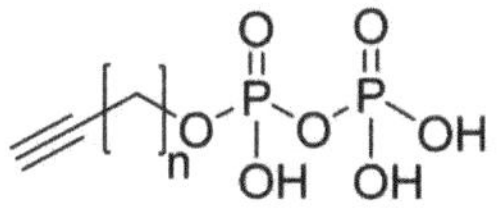

48: n = 1 IC_{50} = 6.7 μM (*A. aeolicus*)
49: n = 2 IC_{50} = 0.45 μM (*A. aeolicus*)
50: n = 3 IC_{50} = 6.5 μM (*A. aeolicus*)

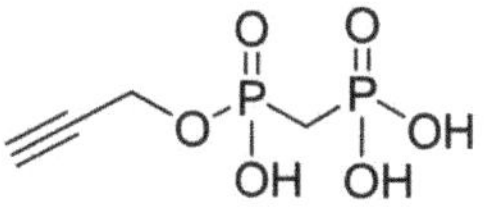

51: IC_{50} > 1mM (*A. aeolicus*)

Figure 9. Compounds **45–51**, tested as inhibitors of IspH from *A. aeolicus*.

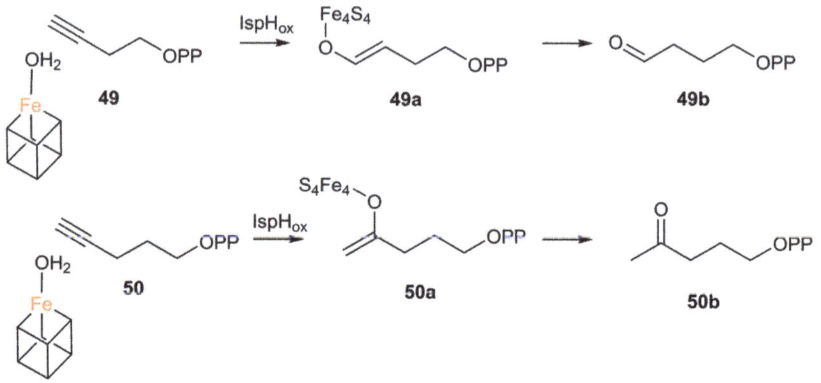

Scheme 5. Reactions of the inhibitors **49** and **50** catalyzed by the enzyme IspH in the oxidized state. The ligand bound to the apical iron could also be an OH.

5.4. Non-Diphosphate Inhibitors

O'Dowd et al. used an in silico approach to screen a series of compounds from ZINC and NCI libraries and after having tested some hits for IspH inhibition, they identified **52** and **53** as two new IspH inhibitors (Figure 10) [84]. In both compounds, it is possible to see a common barbituric acid sub-structure which makes the hits more drug-like, similarly to antibacterial agents developed by Pharmacia Corporation [85]. The compound **52** resulted to be a good inhibitor of IspH from *P. aeruginosa*, with an IC_{50} = 22 µM. On the other hand, the analog **53** inhibits IspH from *E. coli*, showing as result an IC_{50} = 23 µM [84]. After docking experiments of **53** into the *A. aeolicus* IspH structure (PDB ID: 3DNF), the authors proposed a plausible binding mode where the barbiturate might chelate the apical iron of the $[4Fe\text{-}4S]^{2+}$ cluster, similarly to how barbiturate enolates do with Zn^{2+} in other metalloenzymes [86,87]. However, EPR studies for **53** in complex with the reduced IspH from *P. aeruginosa* did show a very low signal intensity, probably due to a change of the redox potential upon ligand binding or to a different relaxation mode [84]. These compounds could be developed as potential drugs for several reasons: they solved the problem related to the lability of the diphosphate moiety present in other inhibitors in vivo [82]; they respect the Lipinski's rule [88] and they are not PAINS compounds [89]. Although they are promising drug-like compounds, further studies to validate the binding mode should be realized.

Potent IspH inhibitors have already been discovered with **26** and **27** showing the best inhibition properties. However, it should be noticed that no inhibitors having an activity on bacteria or apicomplexan parasites have been reported to date.

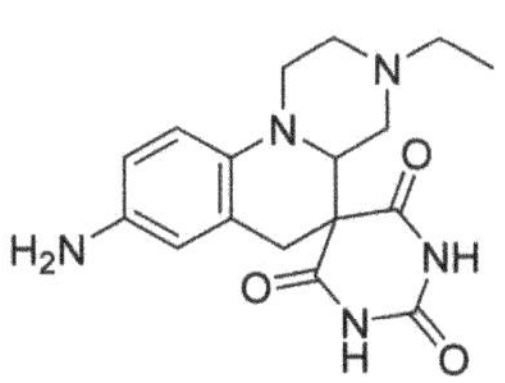

52: IC_{50} = 22 μM (*P. aeruginosa*)

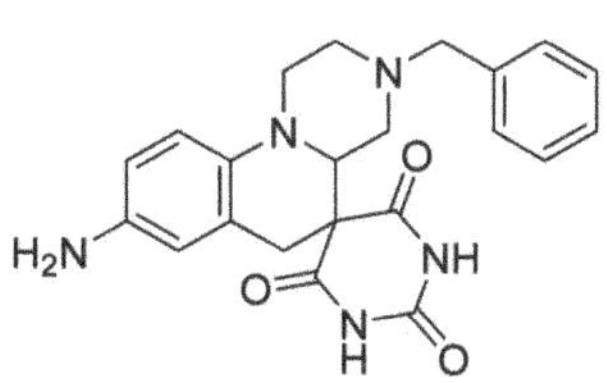

53: IC_{50} = 23 μM (*E. coli*)

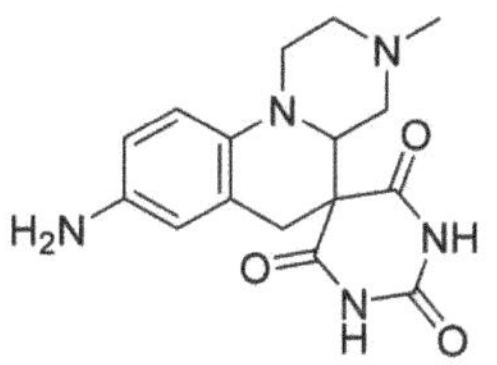

54: IC_{50} = 65 μM

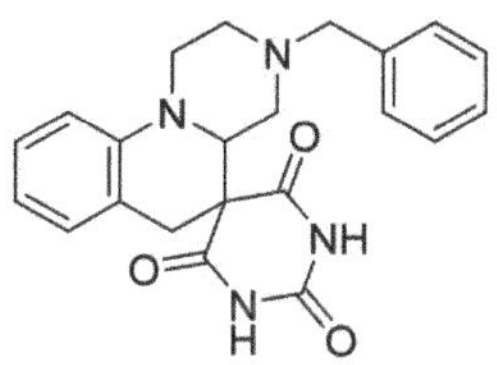

55: IC_{50} = 113 μM

Figure 10. Compounds **52– 55** as inhibitors of IspH. The origin of IspH for inhibitions assays of **54** and **55** was not mentioned by the authors.

6. Conclusions

The oxygen sensitivity of IspH explains why the MEP pathway was overlooked for so many decades and not evidenced in the late 1950s when the mevalonate pathway was discovered by Bloch, Lynen, Cornforth. At that time, the researchers performed their enzymology experiments on the bench and did not have the equipment to work under anaerobic conditions nor the knowledge and spectroscopies needed for the characterization of metalloenzymes. Thanks to the advent of EPR, Mössbauer spectroscopy, and bioinorganic chemistry, it is now clear that IspH contains a peculiar $[4Fe\text{-}4S]^{2+}$ cluster that is the active species and enables the electron transfer towards the substrate HMBPP in concert with a biological reduction system. Crystal structures, (labelled) substrate analogs synthesis and spectroscopic techniques permitted insights into the catalytic mechanism of IspH and led to the discovery of the first examples of bioorganometallic intermediates involving [4Fe-4S] centers. However, the complete mechanism of the IspH-catalyzed reaction is not fully understood. While it is now established that the first step is the binding of the substrate to the $[4Fe\text{-}4S]^{2+}$ cluster, that a rotation of the OH of the substrate occurs upon reduction of the $[4Fe\text{-}4S]^{2+}$ cluster and that the $S = 1/2$ η^3-allyl-$[4Fe\text{-}4S]^{3+}$ intermediate is formed, it is still unclear how the OH or water is removed and how the electron transfer occurs. As most investigations relied on paramagnetic species, no evidence of transient diamagnetic intermediates could be trapped. X-ray crystallography would be an elegant method to trap such intermediates. However, it suffers from the high lability of the apical iron and the obtained structures do not necessarily represent the intermediates present in solution. Obviously, Mössbauer spectroscopy will be a key player in the identification of such intermediates and more sensitive methods such as NRVS will be needed. Further investigations will also rely on the synthesis of new biomimetic models or other innovative molecular tools. The proton source involved in the formation of the two final products is still under discussion. Another question is the ratio of the products formed. What triggers the formation of about six times more of the less thermodynamically favored IPP compared to DMAPP? This raises also the question about the reduction systems. Why do some species use flavodoxin and others ferredoxin to transfer the electrons to the $[4Fe\text{-}4S]^{2+}$ center of IspH? How are these electrons transferred in vivo? One puzzling question is the maintenance of an oxygen sensitive $[4Fe\text{-}4S]^{2+}$ cluster in microorganisms living under aerobic conditions or in plant chloroplasts that produce O_2 in phototropic conditions. Is the cluster continuously repaired? Are protecting enzymes involved? ErpA and NfuA proteins might be such candidates as they were shown to interact with IspH in *E. coli*, and NfuA was proposed to assist ErpA in Fe-S cluster delivery under oxidative stress [90].

Although some questions are still unanswered, most insights acquired about the mechanism of IspH were already exploited to discover promising inhibitors. However, none of these potent inhibitors were reported to be active on cells. Therefore, one of the big remaining challenges will also be the discovery of such molecules as they would represent good starting points to bring to the market innovative drugs with unprecedented modes of action that are urgently needed to fight against multidrug-resistant infectious diseases.

Author Contributions: Writing—original draft preparation, H.J., G.I.B., F.B., P.C., M.S.; writing—review and editing, H.J., G.I.B., F.B., P.C., M.S.; funding acquisition—F.B., M.S. All authors have read and agreed to the published version of the manuscript.

Funding: This project has received funding from the European Union's Horizon 2020 research and innovation programme under the Marie Sklodowska-Curie grant agreement No. 860816 and from the 'Université franco-allemande'. This work of the Interdisciplinary Thematic Institute InnoVec, as part of the ITI program of the University of Strasbourg, CNRS and Inserm, was supported by IdEx Unistra (ANR-10-IDEX-0002), and by SFRI-STRAT'US project (ANR-20-SFRI-0012) under the framework of the French Investments for the Future Program.

Institutional Review Board Statement: Not applicable.

Informed Consent Statement: Not applicable.

Acknowledgments: We are very grateful to our dear colleague and friend Jean-Luc Ferrer who sadly passed away in 2020. We would like to dedicate this manuscript to his memory and to his outstanding contribution in protein crystallography.

Conflicts of Interest: The authors declare no conflict of interest.

References

1. Cane, D.E.; Barton, D.H.R.; Nakanishi, K.; Meth-Cohn, O. (Eds.) *Comprehensive Natural Products Chemistry: Isoprenoids, Including Carotenoids and Steroids*; Elsevier: Oxford, UK, 1999; Volume 2.
2. Sacchettini, J.C.; Poulter, C.D. Creating isoprenoid diversity. *Science* **1997**, *277*, 1788–1789. [CrossRef] [PubMed]
3. Bloch, K. Sterol molecule: Structure, biosynthesis, and function. *Steroids* **1992**, *57*, 378–383. [CrossRef]
4. Rohmer, M. The discovery of a mevalonate-independent pathway for isoprenoid biosynthesis in bacteria, algae and higher plants. *Nat. Prod. Rep.* **1999**, *16*, 565–574. [CrossRef]
5. Eisenreich, W.; Schwarz, M.; Cartayrade, A.; Arigoni, D.; Zenk, M.H.; Bacher, A. The deoxyxylulose phosphate pathway of terpenoid biosynthesis in plants and microorganisms. *Chem. Biol.* **1998**, *5*, R221–R233. [CrossRef]
6. Lichtenthaler, H.K. The 1-deoxy-D-xylulose-5-phosphate pathway of isoprenoid biosynthesis in plants. *Annu. Rev. Plant. Physiol. Plant. Mol. Biol.* **1999**, *50*, 47–65. [CrossRef] [PubMed]
7. Rohmer, M.; Grosdemange-Billiard, C.; Seemann, M.; Tritsch, D. Isoprenoid biosynthesis as a novel target for antibacterial and antiparasitic drugs. *Curr. Opin. Investig. Drugs* **2004**, *5*, 154–162.
8. *Priorization of Pathogens to Guide Discovery, Research and Development of New Antibioitics for Drug-Resistant Bacterial Infections, Including Tuberculosis*; Report No. WHO/EMP/IAU/2017.12; World Health Organization: Geneva, Switzerland, 2017.
9. Miethke, M.; Pieroni, M.; Weber, T.; Brönstrup, M.; Hammann, P.; Halby, L.; Arimondo, P.B.; Glaser, P.; Aigle, B.; Bode, H.B.; et al. Towards the sustainable discovery and development of new antibiotics. *Nat. Rev. Chem.* **2021**, *5*, 726–749. [CrossRef]
10. Mann, F.M.; Xu, M.; Davenport, E.K.; Peters, R.J. Functional characterization and evolution of the isotuberculosinol operon in *Mycobacterium tuberculosis* and related mycobacteria. *Front. Microbiol.* **2012**, *3*, 368. [CrossRef]
11. Brown, A.C.; Kokoczka, R.; Parish, T. LytB1 and LytB2 of *Mycobacterium tuberculosis* are not genetically redundant. *PLoS ONE* **2015**, *10*, e0135638. [CrossRef]
12. Kafai, N.M.; Odom John, A.R. Malaria in children. *Infect. Dis. Clin. North Am.* **2018**, *32*, 189–200. [CrossRef]
13. Mombo-Ngoma, G.; Remppis, J.; Sievers, M.; Zoleko Manego, R.; Endamne, L.; Kabwende, L.; Veletzky, L.; Nguyen, T.T.; Groger, M.; Lötsch, F.; et al. Efficacy and safety of fosmidomycin–piperaquine as nonartemisinin-based combination therapy for uncomplicated falciparum malaria: A single-arm, age de-escalation proof-of-concept study in Gabon. *Clin. Infect. Dis.* **2018**, *66*, 1823–1830. [CrossRef]
14. Hedl, M.; Sutherlin, A.; Wilding, E.I.; Mazzulla, M.; McDevitt, D.; Lane, P.; Burgner, J.W.; Lehnbeuter, K.R.; Stauffacher, C.V.; Gwynn, M.N.; et al. *Enterococcus faecalis* acetoacetyl-coenzyme a thiolase/3-hydroxy-3-methylglutaryl-coenzyme a reductase, a dual-function protein of isopentenyl diphosphate biosynthesis. *J. Bacteriol.* **2002**, *184*, 2116–2122. [CrossRef]
15. Gustafson, C.E.; Kaul, S.; Ishiguro, E.E. Identification of the *Escherichia coli lytB* gene, which is involved in penicillin tolerance and control of the stringent response. *J. Bacteriol.* **1993**, *175*, 1203–1205. [CrossRef]
16. Potter, S.; Yang, X.; Boulanger, M.J.; Ishiguro, E.E. Occurrence of homologs of the *Escherichia coli lytB* gene in gram-negative bacterial species. *J. Bacteriol.* **1998**, *180*, 1959–1961. [CrossRef] [PubMed]

17. Cunningham, F.X.; Lafond, T.P.; Gantt, E. Evidence of a role for LytB in the nonmevalonate pathway of isoprenoid biosynthesis. *J. Bacteriol.* **2000**, *182*, 5841–5848. [CrossRef] [PubMed]
18. Altincicek, B.; Kollas, A.-K.; Sanderbrand, S.; Wiesner, J.; Hintz, M.; Beck, E.; Jomaa, H. GcpE is involved in the 2-C-methyl-D-erythritol 4-phosphate pathway of isoprenoid biosynthesis in *Escherichia coli*. *J. Bacteriol.* **2001**, *183*, 2411–2416. [CrossRef] [PubMed]
19. McAteer, S.; Coulson, A.; McLennan, N.; Masters, M. The *lytB* gene of *Escherichia coli* is essential and specifies a product needed for isoprenoid biosynthesis. *J. Bacteriol.* **2001**, *183*, 7403–7407. [CrossRef] [PubMed]
20. Altincicek, B.; Kollas, A.-K.; Eberl, M.; Wiesner, J.; Sanderbrand, S.; Hintz, M.; Beck, E.; Jomaa, H. LytB, a novel gene of the 2-C-methyl-D-erythritol 4-phosphate pathway of isoprenoid biosynthesis in *Escherichia coli*. *FEBS Lett.* **2001**, *499*, 37–40. [CrossRef]
21. Rohdich, F.; Hecht, S.; Gartner, K.; Adam, P.; Krieger, C.; Amslinger, S.; Arigoni, D.; Bacher, A.; Eisenreich, W. Studies on the nonmevalonate terpene biosynthetic pathway: Metabolic role of IspH (LytB) protein. *Proc. Natl. Acad. Sci. USA* **2002**, *99*, 1158–1163. [CrossRef]
22. Hecht, S.; Eisenreich, W.; Adam, P.; Amslinger, S.; Kis, K.; Bacher, A.; Arigoni, D.; Rohdich, F. Studies on the nonmevalonate pathway to terpenes: The role of the GcpE (IspG) protein. *Proc. Natl. Acad. Sci. USA* **2001**, *98*, 14837–14842. [CrossRef]
23. Adam, P.; Hecht, S.; Eisenreich, W.; Kaiser, J.; Gräwert, T.; Arigoni, D.; Bacher, A.; Rohdich, F. Biosynthesis of terpenes: Studies on 1-hydroxy-2-methyl-2-(*E*)-butenyl 4-diphosphate reductase. *Proc. Natl. Acad. Sci. USA* **2002**, *99*, 12108–12113. [CrossRef]
24. Rohdich, F.; Zepeck, F.; Adam, P.; Hecht, S.; Kaiser, J.; Laupitz, R.; Amslinger, S.; Eisenreich, W.; Bacher, A.; Arigoni, D. The deoxyxylulose phosphate pathway of isoprenoid biosynthesis: Studies on the mechanisms of the reactions catalyzed by IspG and IspH protein. *Proc. Natl. Acad. Sci. USA* **2002**, *100*, 1586–1591. [CrossRef]
25. Altincicek, B.; Duin, E.C.; Reichenberg, A.; Hedderich, R.; Kollas, A.-K.; Hintz, M.; Wagner, S.; Wiesner, J.; Beck, E.; Jomaa, H. LytB protein catalyzes the terminal step of the 2-C-methyl-D-erythritol-4-phosphate pathway of isoprenoid biosynthesis. *FEBS Lett.* **2002**, *532*, 437–440. [CrossRef]
26. Beinert, H. Iron-sulfur proteins: Ancient structures, still full of surprises. *J. Biol. Inorg. Chem.* **2000**, *5*, 2–15. [CrossRef]
27. Seemann, M.; Tse Sum Bui, B.; Wolff, M.; Tritsch, D.; Campos, N.; Boronat, A.; Marquet, A.; Rohmer, M. Isoprenoid biosynthesis through the methylerythritol phosphate pathway: The (*E*)-4-hydroxy-3-methylbut-2-enyl diphosphate synthase (GcpE) is a [4Fe–4S] protein. *Angew. Chem. Int. Ed.* **2002**, *41*, 4337–4339. [CrossRef]
28. Wolff, M.; Seemann, M.; Tse Sum Bui, B.; Frapart, Y.; Tritsch, D.; Estrabot, A.G.; Rodríguez-Concepción, M.; Boronat, A.; Marquet, A.; Rohmer, M. Isoprenoid biosynthesis via the methylerythritol phosphate pathway: The (*E*)-4-hydroxy-3-methylbut-2-enyl diphosphate reductase (lytB/IspH) from *Escherichia coli* is a [4Fe-4S] protein. *FEBS Lett.* **2003**, *541*, 115–120. [CrossRef]
29. Rupp, H.; Rao, K.K.; Hall, D.O.; Cammack, R. Electron spin relaxation of iron-sulphur proteins studied by microwave power saturation. *Biochim. Biophys. Acta Protein Struct.* **1978**, *537*, 255–269. [CrossRef]
30. Ollagnier, S.; Mulliez, E.; Schmidt, P.P.; Eliasson, R.; Gaillard, J.; Deronzier, C.; Bergman, T.; Gräslund, A.; Reichard, P.; Fontecave, M. Activation of the anaerobic ribonucleotide reductase from *Escherichia coli*. *J. Biol. Chem.* **1997**, *272*, 24216–24223. [CrossRef] [PubMed]
31. Petrovich, R.M.; Ruzicka, F.J.; Reed, G.H.; Frey, P.A. Characterization of iron-sulfur clusters in lysine 2,3-aminomutase by electron paramagnetic resonance spectroscopy. *Biochemistry* **1992**, *31*, 10774–10781. [CrossRef]
32. Gräwert, T.; Kaiser, J.; Zepeck, F.; Laupitz, R.; Hecht, S.; Amslinger, S.; Schramek, N.; Schleicher, E.; Weber, S.; Haslbeck, M.; et al. IspH protein of *Escherichia coli*: Studies on iron−sulfur cluster implementation and catalysis. *J. Am. Chem. Soc.* **2004**, *126*, 12847–12855. [CrossRef] [PubMed]
33. Seemann, M.; Janthawornpong, K.; Schweizer, J.; Böttger, L.H.; Janoschka, A.; Ahrens-Botzong, A.; Tambou, E.N.; Rotthaus, O.; Trautwein, A.X.; Rohmer, M.; et al. Isoprenoid biosynthesis via the MEP pathway: In vivo Mössbauer spectroscopy identifies a $[4Fe\text{-}4S]^{2+}$ center with unusual coordination sphere in the LytB protein. *J. Am. Chem. Soc.* **2009**, *131*, 13184–13185. [CrossRef]
34. Xiao, Y.; Chu, L.; Sanakis, Y.; Liu, P. Revisiting the IspH catalytic system in the deoxyxylulose phosphate pathway: Achieving high activity. *J. Am. Chem. Soc.* **2009**, *131*, 9931–9933. [CrossRef] [PubMed]
35. Schünemann, V.; Winkler, H. Structure and dynamics of biomolecules studied by Mössbauer spectroscopy. *Rep. Prog. Phys.* **2000**, *63*, 263–353. [CrossRef]
36. Rekittke, I.; Wiesner, J.; Röhrich, R.; Demmer, U.; Warkentin, E.; Xu, W.; Troschke, K.; Hintz, M.; No, J.H.; Duin, E.C.; et al. Structure of (E)-4-hydroxy-3-methyl-but-2-enyl diphosphate reductase, the terminal enzyme of the non-mevalonate pathway. *J. Am. Chem. Soc.* **2008**, *130*, 17206–17207. [CrossRef] [PubMed]
37. Gräwert, T.; Span, I.; Eisenreich, W.; Rohdich, F.; Eppinger, J.; Bacher, A.; Groll, M. Probing the reaction mechanism of IspH protein by X-ray structure analysis. *Proc. Natl. Acad. Sci. USA* **2010**, *107*, 1077–1081. [CrossRef]
38. Faus, I.; Reinhard, A.; Rackwitz, S.; Wolny, J.A.; Schlage, K.; Wille, H.-C.; Chumakov, A.; Krasutsky, S.; Chaignon, P.; Poulter, C.D.; et al. Isoprenoid biosynthesis in pathogenic bacteria: Nuclear resonance vibrational spectroscopy provides insight into the unusual [4Fe-4S] cluster of the *E. Coli* LytB/IspH protein. *Angew. Chem. Int. Ed.* **2015**, *54*, 12584–12587. [CrossRef] [PubMed]
39. Scheidt, W.; Durbin, S.; Sage, J. Nuclear resonance vibrational spectroscopy? NRVS. *J. Inorg. Biochem.* **2005**, *99*, 60–71. [CrossRef]
40. Xu, W.; Lees, N.S.; Hall, D.; Welideniya, D.; Hoffman, B.M.; Duin, E.C. A closer look at the spectroscopic properties of possible reaction intermediates in wild-type and mutant (*E*)-4-hydroxy-3-methylbut-2-enyl diphosphate reductase. *Biochemistry* **2012**, *51*, 4835–4849. [CrossRef]

41. Röhrich, R.C.; Englert, N.; Troschke, K.; Reichenberg, A.; Hintz, M.; Seeber, F.; Balconi, E.; Aliverti, A.; Zanetti, G.; Köhler, U.; et al. Reconstitution of an apicoplast-localised electron transfer pathway involved in the isoprenoid biosynthesis of *Plasmodium falciparum*. *FEBS Lett.* **2005**, *579*, 6433–6438. [CrossRef] [PubMed]
42. Gräwert, T.; Rohdich, F.; Span, I.; Bacher, A.; Eisenreich, W.; Eppinger, J.; Groll, M. Structure of active IspH enzyme from *Escherichia coli* provides mechanistic insights into substrate reduction. *Angew. Chem. Int. Ed.* **2009**, *48*, 5756–5759. [CrossRef]
43. Rekittke, I.; Olkhova, E.; Wiesner, J.; Demmer, U.; Warkentin, E.; Jomaa, H.; Ermler, U. Structure of the (E)-4-hydroxy-3-methyl-but-2-enyl-diphosphate reductase from *Plasmodium falciparum*. *FEBS Lett.* **2013**, *587*, 3968–3972. [CrossRef] [PubMed]
44. Frank, A.; Groll, M. The methylerythritol phosphate pathway to isoprenoids. *Chem. Rev.* **2017**, *117*, 5675–5703. [CrossRef]
45. Span, I.; Gräwert, T.; Bacher, A.; Eisenreich, W.; Groll, M. Crystal structures of mutant IspH proteins reveal a rotation of the substrate's hydroxymethyl group during catalysis. *J. Mol. Biol.* **2012**, *416*, 1–9. [CrossRef]
46. Span, I.; Wang, K.; Wang, W.; Zhang, Y.; Bacher, A.; Eisenreich, W.; Li, K.; Schulz, C.; Oldfield, E.; Groll, M. Discovery of acetylene hydratase activity of the iron–sulphur protein IspH. *Nat. Commun.* **2012**, *3*, 1042. [CrossRef] [PubMed]
47. Borel, F.; Barbier, E.; Krasutsky, S.; Janthawornpong, K.; Chaignon, P.; Poulter, C.D.; Ferrer, J.-L.; Seemann, M. Further insight into crystal structures of *Escherichia coli* IspH/lytB in complex with two potent inhibitors of the MEP pathway: A starting point for rational design of new antimicrobials. *ChemBioChem* **2017**, *18*, 2137–2144. [CrossRef] [PubMed]
48. Span, I.; Wang, K.; Wang, W.; Jauch, J.; Eisenreich, W.; Bacher, A.; Oldfield, E.; Groll, M. Structures of fluoro, amino, and thiol inhibitors bound to the [Fe_4S_4] protein IspH. *Angew. Chem. Int. Ed.* **2013**, *52*, 2118–2121. [CrossRef]
49. Wang, W.; Wang, K.; Span, I.; Jauch, J.; Bacher, A.; Groll, M.;Oldfield, E. Are free radicals involved in IspH catalysis? An EPR and crystallographic investigation. *J. Am. Chem. Soc.* **2012**, *134*, 11225–11234. [CrossRef]
50. Span, I.; Wang, K.; Eisenreich, W.; Bacher, A.; Zhang, Y.; Oldfield, E.; Groll, M. Insights into the binding of pyridines to the iron–sulfur enzyme IspH. *J. Am. Chem. Soc.* **2014**, *136*, 7926–7932. [CrossRef]
51. Brown, A.C.; Suess, D.L.M. Controlling substrate binding to Fe_4S_4 clusters through remote steric effects. *Inorg. Chem.* **2019**, *58*, 5273–5280. [CrossRef]
52. Blachly, P.G.; Sandala, G.M.; Giammona, D.A.; Bashford, D.; McCammon, J.A.; Noodleman, L. Broken-symmetry DFT computations for the reaction pathway of IspH, an iron–sulfur enzyme in pathogenic bacteria. *Inorg. Chem.* **2015**, *54*, 6439–6461. [CrossRef]
53. Blachly, P.G.; Sandala, G.M.; Giammona, D.A.; Liu, T.; Bashford, D.; McCammon, J.A.; Noodleman, L. Use of broken-symmetry density functional theory to characterize the IspH oxidized state: Implications for IspH mechanism and inhibition. *J. Chem. Theory Comput.* **2014**, *10*, 3871–3884. [CrossRef] [PubMed]
54. Wang, W.; Wang, K.; Liu, Y.-L.; No, J.-H.; Li, J.; Nilges, M.J.; Oldfield, E. Bioorganometallic mechanism of action, and inhibition, of IspH. *Proc. Natl. Acad. Sci. USA* **2010**, *107*, 4522–4527. [CrossRef] [PubMed]
55. Fuss, J.O.; Tsai, C.-L.; Ishida, J.P.; Tainer, J.A. Emerging critical roles of Fe–S clusters in DNA replication and repair. *Bioch. Biophys. Acta Mol. Cell Res.* **2015**, *1853*, 1253–1271. [CrossRef]
56. Ter Beek, J.; Parkash, V.; Bylund, G.O.; Osterman, P.; Sauer-Eriksson, A.E.; Johansson, E. Structural evidence for an essential Fe–S cluster in the catalytic core domain of DNA polymerase ϵ. *Nucleic Acids Res.* **2019**, *47*, 5712–5722. [CrossRef]
57. Citron, C.A.; Brock, N.L.; Rabe, P.; Dickschat, J.S. The stereochemical course and mechanism of the IspH reaction. *Angew. Chem. Int. Ed.* **2012**, *51*, 4053–4057. [CrossRef] [PubMed]
58. Lee, H.-I.; Igarashi, R.Y.; Laryukhin, M.; Doan, P.E.; Dos Santos, P.C.; Dean, D.R.; Seefeldt, L.C.; Hoffman, B.M. An organometallic intermediate during alkyne reduction by nitrogenase. *J. Am. Chem. Soc.* **2004**, *126*, 9563–9569. [CrossRef] [PubMed]
59. Lee, H.-I.; Sørlie, M.; Christiansen, J.; Yang, T.-C.; Shao, J.; Dean, D.R.; Hales, B.J.; Hoffman, B.M. Electron inventory, kinetic assignment (E_n), structure, and bonding of nitrogenase turnover intermediates with C_2H_2 and CO. *J. Am. Chem. Soc.* **2005**, *127*, 15880–15890. [CrossRef] [PubMed]
60. Chaignon, P.; Petit, B.E.; Vincent, B.; Allouche, L.; Seemann, M. Methylerythritol phosphate pathway: Enzymatic evidence for a rotation in the Lytb/IspH catalyzed reaction. *Chem. Eur. J.* **2020**, *26*, 1032–1036. [CrossRef] [PubMed]
61. Pearson, R.G. Hard and soft acids and bases. *J. Am. Chem. Soc.* **1963**, *85*, 7. [CrossRef]
62. Rao, P.V.; Holm, R.H. Synthetic analogues of the active sites of iron−sulfur proteins. *Chem. Rev.* **2004**, *104*, 527–559. [CrossRef]
63. Xiao, Y.; Zhao, Z.K.; Liu, P. Mechanistic studies of IspH in the deoxyxylulose phosphate pathway: Heterolytic C-O bond cleavage at C_4 position. *J. Am. Chem. Soc.* **2008**, *130*, 2164–2165. [CrossRef]
64. Xiao, Y.; Liu, P. IspH protein of the deoxyxylulose phosphate pathway: Mechanistic studies with C_1 -deuterium-labeled substrate and fluorinated analogue. *Angew. Chem. Int. Ed.* **2008**, *47*, 9722–9725. [CrossRef]
65. Abdel-Azeim, S.; Jedidi, A.; Eppinger, J.; Cavallo, L. Mechanistic insights into the reductive dehydroxylation pathway for the biosynthesis of isoprenoids promoted by the IspH enzyme. *Chem. Sci.* **2015**, *6*, 5643–5651. [CrossRef]
66. Walters, E.M.; Garcia-Serres, R.; Jameson, G.N.L.; Glauser, D.A.; Bourquin, F.; Manieri, W.; Schürmann, P.; Johnson, M.K.; Huynh, B.H. Spectroscopic characterization of site-specific [Fe_4S_4] cluster chemistry in ferredoxin:thioredoxin reductase: Implications for the catalytic mechanism. *J. Am. Chem. Soc.* **2005**, *127*, 9612–9624. [CrossRef]
67. Belinskii, M. Spin coupling model for tetrameric iron clusters in ferredoxins. I. Theory, exchange levels, g-factors. *Chem. Phys.* **1993**, *172*, 189–211. [CrossRef]
68. Li, J.; Wang, K.; Smirnova, T.I.; Khade, R.L.; Zhang, Y.; Oldfield, E. Isoprenoid biosynthesis: Ferraoxetane or allyl anion mechanism for IspH catalysis? *Angew. Chem. Int. Ed.* **2013**, *52*, 6522–6525. [CrossRef] [PubMed]

69. Charon, L.; Pale-Grosdemange, C.; Rohmer, M. On the reduction steps in the mevalonate independent 2-C-methyi-D-erythritol 4-phosphate (MEP) pathway for isoprenoid biosynthesis in the bacterium *Zymomonas Mobilis*. *Tetrahedron Lett.* **1999**, *40*, 7231–7234. [CrossRef]
70. Charon, L.; Hoeffler, J.-F.; Pale-Grosdemange, C.; Lois, L.-M.; Campos, N.; Boronat, A.; Rohmer, M. Deuterium-labelled isotopomers of 2-C-methyl-D-erythritol as tools for the elucidation of the 2-C-methyl-D-erythritol 4-phosphate pathway for isoprenoid biosynthesis. *Biochem. J.* **2000**, *346*, 737–742. [CrossRef] [PubMed]
71. Hoeffler, J.-F.; Hemmerlin, A.; Grosdemange-Billiard, C.; Bach, T.J.; Rohmer, M. Isoprenoid biosynthesis in higher plants and in *Escherichia coli*: On the branching in the methylerythritol phosphate pathway and the independent biosynthesis of isopentenyl diphosphate and dimethylallyl diphosphate. *Biochem. J.* **2002**, *366*, 573–583. [CrossRef]
72. Giner, J.-L.; Jaun, B.; Arigoni, D. Biosynthesis of isoprenoids in *Escherichia coli*: The fate of the 3-H and 4-H atoms of 1-deoxy-d-xylulose. *Chem. Commun.* **1998**, 1857–1858. [CrossRef]
73. Rieder, C.; Jaun, B.; Arigoni, D. On the early steps of cineol biosynthesis in *Eucalyptus globulus*. *Helv. Chim. Acta* **2000**, *83*, 2504–2513. [CrossRef]
74. Arigoni, D.; Eisenreich, W.; Latzel, C.; Sagner, S.; Radykewicz, T.; Zenk, M.H.; Bacher, A. Dimethylallyl pyrophosphate is not the committed precursor of isopentenyl pyrophosphate during terpenoid biosynthesis from 1-deoxyxylulose in higher plants. *Proc. Natl. Acad. Sci. USA* **1999**, *96*, 1309–1314. [CrossRef]
75. Laupitz, R.; Gräwert, T.; Rieder, C.; Zepeck, F.; Bacher, A.; Arigoni, D.; Rohdich, F.; Eisenreich, W. Stereochemical studies on the making and unmaking of isopentenyl diphosphate in different biological systems. *Chem. Biodivers.* **2004**, *1*, 1367–1376. [CrossRef]
76. Chang, W.; Xiao, Y.; Liu, H.; Liu, P. Mechanistic studies of an IspH-catalyzed reaction: Implications for substrate binding and protonation in the biosynthesis of isoprenoids. *Angew. Chem. Int. Ed.* **2011**, *50*, 12304–12307. [CrossRef] [PubMed]
77. Masini, T.; Kroezen, B.S.; Hirsch, A.K.H. Druggability of the enzymes of the non-mevalonate-pathway. *Drug Discov. Today* **2013**, *18*, 1256–1262. [CrossRef]
78. Janthawornpong, K.; Krasutsky, S.; Chaignon, P.; Rohmer, M.; Poulter, C.D.; Seemann, M. Inhibition of IspH, a $[4Fe–4S]^{2+}$ enzyme involved in the biosynthesis of isoprenoids via the methylerythritol phosphate pathway. *J. Am. Chem. Soc.* **2013**, *135*, 1816–1822. [CrossRef] [PubMed]
79. Ahrens-Botzong, A.; Janthawornpong, K.; Wolny, J.A.; Tambou, E.N.; Rohmer, M.; Krasutsky, S.; Poulter, C.D.; Schünemann, V.; Seemann, M. Biosynthesis of isoprene units: Mössbauer spectroscopy of substrate and inhibitor binding to the [4Fe-4S] cluster of the LytB/IspH enzyme. *Angew. Chem. Int. Ed.* **2011**, *50*, 11976–11979. [CrossRef]
80. Van Hoof, S.; Lacey, C.J.; Röhrich, R.C.; Wiesner, J.; Jomaa, H.; Van Calenbergh, S. Synthesis of analogues of (*E*)-1-hydroxy-2-methylbut-2-enyl 4-diphosphate, an isoprenoid precursor and human Γδ T cell activator. *J. Org. Chem.* **2008**, *73*, 1365–1370. [CrossRef]
81. Wang, K.; Wang, W.; No, J.-H.; Zhang, Y.; Zhang, Y.; Oldfield, E. Inhibition of the Fe_4S_4 -cluster-containing protein IspH (lytB): Electron paramagnetic resonance, metallacycles, and mechanisms. *J. Am. Chem. Soc.* **2010**, *132*, 6719–6727. [CrossRef]
82. Elliott, T.S.; Slowey, A.; Ye, Y.; Conway, S.J. The use of phosphate bioisosteres in medicinal chemistry and chemical biology. *Med. Chem. Commun.* **2012**, *3*, 735. [CrossRef]
83. Wang, W.; Li, J.; Wang, K.; Smirnova, T.I.; Oldfield, E. Pyridine inhibitor binding to the 4Fe-4S protein, *A. Aeolicus* IspH (LytB): A HYSCORE Investigation. *J. Am. Chem. Soc.* **2011**, *133*, 6525–6528. [CrossRef]
84. O'Dowd, B.; Williams, S.; Wang, H.; No, J.H.; Rao, G.; Wang, W.; McCammon, J.A.; Cramer, S.P.; Oldfield, E. Spectroscopic and computational investigations of ligand binding to IspH: Discovery of non-diphosphate inhibitors. *ChemBioChem* **2017**, *18*, 914–920. [CrossRef]
85. Miller, A.A.; Bundy, G.L.; Mott, J.E.; Skepner, J.E.; Boyle, T.P.; Harris, D.W.; Hromockyj, A.E.; Marotti, K.R.; Zurenko, G.E.; Munzner, J.B.; et al. Discovery and characterization of QPT-1, the progenitor of a new class of bacterial topoisomerase inhibitors. *Antimicrob. Agents Chemother.* **2008**, *52*, 2806–2812. [CrossRef] [PubMed]
86. Brandstetter, H.; Grams, F.; Glitz, D.; Lang, A.; Huber, R.; Bode, W.; Krell, H.-W.; Engh, R.A. The 1.8-Å crystal structure of a matrix metalloproteinase 8-barbiturate inhibitor complex reveals a previously unobserved mechanism for collagenase substrate recognition. *J. Biol. Chem.* **2001**, *276*, 17405–17412. [CrossRef] [PubMed]
87. Dunten, P.; Kammlott, U.; Crowther, R.; Levin, W.; Foley, L.H.; Wang, P.; Palermo, R. X-ray structure of a novel matrix metalloproteinase inhibitor complexed to stromelysin. *Protein Sci.* **2001**, *10*, 923–926. [CrossRef]
88. Lipinski, C.A.; Lombardo, F.; Dominy, B.W.; Feeney, P.J. Experimental and computational approaches to estimate solubility and permeability in drug discovery and development settings. *Adv. Drug Del. Rev.* **2001**, *46*, 3–26. [CrossRef]
89. Baell, J.B.; Holloway, G.A. New substructure filters for removal of pan assay interference compounds (PAINS) from screening libraries and for their exclusion in bioassays. *J. Med. Chem.* **2010**, *53*, 2719–2740. [CrossRef] [PubMed]
90. Py, B.; Gerez, C.; Huguenot, A.; Vidaud, C.; Fontecave, M.; Ollagnier de Choudens, S.; Barras, F. The ErpA/NfuA complex builds an oxidation-resistant Fe-S cluster delivery pathway. *J. Biol. Chem.* **2018**, *293*, 7689–7702. [CrossRef]

Article

Model Systems for Evidencing the Mediator Role of Riboflavin in the UVA Cross-Linking Treatment of Keratoconus

Mihaela Monica Constantin [1], Cătălina Gabriela Corbu [1,2], Sorin Mocanu [3], Elena Irina Popescu [3], Marin Micutz [4,*], Teodora Staicu [4], Raluca Şomoghi [5,6], Bogdan Trică [6], Vlad Tudor Popa [3], Aurica Precupas [3,*], Iulia Matei [3] and Gabriela Ionita [3,*]

1 Oftaclinic Clinic, Bd. Marasesti 2B, 040254 Bucharest, Romania; mihaelamonicaconstantin@yahoo.com (M.M.C.); carol_corbu@yahoo.com (C.G.C.)

2 Clinical Hospital of Ophthalmologic Emergencies, Alexandru Lahovari 1 Square, 010464 Bucharest, Romania

3 "Ilie Murgulescu" Institute of Physical Chemistry of the Romanian Academy, Splaiul Independentei 202, 060021 Bucharest, Romania; smocanu@icf.ro (S.M.); elenairinapopescu@gmail.com (E.I.P.); vtpopa@icf.ro (V.T.P.); iulia.matei@yahoo.com (I.M.)

4 Department of Physical Chemistry, Faculty of Chemistry, University of Bucharest, Bd. Regina Elisabeta 4-12, 030018 Bucharest, Romania; teos@gw-chimie.math.unibuc.ro

5 Chemistry Department, Faculty of Petroleum Technology and Petrochemistry, Petroleum-Gas University of Ploiesti, Bd. Bucuresti 39, 100680 Ploiesti, Romania; r.somoghi@gmail.com

6 Department of Bioresources, National Institute for Research & Development in Chemistry and Petrochemistry—ICECHIM, Splaiul Independentei nr. 202, Sector 6, 060021 Bucharest, Romania; trica.bogdan@gmail.com

* Correspondence: micutz@gw-chimie.math.unibuc.ro (M.M.); aprecupas@icf.ro (A.P.); ige@icf.ro (G.I.)

Abstract: Riboflavin under UVA radiation generates reactive oxygen species (ROS) that can induce various changes in biological systems. Under controlled conditions, these processes can be used in some treatments for ocular or dermal diseases. For instance, corneal cross-linking (CXL) treatment of keratoconus involves UVA irradiation combined with riboflavin aiming to induce the formation of new collagen fibrils in cornea. To reduce the damaging effect of ROS formed in the presence of riboflavin and UVA, the CXL treatment is performed with the addition of polysaccharides (dextran). Hyaluronic acid is a polysaccharide that can be found in the aqueous layer of the tear film. In many cases, keratoconus patients also present dry eye syndrome that can be reduced by the application of topical solutions containing hyaluronic acid. This study presents physico-chemical evidence on the effect of riboflavin on collagen fibril formation revealed by the following methods: differential scanning microcalorimetry, rheology, and STEM images. The collagen used was extracted from calf skin that contains type I collagen similar to that found in the eye. Spin trapping experiments on collagen/hyaluronic acid/riboflavin solutions evidenced the formation of ROS species by electron paramagnetic resonance measurements.

Keywords: collagen; riboflavin; hyaluronic acid; EPR spectroscopy; keratoconus; STEM

Citation: Constantin, M.M.; Corbu, C.G.; Mocanu, S.; Popescu, E.I.; Micutz, M.; Staicu, T.; Şomoghi, R.; Trică, B.; Popa, V.T.; Precupas, A.; et al. Model Systems for Evidencing the Mediator Role of Riboflavin in the UVA Cross-Linking Treatment of Keratoconus. *Molecules* **2022**, *27*, 190. https://doi.org/10.3390/molecules27010190

Academic Editor: Chryssostomos Chatgilialoglu

Received: 11 November 2021
Accepted: 26 December 2021
Published: 29 December 2021

Publisher's Note: MDPI stays neutral with regard to jurisdictional claims in published maps and institutional affiliations.

1. Introduction

Keratoconus (KC) is an ocular disease with a relatively high prevalence (approximately 1:2000) [1] usually diagnosed during the second or third decade of life that causes an irregularly shaped cornea leading to severe impairment of vision. This disease is characterized by progressive thinning of the cornea, giving rise to a cone-shaped cornea instead of the normal spherical shape [1,2]. Hence, understanding the pathogenesis of this disease and controlling its evolution have attracted the attention of specialists in ophthalmology, and their efforts were accompanied by those of researchers in other fields, such as biochemistry, physics, or chemistry. Dry eye is also a symptom and a disease accompanying KC, with some studies including it in the multifactorial etiology of KC [3].

A relatively new technique introduced in 2003 by Wollensak et al. [4] is the corneal collagen cross-linking (CXL), based on the combined use of the photosensitizer riboflavin and UVA light of 370 nm. This method exploits the property of riboflavin to generate, under UVA irradiation, reactive oxygen species that further interact with biological assemblies [5]. The CXL method is also applied for the treatment of other diseases associated with the extracellular matrix involving various veins [6], degraded dentine [7], or the eye (pellucid marginal corneal degeneration, post-LASIK ectasia, infectious keratitis, bullous keratopathy) [8–11]. Although numerous clinical trials involving patients with KC are dedicated to the analysis of the efficacy and effects of CXL, literature data are scarce on the physico-chemical aspects behind this treatment. In a recent study, we investigated the changes in the protein profile in tears collected from patients with dry eye syndrome and KC during treatment and explored the possible correlation of the electron paramagnetic resonance (EPR) parameters of spin probes that bind to some tear proteins with the composition of tears or ophthalmic parameters [12]. Lactoferrin (LF), lysozyme (LYZ) and lipocalin (a prealbumin protein) are among the proteins most abundant in tears [13–15]. The level of human serum albumin (HSA) may increase in tear secretions of patients suffering from dry eye syndrome and KC [16]. With this study, we aimed to find evidence of changes in collagen properties induced by riboflavin in the presence of UVA, and to identify the type of reactive species formed in various systems relevant for the CXL treatment of KC. Fibrous collagenous structures (type I, II and III) are present in the connective tissues of the eye, such as the cornea, sclera, vitreous body, retina, and also in other tissues like the skin, bones, and tendons [17]. To get information on how riboflavin influences the properties of collagen and the process of fibrillogenesis, collagen extracted from calf skin was used. This extract has type I collagen as the main component [18]. Also, we analyzed the influence of hyaluronic acid on fibrillogenesis. The present study involved rheological, microcalorimetric, electron microscopy, and electron paramagnetic resonance measurements. The rheological data evidenced the mechanical properties of the systems, while the microcalorimetric measurements provided information on the thermodynamic stability of collagen and collagen fibrils. Using three spin trapping agents, DMPO, CPH, and TEMP, EPR spectroscopy evidenced the reactive radical species formed in samples containing collagen, in solutions of proteins present in tears (LF, LYZ, HSA), and in tears collected from patients during CXL treatment.

2. Results and Discussion

The collagen systems investigated in this study are described in Table 1. These systems were characterized by rheology, STEM, and microDSC methods at a pH value of 3.5 or 7.5 in order to get bulk information and microscopic information on collagen systems.

Table 1. The composition of collagen samples investigated in this study, in the absence and in the presence of UVA light.

Sample	Collagen 0.25%	Riboflavin 0.1%	Hyaluronic Acid 0.1%	UVA Exposure
1	x	-	-	-
1a	x	-	-	x
2	x	x	-	-
2a	x	x	-	x
3	x	-	x	-
3a	x	-	x	x
4	x	x	x	-
4a	x	x	x	x

Fibril formation in collagen solutions was initiated following the procedure described in [16–25]. Fibril formation in all collagen systems under investigation occurs only at pH 7.5. For instance, Figure 1 presents the gels formed in the absence of riboflavin (Figure 1a), in the presence of riboflavin (Figure 1b), and by the irradiated collagen/riboflavin sample (Figure 1c).

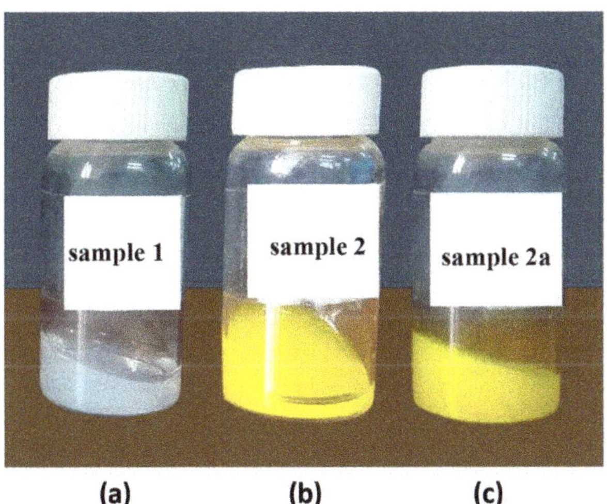

Figure 1. Collagen gels formed at pH 7.5 in the absence of riboflavin (**a**) and in the presence of riboflavin prior to (**b**) and after (**c**) UVA irradiation.

The spin trapping experiments were performed for samples **1a**–**4a**, for solutions of proteins that can be found in tears, for a pharmaceutic solution containing riboflavin and dextran (Peschke solution) used in CLX treatment and for tears collected from patients during the CLX treatment.

2.1. Rheological Measurements

Rheological data for samples prepared at pH 7.5 were acquired at four temperatures in the range 10–37 °C: 10, 20, 25, and 37 °C. As the process of collagen fibril formation is favored at 37 °C, rheometric measurements at this temperature were performed at different times from the onset of gelation (1, 2 and 3 h). The rheological behavior of the samples described in Table 1 is similar. For instance, in the case of sample **1**, the variation of the rheological parameters (storage modulus G′, loss modulus G″, dynamic viscosity η_{dyn} = G″/ω and loss tangent tgδ = G″/G′, where ω is angular frequency in rad s^{-1}) as a function of frequency (in Hz) is illustrated in Figure 2a–d, while the change in the dynamic viscosity corresponding to a frequency of 1 Hz as a function of temperature is shown in Figure 2e. The rheological behavior of samples **2**–**4** and **1a**–**4a** is presented in the supplementary material file (Figures S1–S7). In the temperature range of 10–25 °C, all samples behave as viscoelastic liquids, while at the temperature of 37 °C the rheological parameters correspond to a gel system as the fibril formation occurs. At 37 °C, G′ >> G″ over the whole frequency domain, and the values of G′, G″ and η_{dyn} are at least an order of magnitude higher than at lower temperatures, depending on the time the system has been maintained at the gelation temperature. At the gelation temperature, the loss tangent has sub-unitary values over the frequency domain. Figure 2e shows that the dynamic viscosity η_{dyn} corresponding to 1 Hz increases suddenly at the gelation temperature after incubation from 1 to 3 h.

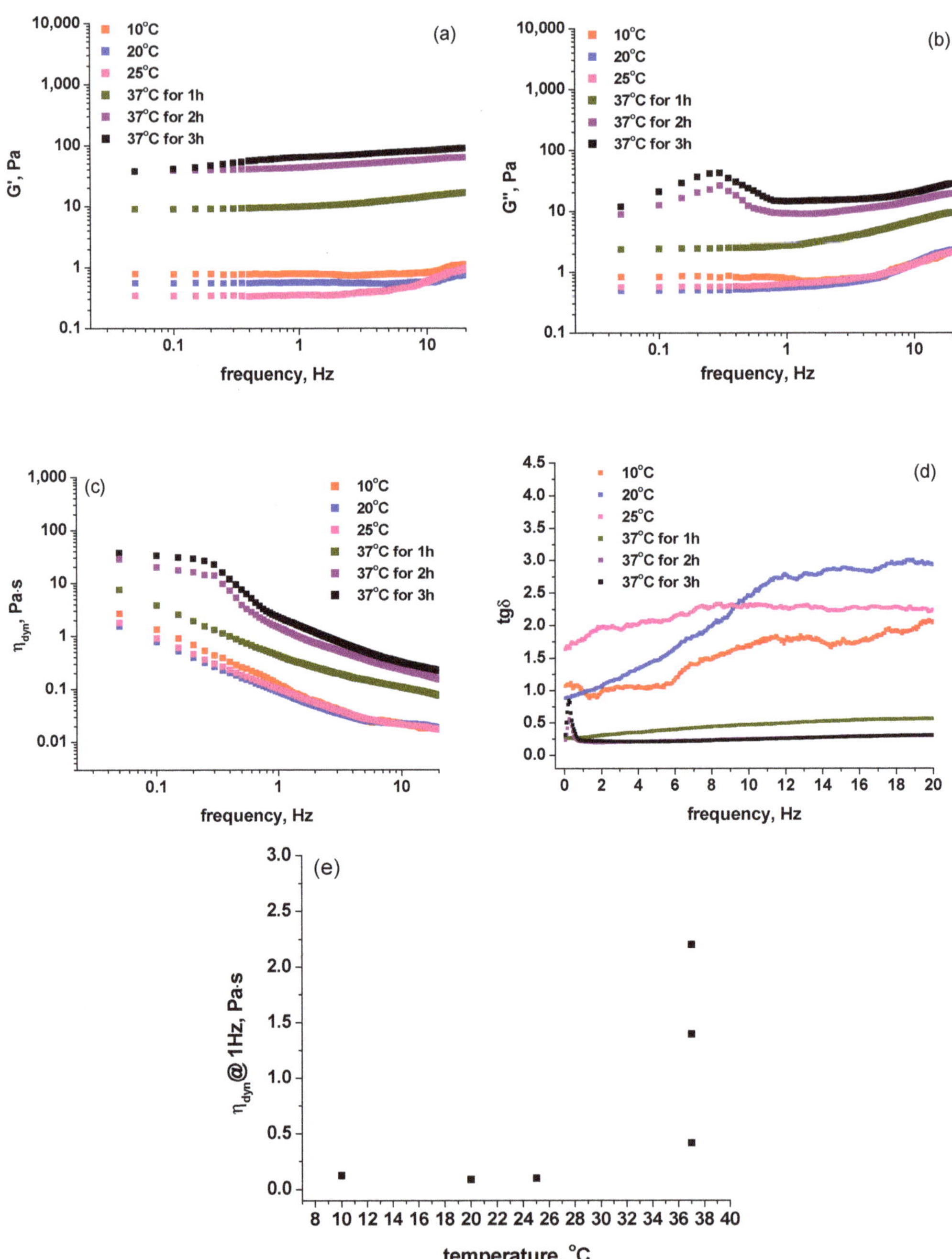

Figure 2. The variation of (**a**) storage modulus (G′), (**b**) loss modulus (G″), (**c**) dynamic viscosity (η_{dyn} = G″/ω) and (**d**) loss tangent (tgδ = G″/G′) with frequency, and the variation of η_{dyn} with temperature (**e**) for sample **1**.

The influence of riboflavin, hyaluronic acid and UVA treatment on fibril formation can be evidenced by changes in the G′ value, as indicated in Table 2 for an angular frequency of 1 Hz and the temperatures indicated above. In the case of sample **1a** corresponding to a collagen sample exposed to UVA radiation, it can be noticed that the steep increase in dynamic viscosity at 37 °C is more pronounced (Figure S1d). In the case of sample **2** containing riboflavin, higher values of G′, G″ and η_{dyn} (at 37 °C, after 1 h and 2 h of equilibration) were noticed as compared to sample **1** (Figure S2a–d). This indicates that gelation occurs faster as it is favored by the presence of riboflavin.

Table 2. The G′ values (in Pa) for samples **1–4** and **1a–4a** corresponding to a frequency value of 1 Hz.

Sample	10 °C	20 °C	25 °C	37 °C (1 h)	37 °C (2 h)	37 °C (3 h)
1	0.79	0.57	0.36	9.90	43.43	63.88
1a	0.66	0.66	0.53	14.16	77.04	149.10
2	0.77	0.49	0.34	40.08	72.16	113.26
2a	0.60	0.39	0.47	1.38	19.36	45.00
3	0.67	0.66	0.42	62.39	104.70	137.52
3a	0.51	0.51	0.43	41.11	75.05	94.90
4	0.59	0.49	0.24	49.88	50.11	52.88
4a	0.61	0.52	0.43	8.28	46.73	71.83

The UVA irradiation of the collagen-riboflavin system (sample **2a**, Figure S3) determines a different rheological behavior by comparison to the non-irradiated system. Gelation during collagen fibril formation is slower, and the completeness of gel formation after 3 h of fibril formation led to a much weaker physical gel. For instance, the G′, G″ and η_{dyn} values of sample **2a** are lower compared to those of sample **2** (see Figures S2 and S3). This behavior suggests that UVA exposure in sample **2a** impacted the processes occurring in the collagen-riboflavin system. Under UVA radiation, riboflavin undergoes photo degradation processes involving intramolecular photodealkylation and/or photo addition corresponding to the excited singlet state, or intramolecular photo reduction corresponding to the excited triplet state [26]. Free radical formation accompanying the photodegradation of riboflavin may initiate the intermolecular cross-linking of collagen prior to the onset of the collagen fibril formation process.

Samples **3** and **3a** are characterized by the formation of well-defined strong gels at 37 °C even for the case of 1 h equilibration time (Figures S4 and S5). This stands for a physical interaction between collagen and hyaluronic acid that adds to the fibril formation process. The rheograms almost overlap for the equilibration times of 2 h and 3 h at 37 °C (Figures S4a–c and S5a–c), which indicates that, in the presence of hyaluronic acid, the process of fibril formation is completed faster. This behavior is in good agreement with the favorable role played by hyaluronic acid in strengthening and improving the elasticity of the collagen gel resulting during fibrillogenesis [27]. Literature data showed that hyaluronic acid may experience photodegradation under the action of UV light that gradually affects its molecular weight, leading to the decrease of its intrinsic viscosity observed in saline solution (0.2 M NaCl) at 25 °C [28]. In the case of sample **3a**, the gels have a slightly lower consistency (Figure S5b–d) compared to the non-irradiated sample **3**.

Sample **4** reaches the gel state equilibrium in a shorter time, as dependences of G′ and G″ as a function of frequency at 37 °C at different equilibration times are almost overlapped at applied frequencies higher than or equal to 1 Hz (Figure S6). The exposure of this tricomponent system to UVA light (sample **4a**) determines a different behavior compared to sample **4**. The behavior of sample **4a** is similar to that of sample **2a** (see Figures S3 and S7), and this result emphasizes the preponderant role played by riboflavin relative to hyaluronic acid in controlling the rheological properties of these mixed systems.

2.2. STEM Images

To evidence the role of riboflavin and hyaluronic acid on collagen fibrillogenesis, the samples have been examined after performing the negative staining procedure described in the experimental section. Figure 3 presents the STEM images obtained for sample **1** and sample **2**. The STEM images revealed a random formation of the fibrils in the case of sample **1**, which contains only collagen (Figure 3), while in the case of the collagen solutions containing riboflavin and/or hyaluronic acid, the fibrils display a clear axial and a regular transverse D-periodic banding pattern (Figure 3b). The distances between ridges are in the range of 65–70 nm, a value close to that of the native fibrillar structure of type I collagen [29,30]. This latter aspect proves the regulatory effect in collagen fibril formation displayed by both riboflavin and hyaluronic acid.

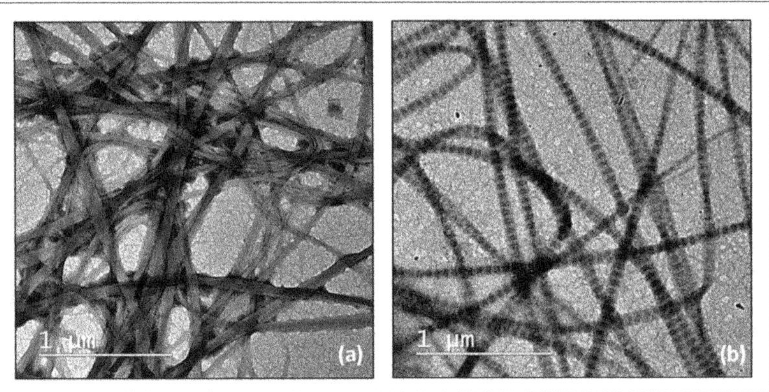

Figure 3. STEM images for collagen fibrils in the absence (**a**) and presence (**b**) of riboflavin, pH 7.5.

Rheological measurements and STEM investigation evidenced that both riboflavin and hyaluronic acid influence the fibrillogenesis of collagen. We further investigated by microDSC measurements the effect of riboflavin, hyaluronic acid, and UVA exposure of collagen samples on the thermal stability of the protein assemblies.

2.3. MicroDSC Measurements

In collagen samples, an increase in temperature determines the breaking of hydrogen bonds and induces irreversible unfolding of the triple helix to random coils [31]. The DSC signal obtained for collagen thermal denaturation displays one endothermic peak that can be decomposed into components that correspond to pre-, post-, and major transition. Literature data report the presence in the collagen DSC curve of either a main peak, attributed to triple helix melting, with a shoulder at lower temperature associated with the fibrillation of collagen [32], or one main peak located between two shoulders [33]. The number of transition peaks assigned to fibril denaturation varies from one peak to three peaks [34,35], the high-temperature peak corresponding to the melting of the fibrillar form of collagen, while the low temperature transition is related to the melting of monomeric collagen [36].

The DSC thermograms for collagen thermal denaturation in different systems are discussed in the following. Figure 4 displays the influence of pH on collagen thermal denaturation and fibril formation. The DSC signal of sample 1 (in solution at pH 3.5 or 7.5) can be decomposed in three peaks (Figure S8). The major peak is attributed to the melting of the triple helix, while the minor peaks correspond to fibril denaturation [33].

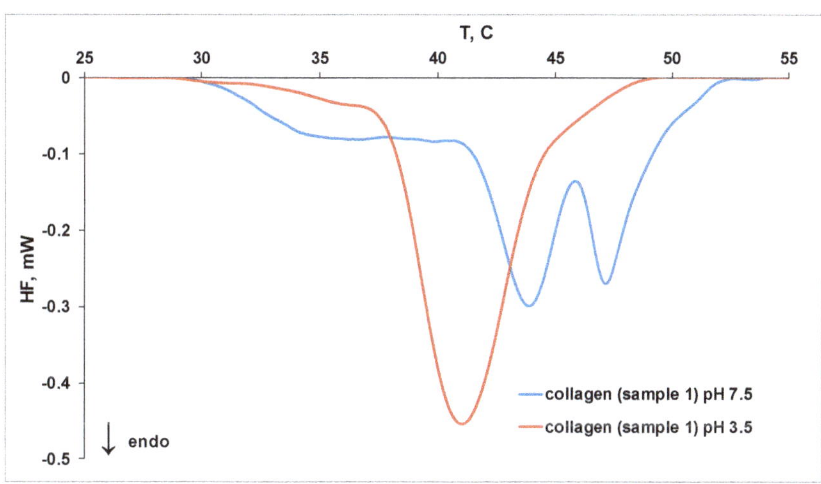

Figure 4. The DSC thermograms for thermal denaturation of sample **1** at pH 3.5 and 7.5.

Higher values of peak temperatures for collagen thermal denaturation (Table 3) were obtained at pH 7.5 and suggest a higher thermal stability of triple helix (T_{peak2}) and fibrillar collagen (T_{peak1} and T_{peak3}).

Table 3. Peak temperatures and denaturation heat obtained for collagen at pH 3.5 and pH 7.5.

Sample	T_{peak1}, °C	T_{peak2}, °C	T_{peak3}, °C	Denaturation Heat, J/g
1 (pH 3.5)	36.10	41.06	45.82	0.27
1 (pH 7.5)	36.74	43.93	47.15	0.29

The influence of riboflavin and UVA irradiation on collagen thermal denaturation and fibril formation at different pH values is displayed in Figure 5. At pH 3.5, the collagen thermal stability is reduced in the presence of riboflavin (sample **2**) and in the presence of riboflavin and UVA irradiation (sample **2a**), as the peaks are shifted to lower values. An opposite effect was obtained at pH 7.5: riboflavin induces the formation of new collagen fibrils that show a better thermal stability (T_{peak1} and T_{peak3} increase), while the thermal stability of the triple helix structure remains unchanged.

The exposure of collagen/riboflavin solution to UVA irradiation for 30 min (sample **2a**) shifts the first and third peaks to higher temperature values, indicating a higher thermal stability of collagen fibrils. Furthermore, an additional peak at higher temperature (T_{peak4}) was observed, suggesting the formation of new collagen fibrils. Partial defibrillation of collagen under riboflavin/UVA radiation action might be an initial step that enables the formation of radicals on the protein chain, which further ensures a packing of the fibrils into a stronger network.

Figure 6 presents the influence of hyaluronic acid and riboflavin (samples **3** and **4**) on collagen thermal stability at different pH values. At pH 3.5, the presence of hyaluronic acid and riboflavin (sample **4**) shifts the DSC peaks towards lower temperatures, indicating that these compounds favor defibrillation and denaturation of collagen. At pH 7.5, the destabilization effect induced by hyaluronic acid on collagen is reduced in the presence of riboflavin. Moreover, a fourth peak could be noticed at higher temperatures, suggesting that additional intermolecular cross-linking of collagen fibers occurs. Cross-linking stabilizes the fibrous structure of collagen by reducing the separation of the individual molecules [37]. Riboflavin also protects the new fibrillar collagen formed at higher temperature, as evidenced by the corresponding (higher value) denaturation peak (T_{peak4}).

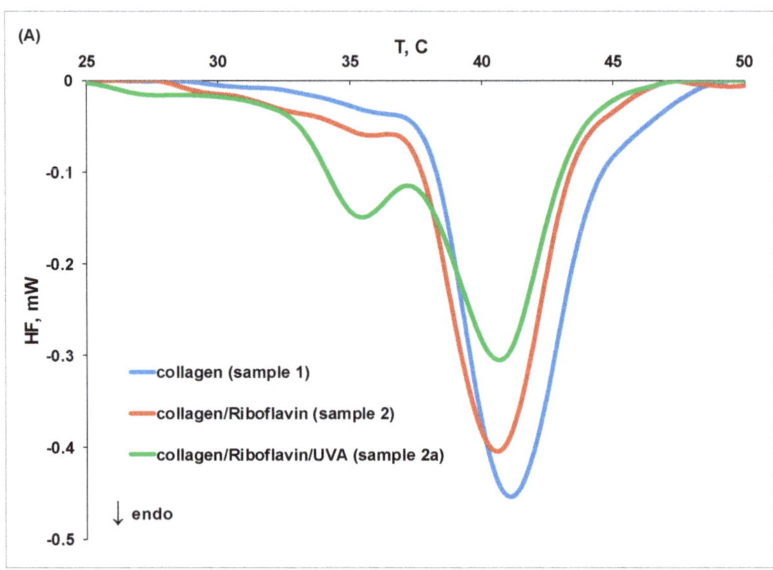

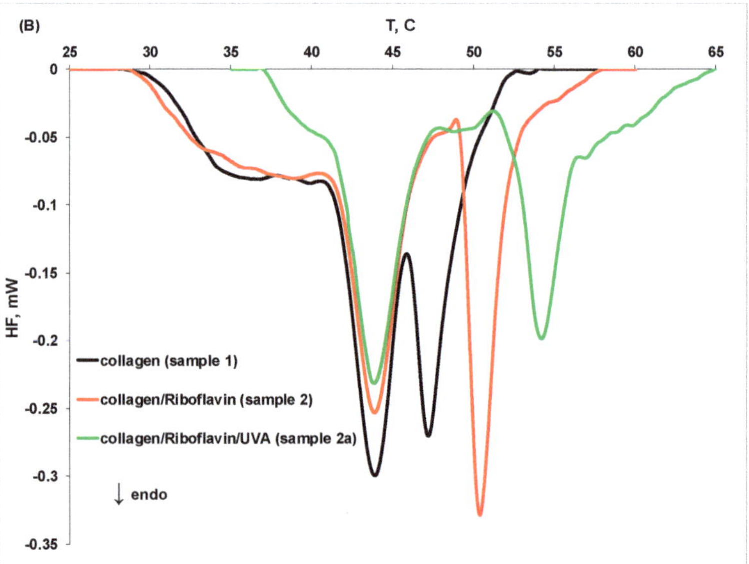

Figure 5. The DSC thermograms for thermal denaturation of collagen (sample **1**), collagen in the presence of riboflavin (sample **2**), and collagen in the presence of riboflavin after UVA exposure (sample **2a**) at (**A**) pH 3.5 and (**B**) pH 7.5.

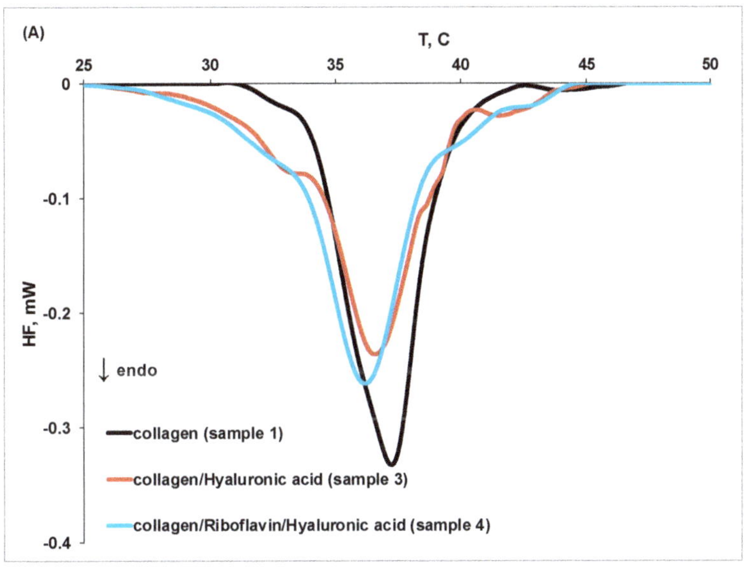

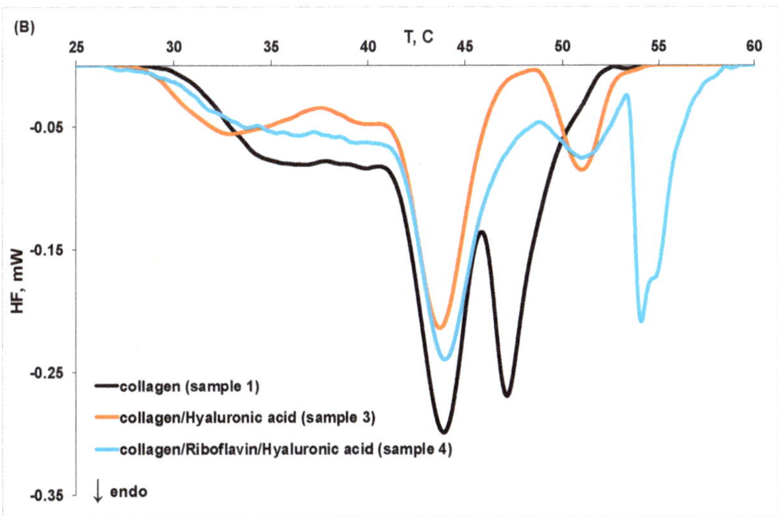

Figure 6. The DSC thermograms for thermal denaturation of collagen (sample **1**), collagen in the presence of hyaluronic acid (sample **3**), and collagen in the presence of hyaluronic acid and riboflavin (sample **4**) at (**A**) pH 3.5 and (**B**) pH 7.5.

Figure 7 displays the effect of UVA irradiation on collagen thermal denaturation in different systems at pH 7.5, corresponding to samples **1a**, **2a,** and **4a**. The exposure of collagen/riboflavin solution to UVA irradiation (sample **2a**) for 30 min shifts the peaks to higher temperatures and induces the presence of a fourth peak, suggesting the formation of new collagen fibrils with better thermal stability. Hyaluronic acid destabilizes the collagen, and the thermal stability of fibrils is reduced (sample **4a**). The influence of riboflavin, hyaluronic acid, and UVA irradiation on collagen thermal denaturation at pH 7.5 is summarized in Table 4.

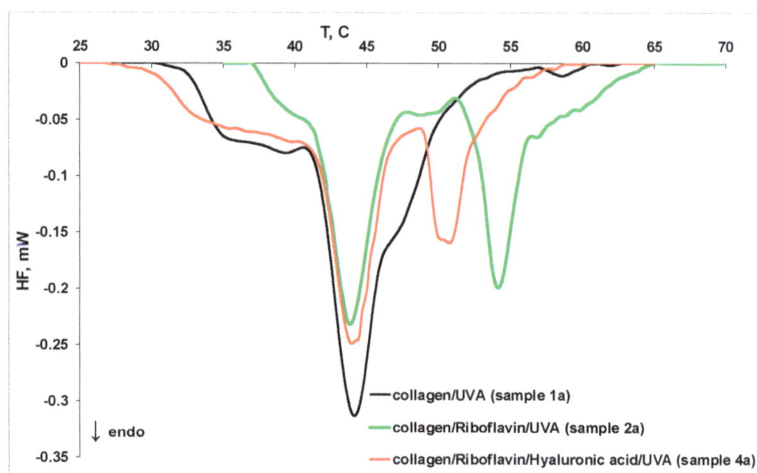

Figure 7. The DSC thermograms for thermal denaturation of collagen (sample **1a**), collagen/riboflavin (sample **2a**), and collagen/riboflavin/hyaluronic acid (sample **4a**) under UVA irradiation at pH 7.5.

Table 4. Peak temperatures and denaturation heat of collagen in different systems at pH 7.5.

Sample	T_{peak1}, °C	T_{peak2}, °C	T_{peak3}, °C	T_{peak4}, °C	Denaturation Heat, J/g
1	36.74	43.93	47.15	-	0.23
1a	37.22	44.15	47.42	-	0.27
2	37.60	43.89	50.32	-	0.29
2a	42.44	43.88	54.24	57.33	0.24
3	32.96	40.31	43.71	50.95	0.17
3a	40.80	43.70	50.63	54.04	0.13
4	37.46	43.93	51.43	54.15	0.27
4a	38.73	44.04	50.43	-	0.26

Higher values of denaturation heat were obtained at pH 7.5 for all systems, indicating a higher content of fibrillar collagen. Riboflavin induces a stabilization effect on collagen fibrils, evidenced by the higher values of denaturation heat at both pH values. The exposure of collagen/riboflavin solution to UVA irradiation for 30 min decreases the denaturation heat at pH 7.5, indicating that a process of collagen partial defibrillation occurs. A significant destabilization effect on collagen is induced by hyaluronic acid, especially at pH 3.5.

The averages for the denaturation heat of collagen in different systems at pH 3.5 and 7.5 are 0.152 J/g and 0.241 J/g, respectively. These average values were used as references for the histograms presented in Figure 8. While at pH 7.5 there are 5 systems with denaturation heat above the average, at pH 3.5, the trend is opposite: only 3 systems have a higher-than-average denaturation heat. These results confirm the higher fibrillar content of collagen at pH 7.5.

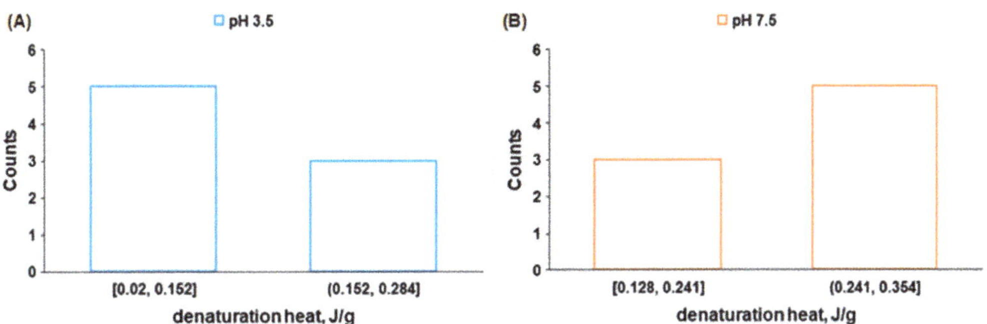

Figure 8. Denaturation heat histograms of collagen systems at pH 3.5 (**A**) and 7.5 (**B**).

The rheological and thermodynamic changes observed in the collagen systems exposed to riboflavin and UVA are the results of interactions between the reactive oxygen species (ROS) generated by the interaction of riboflavin with UVA light. In the next section, the information obtained by spin trapping experiments in various systems that contain riboflavin exposed to UVA and components of tear films (proteins, hyaluronic acid) are presented.

2.4. Radical Species Formed in Collagen/Riboflavin/UVA Systems

Under UVA light, riboflavin generates ROS, such as hydroxyl ($HO^{\bullet}$), superoxide ($O_2^{\bullet -}$), hydroperoxide ($HOO^{\bullet}$) radicals, and singlet oxygen ($^1O_2$), following two mechanisms [38,39]. ROS generated by photodegradation of riboflavin mediate the formation of intra or interfibrillar bonds in collagen during CXL treatment, but their side effects cannot be omitted as free radicals may induce the death of endothelial cells, affecting the crystalline lens or retina. However, these effects are reduced by the presence in the composition of the tear film of proteins like lactoferrin (LF) and lysozyme (LYZ), which have the role of removing these free radicals. In addition, the Peschke solution used for CXL treatment contains dextran, which limits the concentration of oxygen-centered radicals generated by riboflavin and ensures a certain viscosity of the solution. Although the main process in CXL treatment regards the cross-linking of collagen, the interaction of the free radicals with components of the tear film also occurs. It is known that the keratoconus disease is often associated with dry eye syndrome (DES), which causes a low content of LF and LYZ in the tear film [3,15]. The treatment of DES involves different formulations for artificial tears that have hyaluronic acid as the main component, which can in fact be found in the mucin layer of the tear film.

By EPR measurements, we aim to highlight the short-lived free radicals (especially those centered on oxygen) formed in various systems exposed to riboflavin and UVA radiation. For this purpose, three spin traps have been used: 5,5-dimethyl-1-pyrroline N-oxide (DMPO), used to identify $HO^{\bullet}$ radicals, 1-hydroxy-3-carboxy-2,2,5,5-tetramethylpyrrolidine (CPH), used to identify $O_2^{\bullet -}$ and $HO^{\bullet}$ radicals, and 2,2,6,6-tetramethylpiperidin-4-one (TEMP) used to identify $^1O_2$ (Figure S9). CPH is a radical scavenger with higher sensitivity than DMPO in identifying $HO^{\bullet}$ and $HOO^{\bullet}$ [40]. The total protein content of the tear film is about 15 mg/mL, with LF and LYZ in concentrations up to 3 mg/mL. Human serum albumin (HSA) can be found in high concentration in tears only upon the rupture of blood vessels in the conjunctiva [15]. Therefore, the concentration of each protein in the solutions examined was 3 mg/mL. Determinations were also carried out in a mixture of the three proteins containing 0.1% hyaluronic acid.

To evidence the role of UVA irradiation, the EPR spectra of riboflavin solution were firstly recorded in the dark, noting the absence of a signal (the black spectrum in Figure S10).

After exposure to UVA radiation, the 4-line EPR signal characteristic to the •DMPO-OH adduct (hyperfine splitting constants a_N = 14.9 G and a_H = 14.7 G) is observed (cyan spectrum in Figure S10). The spin adducts were evidenced in the following systems: solution of LF, LYZ, and HSA in the presence of riboflavin 0.1%, mixture of the three proteins and hyaluronic acid, Peschke solution, sample **1a**, sample **4a**, and in tears collected from patients immediately after the CXL procedure.

The EPR spectra of the spin adducts formed in the protein systems in the presence of riboflavin and UVA are a sum of two main components. One component corresponds to the •DMPO-OH adduct, and the other to a DMPO adduct with a carbon-centered radical (hyperfine splitting constants a_N = 16.1 G and a_H = 23.3 G) formed probably by the interaction of oxygen-centered reactive radicals generated by riboflavin with the proteins. The nitroxide resulted by DMPO degradation is also evidenced but its contribution is less than 10% (see Table S1). The lines corresponding to each adduct are marked with a red dot for •DMPO-OH and with a blue dot for the carbon-centered radical adduct (Figure 9). In the presence of hyaluronic acid or dextran, it was found that the formation of carbon-centered radicals is reduced. Thus, we can affirm that hyaluronic acid plays a role not only in restoring the integrity of the tear film, but also an antioxidant role against highly reactive radicals. We also observed the presence of HO• and carbon-centered radicals in the solution containing collagen and riboflavin (sample **1a**) after exposure to UVA radiation (Figure 10a). The presence of hyaluronic acid (Figure 10b) and of the three proteins and hyaluronic acid (Figure 10c) reduces the contribution of the carbon-centered radical.

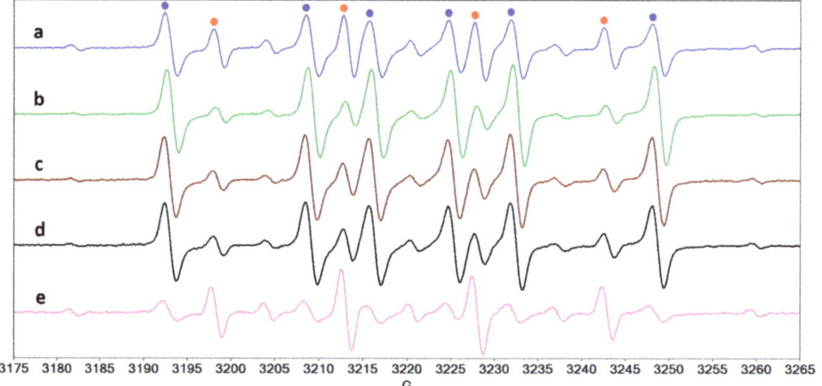

Figure 9. The EPR spectra of DMPO adducts formed after UVA exposure, in the presence of riboflavin, of a solution of (**a**) HSA, (**b**) LF, (**c**) LYZ, and of the mixture of the three proteins in the absence (**d**) and in the presence (**e**) of hyaluronic acid.

In addition to the HO• radical, irradiation of riboflavin-containing solutions can lead to the formation of the $O_2^{\bullet -}$ radical that, although having lower reactivity compared to the HO• radical due to low-rate constant values [41], can be responsible for the generation of highly-reactive secondary species with high biological toxicity [42]. Such species may, in theory, have greater negative side effects in the treatment of CXL. The CPH spin trap can evidence the formation of this radical by use of superoxide dismutase (SOD). Figure 11 shows the EPR spectra of CPH adducts formed in collagen solution in the absence (Figure 11a) and in the presence (Figure 11b) of SOD. It can be observed that the intensity of the spectrum in the presence of SOD is slightly lower, which may lead to the conclusion that the $O_2^{\bullet -}$ radical is generated in collagen solution. In the case of the solution containing collagen and hyaluronic acid, the intensities of the two spectra are similar and lower (Figure 11c,d). This result highlights the protective role that hyaluronic acid may play during CXL treatment. However, it should be underlined that we did not perform quantitative measurements to determine the concentration of the radical species formed; therefore the formation of the $O_2^{\bullet-}$ species in this system is not certain.

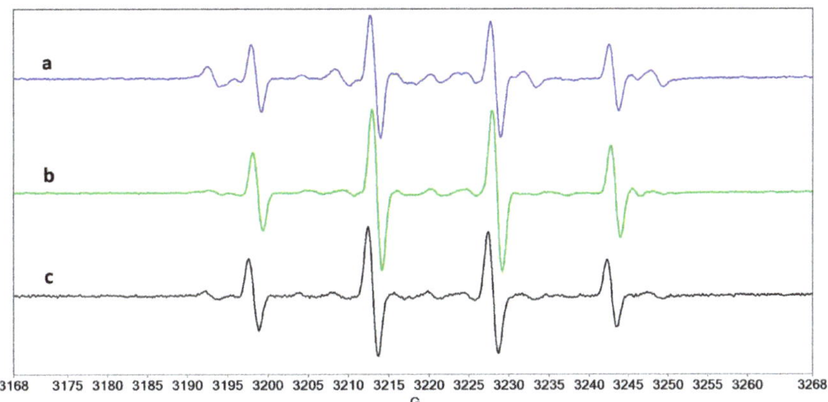

Figure 10. The EPR spectra of DMPO adducts formed after UVA exposure, in the presence of riboflavin, of collagen solution (**a**) in the absence of hyaluronic acid, (**b**) in the presence of hyaluronic acid, and (**c**) in the presence of the mixture of the three proteins and hyaluronic acid.

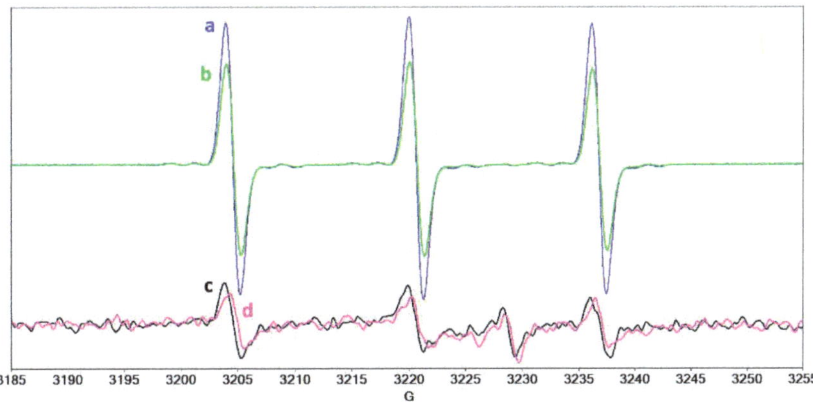

Figure 11. The EPR spectra of CPH adducts formed by UVA irradiation of a solution containing riboflavin and (**a**) collagen, (**b**) collagen/SOD, (**c**) collagen/hyaluronic acid, and (**d**) collagen/hyaluronic acid/SOD.

The next step was to examine the possible formation of $^1O_2$ that poses the highest risk of eye tissue destruction following CXL treatment. Several systems have been evaluated to ascertain the presence of this radical by generating TEMPONE (4-oxo-TEMPO). Figure 12 shows that TEMPONE can be generated in riboflavin solution, in collagen solution containing riboflavin, in collagen solution, but also in the Peschke solution containing riboflavin and dextran. In contrast, TEMPONE was not present in tears, in a mixture of collagen and hyaluronic acid, or in a mixture of collagen, hyaluronic acid, and protein. This set of determinations emphasizes, once again, the protective role of hyaluronic acid and of the proteins present in tears.

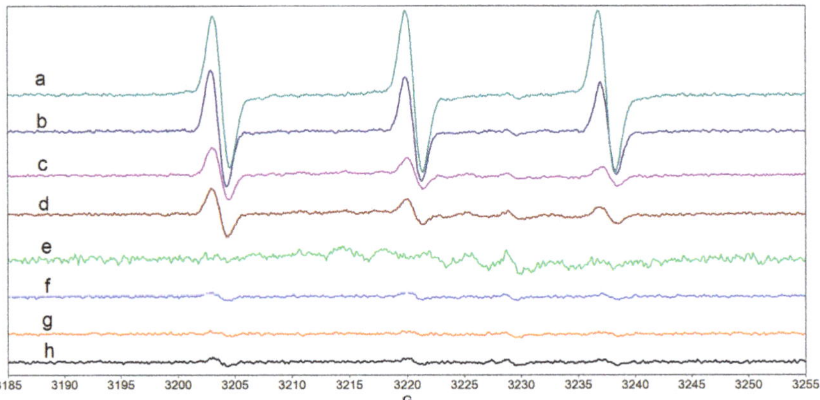

Figure 12. The EPR spectra of TEMPONE formed after UVA irradiation of the following systems: (**a**) water, (**b**) collagen/riboflavin, (**c**) collagen, (**d**) Peschke solution, (**e**) tears, (**f**) collagen/hyaluronic acid, (**g**) collagen/hyaluronic acid/HSA/LF, (**h**) HSA/LF/LYZ.

3. Materials and Methods

5,5-dimethyl-1-pyrroline-N-oxide (DMPO), 2,2,6,6-tetramethylpiperidine (TEMP), superoxide dismutase (SOD), lysozyme, lactoferrin, and human serum albumin were purchased from Sigma Aldrich. 1-hydroxy-3-carboxy-2,2,5,5-tetramethylpyrrolidine (CPH) was obtained from Enzo. Riboflavin and hyaluronic acid were purchased from Roth and Acros Organics, respectively. The Peschke D solution was obtained from Lightmed.

3.1. Collagen Extraction and Purification

Collagen type I was extracted from calf skin following a modified procedure adapted from refs. [43–50], which is described in detail in the supplementary material file.

3.2. Sample Preparation

Collagen samples for STEM visualization of fibril formation in vitro, in the absence or in the presence of riboflavin and hyaluronic acid, were prepared in accordance to the procedure described in [19–25], as follows: 5 mL of phosphate-buffered saline (PBS) buffer solution were added (under vigorous magnetic stirring on ice) into the same volume of acetic solution of collagen or of its acetic mixtures with the same initial collagen concentration of 0.5%. The final pH of the so-prepared systems was adjusted to 7.5 using a concentrated aqueous NaOH solution (40%). Similar samples containing riboflavin and/or hyaluronic acid in concentration of 0.1% were prepared. The composition of PBS consisted of 270 mM NaCl, 60 mM Na_2HPO_4, and 60 mM N-[tris(hydroxymethyl)methyl]-2-aminoethansulfonic acid (TES) [19], with a final pH set to 8.50 by adding concentrated NaOH solution. The collagen fibrillogenesis was triggered by incubating the mixtures at 37 °C for 4 h. After that, the process of fibril formation was assessed visually when turbid and physical cross-linked gels resulted. The collagen samples were studied at pH 3.5 and pH 7.5 by rheometry and microDSC. For evidencing the reactive species formed in collagen during UVA irradiation, tear secretions from patients were collected during exposure to UVA, and solutions of collagen in the absence and in the presence of hyaluronic acid and/or riboflavin were prepared.

For EPR measurements, stock solutions of each spin trapping agent were prepared in a concentration of 50 mM. The concentration of lysozyme, lactoferrin, and human serum albumin was 3 mg/mL while the concentration of collagen was 0.25%. For samples containing riboflavin and/or hyaluronic acid, the concentration was 0.1%. The tear samples were collected from patients during CXL treatment and mixed immediately with an equal volume of spin trapping agent solution, then the sample was transferred to a capillary that was introduced in liquid nitrogen prior to EPR measurement.

3.3. Instruments

3.3.1. STEM

To visualize the fibril structure of collagen, the grids were exposed to the electron beam of the instrument (TEM Tecnai™ G2 F20 TWIN Cryo-TEM, FEI Company™) at an accelerating voltage of 200 kV. TEM samples were prepared in accordance with adapted protocols applied to collagen fibrils [19,51–53]. Thus, an aliquot extracted from each of the individual turbid hydrogels resulted after collagen fibrillogenesis was dispersed into PBS of pH 7.5 to give a final fibril suspension cca. 20-fold diluted. Then, a droplet of diluted suspension was poured onto a copper grid (holey carbon coated copper grid) for 5 min. The excess of solution was carefully removed by means of a piece of filter paper placed at the grid edge. The fibrils adsorbed onto the carbon-coated grid were negatively stained with a droplet of aqueous phosphotungstic acid solution (2–3%, pH 7.4) for 1–2 min. Then, the treated copper grid was rinsed with 3-4 droplets of distilled water and eventually air dried.

3.3.2. Rheometry

The viscoelastic behavior of collagen and collagen-based systems and gel formation during collagen fibrillogenesis were investigated by dynamic rheometry. To this end, an MFR 2100 instrument (GBC, Australia) was employed, as previously described [54–56]. The rheological measurements were carried out at the following temperature values: 10, 20, 25, and 37 °C. A small volume of initial viscoelastic sample was rapidly sandwiched between the two chilled (5 °C) parallel plates of the rheometer equipped with a home-made temperature control jacket connected to a circulating water bath Lauda E100. Prior to performing the measurements, every single sample was equilibrated at 10, 20, and 25 °C for 15 min, except for 37 °C where the equilibration times were 1, 2, and 3 h. The values of storage and loss moduli (G′, G″), dynamic viscosity (η_{dyn}), and loss tangent (tgδ) were determined using the following instrument parameters: pseudorandom noise oscillation, gap between plates 300 μm, oscillation amplitude 0.04 μm, frequency domain 0.05–20.00 Hz, and 30 scans per rheogram. To avoid water evaporation from the sample during fibrillogenesis (over 3 h), a tiny amount of low-viscous polydimethylsiloxane (viscosity of 5 mPa·s at 25 °C) was used to seal the system to be measured.

3.3.3. MicroDSC

Thermal denaturation of collagen at pH 3.5 and 7.5 was studied using differential scanning microcalorimetry (MicroDSC). The measurements were carried out with a SETARAM MicroDSC III calorimeter in the temperature range 15–85 °C at a 1 °C min^{-1} heating rate. Experimental DSC data were analyzed using the Calisto v.1077 software package; thus, the denaturation temperature (T_{peak}) and the denaturation heat were obtained. PeakFit 4.12 software was used to decompose the endothermic peak corresponding to the collagen thermal denaturation.

3.3.4. EPR Spectroscopy

The EPR spectra of spin adducts were recorded on a JEOL FA 100 spectrometer equipped with a cylindrical-type resonator TE011, with a frequency modulation of 100 kHz, microwave power of 0.998 mW, sweep time of 480 s, modulation amplitude of 1 G, time constant of 0.3 s, and a magnetic field scan range of 100 G. For the spin trapping experiments, the following settings were used: sweep field 150 G, frequency 100 kHz, gain 800, sweep time 240 s, time constant 0.1 s, modulation width 1 G, and microwave power 1 mW. The collagen samples were exposed to UVA radiation (370 nm) using an UV irradiation device directly in the spectrometer's cavity, as follows. The UV light generated by a mercury arc lamp (500 W, LOT-Quantum Design, Darmstadt, Germany) was passed through an UV irradiation accessory (ES-UVAT1, JEOL Resonance Inc., Tokyo, Japan; 370 nm HOYA colored optical glass filter ES-13020FL) connected to the spectrometer cavity via a condenser lens (ES-UVLL/UVLS, JEOL Resonance Inc.). The time of irradiation was 30 min. The EPR spectral simulations of spin adducts were performed using the WinSim program [57–59].

4. Conclusions

In conclusion, we can state that the results of our physicochemical determinations on systems containing collagen and hyaluronic acid exposed to the action of riboflavin and UVA irradiation support the roles that riboflavin and hyaluronic acid play in the treatment of CXL. Riboflavin, by generating free radicals, ensures the formation of interfibrillary bonds that lead to an increased mechanical strength of the cornea, while hyaluronic acid has the role of regulating or neutralizing some of the free radicals formed. Dextran has similar role to that of hyaluronic acid.

Supplementary Materials: The following are available online. Figure S1: Rheograms for UVA-irradiated collagen (sample **1a**) at the indicated temperatures (collagen 0.25%, pH 7.5), Figure S2: Rheograms for the collagen-riboflavin system (sample **2**) at the indicated temperatures (collagen 0.25%, riboflavin 0.1%, pH 7.5), Figure S3: Rheograms for the UVA-irradiated collagen-riboflavin system (sample **2a**) at the indicated temperatures (collagen 0.25%, riboflavin 0.1%, pH 7.5), Figure S4: Rheograms for the collagen-hyaluronic acid system (sample **3**) at the indicated temperatures (collagen 0.25%, hyaluronic acid 0.1%, pH 7.5), Figure S5: Rheograms for the UVA-irradiated collagen-hyaluronic acid system (sample **3a**) at the indicated temperatures (collagen 0.25%, hyaluronic acid 0.1%, pH 7.5), Figure S6: Rheograms for the collagen-riboflavin-hyaluronic acid system (sample **4**) at the indicated temperatures (collagen 0.25%, riboflavin 0.1%, hyaluronic acid 0.1%, pH 7.5), Figure S7: Rheograms for the UVA-irradiated collagen-riboflavin-hyaluronic acid system (sample **4a**) at the indicated temperatures (collagen 0.25%, riboflavin 0.1%, hyaluronic acid 0.1%, pH 7.5), Figure S8: PeakFit decomposition of DSC thermograms for collagen at (A) pH 3.5 and (B) pH 7.5, Figure S9: Molecular structures and spin trapping mechanisms of the spin traps used, Figure S10: The spectrum of riboflavin/DMPO solution in the absence (black) and in the presence (cyan) of UVA light. Collagen extraction and purification procedure [60,61]. Table S1: Contributions of DMPO spin adducts detected in solutions of tear proteins, in the presence of riboflavin, after UVA exposure.

Author Contributions: Conceptualization: G.I.; methodology, G.I., M.M.C., M.M. and I.M.; tear samples collection: M.M. and C.G.C.; EPR measurements: S.M., E.I.P. and I.M.; DSC measurements: A.P. and V.T.P.; sample preparation for TEM: M.M.; TEM images: B.T. and R.Ş.; rheology: T.S. and M.M.; writing—original draft preparation: all authors.; writing—review and editing: I.M. and G.I.; supervision: G.I.; project administration: G.I.; funding acquisition: G.I. All authors have read and agreed to the published version of the manuscript.

Funding: This research was partially funded by the Romanian National Authority for Scientific Research, CNCS–UEFISCDI, grant number PN-III-P2-2.1-PED-2016-0187.

Institutional Review Board Statement: Not applicable.

Informed Consent Statement: Patients that became subjects of this study were recruited during the visits to the "Oftaclinic Grup" clinic between January 2017 and July 2017. Informed consent was obtained from all subjects prior to the start of the study, in adherence to the Declaration of Helsinki. The study followed the institutional ethics guidelines and was approved by the ethics committee of the University of Medicine and Pharmacy "Carol Davila" Bucharest (PO-35-F-03, no. 130/7.07.2017).

Data Availability Statement: Not applicable.

Acknowledgments: This work was performed in the frame of the "Ilie Murgulescu" Institute of Physical Chemistry research plan "EPR and fluorescence studies on supramolecular interactions in inhomogeneous systems".

Conflicts of Interest: The authors declare no conflict of interest.

Sample Availability: Samples are available from the authors.

References

1. Kennedy, R.H.; Bourne, W.M.; Dyer, J.A. A 48-year clinical and epidemiologic study of keratoconus. *Am. J. Ophthalmol.* **1986**, *101*, 267–273. [CrossRef]
2. Balasubramanian, S.A.; Mohan, S.; Pye, D.C.; Willcox, M.D.P. Proteases, proteolysis and inflammatory molecules in the tears of people with keratoconus. *Acta Ophthalmol.* **2012**, *90*, 303–309. [CrossRef] [PubMed]
3. Carracedo, G.; Recchioni, A.; Alejandre-Alba, N.; Martin-Gil, A.; Crooke, A.; Morote, I.J.-A.; Pintor, J. Signs and symptoms of dry eye in keratoconus patients: A pilot study. *Curr. Eye Res.* **2015**, *40*, 1088–1094. [CrossRef] [PubMed]
4. Wollensak, G.; Sperl, E.; Seiler, T. Riboflavin/ultraviolet-A–induced collagen crosslinking for the treatment of keratoconus. *Am. J. Ophathalmol.* **2003**, *135*, 620–627. [CrossRef]
5. O'Brart, D.P.S. Corneal collagen cross-linking: A review. *J. Optom.* **2014**, *7*, 113–124. [CrossRef] [PubMed]
6. Frullini, A.; Manetti, L.; Di Cicco, E.; Fortuna, D. Photoinduced collagen cross-linking: A new approach to venous insufficiency. *Dermatol. Surg.* **2011**, *37*, 1113–1118. [CrossRef] [PubMed]
7. Cai, J.; Palamara, J.E.A.; Burrow, M.F. Effects of collagen crosslinkers on dentine: A literature review. *Calcif. Tissue Int.* **2018**, *102*, 265–279. [CrossRef]
8. Spoerl, E.; Huhle, M.; Seiler, T. Induction of cross-links in corneal tissue. *Exp. Eye Res.* **1998**, *66*, 97–103. [CrossRef]
9. Wollensak, G.; Iomdina, E.; Dittert, D.D.; Salamatina, O.; Stoltenburg, G. Cross-linking of scleral collagen in the rabbit using riboflavin and UVA. *Acta Ophthalmol. Scand.* **2005**, *83*, 477–482. [CrossRef]
10. Spoerl, E.; Wollensak, G.; Seiler, T. Increased resistance of crosslinked cornea against enzymatic digestion. *Curr. Eye Res.* **2004**, *29*, 35–40. [CrossRef]
11. Hsu, K.M.; Sugar, J. Keratoconus and other corneal diseases: Pharmacologic cross-linking and future therapy. *Pharm. Log. Ther. Ocul. Dis.* **2017**, *242*, 137–161. [CrossRef]
12. Constantin, M.M.; Corbu, C.G.; Tanase, C.; Codrici, E.; Mihai, S.; Popescu, I.D.; Enciu, A.M.; Mocanu, S.; Matei, I.; Ionita, G. Spin probe method of electron paramagnetic resonance spectroscopy-a qualitative test for measuring the evolution of dry eye syndrome under treatment. *Anal. Methods* **2019**, *11*, 965–972. [CrossRef]
13. Balasubramanian, S.A.; Pye, D.C.; Willcox, M.D.P. Levels of lactoferrin, secretory IgA and serum albumin in the tear film of people with keratoconus. *Exp. Eye Res.* **2012**, *96*, 132–137. [CrossRef]
14. Glasgow, B.J.; Marshall, G.; Gasymov, O.K.; Abduragimov, A.R.; Yusifov, T.N.; Knobler, C.M. Tear lipocalins: Potential lipid scavengers for the corneal surface. *Investig. Ophthalmol. Vis. Sci.* **1999**, *40*, 3100–3107.
15. Ohashi, Y.; Dogru, M.; Tsubota, K. Laboratory findings in tear fluid analysis. *Clin. Chim. Acta* **2006**, *369*, 17–28. [CrossRef] [PubMed]
16. Bron, A.J.; Mengher, L.S. The ocular surface in keratoconjunctivitis sicca. *Eye* **1989**, *3*, 428–437. [CrossRef]
17. Bailey, A.J. Structure, function and ageing of the collagens of the eye. *Eye* **1987**, *1*, 175–183. [CrossRef] [PubMed]
18. Wang, Y.; Zheng, M.; Liu, X.; Yue, O.; Wang, X.; Jiang, H. Advanced collagen nanofibers-based functional bio-composites for high-value utilization of leather: A review. *J. Sci. Adv. Mater. Devices* **2021**, *6*, 153–166. [CrossRef]
19. Williams, B.R.; Gelman, R.A.; Poppke, D.C.; Piez, K.A. Collagen fibril formation: Optimal in vitro conditions and preliminary kintic results. *J. Biol. Chem.* **1978**, *253*, 6578–6585. [CrossRef]
20. Silver, F.H. Type I collagen fibrillogenesis in vitro: Additional evidence for the assembly mechanism. *J. Biol. Chem.* **1981**, *256*, 4973–4977. [CrossRef]
21. Silver, F.H.; Birk, D.E. Kinetic analysis of collagen fibrillogenesis: I. Use of turbidity-time data. *Collagen Rel. Res.* **1983**, *3*, 393–405. [CrossRef]
22. Birk, D.E.; Silver, F.H. Kinetic analysis of collagen fibrillogenesis: II. Corneal and scleral type I collagen. *Collagen Rel. Res.* **1984**, *4*, 265–277. [CrossRef]
23. Hansen, U.; Bruckner, P. Macromolecular specificity of collagen fibrillogenesis: Fibrils of collagens I and XI contain a heteritypic alloyed core and a collagen I sheath. *J. Biol. Chem.* **2003**, *278*, 37352–37359. [CrossRef]
24. Gobeaux, F.; Mosser, G.; Anglo, A.; Panine, P.; Davidson, P.; Giraud-Guille, M.M.; Belamie, E. Fibrillogenesis in dense collagen solutions: A physicochemical study. *J. Biol. Chem.* **2008**, *376*, 1509–1522. [CrossRef] [PubMed]
25. Han, S.; McBride, D.J.; Losert, W.; Leikin, S. Segregation of type I collagen homo- and heterotrimers in fibrils. *J. Biol. Chem.* **2008**, *383*, 122–132. [CrossRef]
26. Sheraz, M.A.; Kazi, S.H.; Ahmed, S.; Anwar, Z.; Ahmad, I. Photo, thermal and chemical degradation of riboflavin. *Beilstein J. Org. Chem.* **2014**, *10*, 1999–2012. [CrossRef]
27. Balazs, E.A. Interaction between cells, hyaluronic acid and collagen. *Upsala J. Med. Sci.* **1977**, *82*, 94. [CrossRef]
28. Lapcik, L., Jr.; Omelka, L.; Kubena, K.; Galatik, A.; Kello, V. Photodegradation of hyaluronic acid and of the vitreous body. *Gen. Physiol. Biophys.* **1990**, *9*, 419–429. [PubMed]
29. Marchini, M.; Morocutti, M.; Ruggeri, A.; Koch, M.H.J.; Bigi, A.; Roveri, N. Differences in the fibril structure of corneal and tendon collagen. An electron microscopy and X-ray diffraction investigation. *Connect. Tissue Res.* **1986**, *15*, 269–281. [CrossRef]
30. Su, H.N.; Ran, L.Y.; Chen, Z.H.; Qin, Q.L.; Shi, M.; Song, X.Y.; Chen, X.L.; Zhang, Y.Z.; Xie, B.B. The ultrastructure of type I collagen at nanoscale: Large or small D-spacing distribution? *Nanoscale* **2014**, *6*, 8134–8139. [CrossRef]
31. Komsa-Penkova, R.; Koynova, R.; Kostov, G.; Tenchov, B.G. Thermal stability of calf skin collagen type I in salt solutions. *BBA Protein Struct. Mol. Enzymol.* **1996**, *1297*, 171–181. [CrossRef]

32. Mu, C.; Li, D.; Lin, W.; Ding, Y.; Zhang, G. Temperature induced denaturation of collagen in acidic solution. *Biopolymers* **2007**, *86*, 282–287. [CrossRef] [PubMed]
33. Flandin, F.; Buffevant, C.; Herbage, D. A differential scanning calorimetry analysis of the age-related changes in the thermal stability of rat skin collagen. *Biochim. Biophys. Acta.* **1984**, *791*, 205–211. [CrossRef]
34. Wallace, D.G.; Condell, R.A.; Donovan, J.W.; Paivinen, A.; Rhee, W.M.; Wade, S.B. Multiple denaturational transitions in fibrillar collagen. *Biopolymers* **1986**, *25*, 1875–1895. [CrossRef]
35. He, L.; Mu, C.; Li, D.; Lin, W. Revisit the pre-transition of type I collagen denaturation in dilute solution by ultrasensitive differential scanning calorimetry. *Thermochim. Acta* **2012**, *548*, 1–5. [CrossRef]
36. Tiktopulo, E.I.; Kajava, A.V. Denaturation of type I collagen fibrils is an endothermic process accompanied by a noticeable change in the partial heat capacity. *Biochemistry* **1998**, *37*, 8147–8152. [CrossRef]
37. Miles, C.A.; Avery, N.C.; Rodin, V.V.; Bailey, A.J. The increase in denaturation temperature following cross-linking of collagen is caused by dehydration of the fibres. *J. Mol. Biol.* **2005**, *346*, 551–556. [CrossRef]
38. Choe, E.; Huang, R.; Min, D.B. Chemical reactions and stability of riboflavin in foods. *J. Food Sci.* **2005**, *70*, R28–R36. [CrossRef]
39. Ionita, G.; Matei, I. Application of riboflavin photochemical properties in hydrogel synthesis. In *Biophysical Chemistry—Advance Applications*; Khalid, M.A.A., Ed.; IntechOpen: London, UK, 2019; pp. 1–14. [CrossRef]
40. Dikalov, S.; Grigorev, I.A.; Voinov, M.; Bassenge, E. Detection of superoxide radicals and peroxynitrite by 1-hydroxy-4-phosphonooxy-2,2,6,6-tetramethylpiperidine: Quantification of extracellular superoxide radicals formation. *Biochem. Biophys. Res. Commun.* **1998**, *248*, 211. [CrossRef]
41. Bielski, B.H.J.; Cabelli, D.E.; Arudi, R.L.; Ross, A.B. Reactivity of HO_2/O_2 radicals in aqueous solution. *J. Phys. Chem. Ref. Data* **1985**, *14*, 1041–1100. [CrossRef]
42. Collin, F. Chemical basis of reactive oxygen species reactivity and involvement in neurodegenerative diseases. *Int. J. Mol. Sci.* **2019**, *20*, 2407. [CrossRef]
43. Fujii, T.; Kühn, K. Isolation and characterization of pepsin-treated type III collagen from calf skin. *Hoppe-Seyler's Z. Phys. Chem.* **1975**, *356*, 1793–1802. [CrossRef] [PubMed]
44. Seyer, J.M.; Hutcheson, E.T.; Kang, A.H. Collagen polymorphism in normal and cirrhotic human liver. *J. Clin. Investig.* **1977**, *59*, 241–248. [CrossRef] [PubMed]
45. Seyer, J.M.; Hutcheson, E.T.; Kang, A.H. Collagen polymorphism in idiopathic chronic pulmonary fibrosis. *J. Clin. Investig.* **1976**, *57*, 1498–1507. [CrossRef]
46. Siddiqui, Y.D.; Arief, E.M.; Yusoff, A.; Hamid, S.S.A.; Norani, T.Y.; Abdullah, M.Y.S. Extraction, purification and physical characterization of collagen from body wall of sea cucumber *Bohadschia bivitatta. Health Environ. J.* **2013**, *4*, 53–65.
47. Gao, L.; Wang, Z.; Li, Z.; Zhang, C.; Zhang, D. The characterization of acid and pepsin soluble collagen from ovine bones (*Ujumuqin sheep*). *J. Integr. Agric.* **2018**, *17*, 704–711. [CrossRef]
48. Blanco, M.; Vázquez, J.A.; Pérez-Martín, R.I.; Sotelo, C.G. Collagen extraction optimization from the skin of the small-spotted catshark (*S. canicula*) by response surface methodology. *Mar. Drugs* **2019**, *17*, 40. [CrossRef]
49. Noorzai, S.; Verbeek, C.J.R.; Lay, M.C.; Swan, J. Collagen extraction from various waste bovine hide sources. *Waste Biomass Valori.* **2020**, *11*, 5687–5698. [CrossRef]
50. Bachinger, H.P.; Mizuno, K.; Vranka, J.A.; Boudko, S.P. Collagen formation and structure. In *Comprehensive Natural Products II. Chemistry and Biology, Aminoacids, Peptides and Proteins*; Liu, H.-W., Mander, L., Eds.; Elsevier Ltd.: Amsterdam, The Netherlands, 2010; Volume 5, pp. 481–482.
51. Chandrasekhar, S.; Laurie, G.W.; Cannon, F.B.; Martin, G.R.; Kleinman, H.K. In vitro regulation of cartilage matrix assembly by a Mr 54,000 collagen-binding proteins. *Proc. Natl. Acad. Sci. USA* **1986**, *83*, 5126–5130. [CrossRef]
52. Birk, D.E.; Fitch, J.M.; Babiarz, J.P.; Doane, K.J.; Linsenmayer, T.F. Collagen fibrillogenesis in vitro: Interaction of types I and V collagen regulates fibril diameter. *J. Cell Sci.* **1990**, *95*, 649–657. [CrossRef]
53. Li, Y.; Asadi, A.; Monroe, M.R.; Douglas, E.P. pH effects on collagen fibrillogenesis in vitro: Electrostatic interactions and phosphate binding. *Mater. Sci. Eng. C* **2009**, *29*, 1643–1649. [CrossRef]
54. Micutz, M.; Matalon, E.; Staicu, T.; Angelescu, D.G.; Ariciu, A.M.; Rogozea, A.; Turcu, I.M.; Ionita, G. The influence of hydroxypropyl β-cyclodextrin on the micellar to gel transition in F127 solutions investigated at macro and nanoscale levels. *New J. Chem.* **2014**, *38*, 2801–2812. [CrossRef]
55. Neacsu, M.V.; Matei, I.; Micutz, M.; Staicu, T.; Precupas, A.; Popa, V.T.; Salifoglou, A.; Ionita, G. Interaction between albumin and Pluronic F127 block copolymer by global and local physicochemical profiling. *J. Phys. Chem. B* **2016**, *120*, 4258–4267. [CrossRef] [PubMed]
56. Staicu, T.; Ilis, M.; Circu, V.; Micutz, M. Influence of hydrocarbon moieties of partially fluorinated N-benzoyl thiourea compounds on their gelation properties. A detailed rheological study of complex viscoelastic behavior of decanol/N-benzoyl thiourea mixtures. *J. Mol. Liq.* **2018**, *255*, 297–312. [CrossRef]
57. Staicu, T.; Circu, V.; Ionita, G.; Ghica, C.; Popa, V.T.; Micutz, M. Analysis of bimodal thermally-induced denaturation of type I collagen extracted from calfskin. *RSC Adv.* **2015**, *5*, 38391–38406. [CrossRef]
58. Duling, D.R. Simulation of multiple isotropic spin-trap EPR spectra. *J. Magn. Reson. B* **1994**, *104*, 105–110. [CrossRef] [PubMed]
59. Duling, D.R. *PEST Winsim*; Version 0.96; National Institute of Environmental Health Sciences: Triangle Park, NC, USA, 1996.

60. Kühn, K. The classical collagens: Types I, II, and III. In *Structure and Function of Collagen Types*; Mayne, R., Burgeson, R.E., Eds.; Academic Press: Orlando, Fl, USA, 1987; pp. 1–42.
61. Ventre, M.; Padovani, M.; Covington, A.D.; Netti, P.A. Composition, structure and physical properties of foetal calf skin. In Proceedings of the II IULTCS Eurocongress: Innovation and New Technologies for the Future of Global Leather Industry, Istanbul, Turkey, 24–27 May 2006.

MDPI

Review

Photo- and Radiation-Induced One-Electron Oxidation of Methionine in Various Structural Environments Studied by Time-Resolved Techniques

Bronislaw Marciniak [1,*] and Krzysztof Bobrowski [2,*]

[1] Center for Advanced Technology, and Faculty of Chemistry, Adam Mickiewicz University, Uniwersytetu Poznanskiego 10, 61-712 Poznan, Poland
[2] Institute of Nuclear Chemistry and Technology, Dorodna 16, 03-195 Warsaw, Poland
* Correspondence: marcinia@amu.edu.pl (B.M.); kris@ichtj.pl (K.B.)

Abstract: Oxidation of methionine (Met) is an important reaction that plays a key role in protein modifications during oxidative stress and aging. The first steps of Met oxidation involve the creation of very reactive and short-lived transients. Application of complementary time-resolved radiation and photochemical techniques (pulse radiolysis and laser flash photolysis together with time-resolved CIDNP and ESR techniques) allowed comparing in detail the one-electron oxidation mechanisms initiated either by $^{\bullet}$OH radicals and other one-electron oxidants or the excited triplet state of the sensitizers e.g., 4-,3-carboxybenzophenones. The main purpose of this review is to present various factors that influence the character of the forming intermediates. They are divided into two parts: those inextricably related to the structures of molecules containing Met and those related to external factors. The former include (i) the protection of terminal amine and carboxyl groups, (ii) the location of Met in the peptide molecule, (iii) the character of neighboring amino acid other than Met, (iv) the character of the peptide chain (open vs cyclic), (v) the number of Met residues in peptide and protein, and (vi) the optical isomerism of Met residues. External factors include the type of the oxidant, pH, and concentration of Met-containing compounds in the reaction environment. Particular attention is given to the neighboring group participation, which is an essential parameter controlling one-electron oxidation of Met. Mechanistic aspects of oxidation processes by various one-electron oxidants in various structural and pH environments are summarized and discussed. The importance of these studies for understanding oxidation of Met in real biological systems is also addressed.

Keywords: methionine; oxidation; neighboring group effect; hydroxyl radical; triplet state of carboxybenzophenone; one-electron oxidants; pulse radiolysis; laser flash photolysis; peptides; proteins

Citation: Marciniak, B.; Bobrowski, K. Photo- and Radiation-Induced One-Electron Oxidation of Methionine in Various Structural Environments Studied by Time-Resolved Techniques. *Molecules* **2022**, *27*, 1028. https://doi.org/10.3390/molecules27031028

Academic Editor: Danilo Roccatano

Received: 27 December 2021
Accepted: 27 January 2022
Published: 2 February 2022

1. Introduction

Methionine (Met) is an important sulfur-containing amino acid often playing a protective role in a protein oxidation due to the fact that the thioether group is easily oxidized by many reactive species [1,2]. These processes play a key role in protein oxidation during oxidative stress and biological aging as certain conditions promote the conversion of sulfur radical cations into sulfoxide [3,4]. The first steps of oxidation mechanisms leading to proteome modifications are extremely fast, i.e., they take place within sub-microsecond and microsecond time domains [5]. They involve creation of the very reactive transients such as radicals and radical cations, which are responsible for the subsequent protein damage. For these reasons, it is important to characterize spectrally and kinetically these very first steps. Radiolysis of water and photo-excitation of benzophenone carboxyl derivatives in aqueous solutions provide very convenient source of hydroxyl radicals HO$^{\bullet}$ and excited triplet states (CB*), respectively, which were very useful for studying one-electron oxidation reactions of Met [5,6]. The transient nature of free radicals requires specific time-resolved methods. Therefore, these reactions were followed either in radiation- or photochemical studies by means of pulse radiolysis [7–14], laser flash photolysis [15–22], CIDNP [23–25], and ESR techniques [26]. They were the subject of several review articles [6,27–30], and also chapters in books [31–35], presenting free radical chemistry in amino acids, peptides, and proteins in a more general manner.

A wealth of knowledge has been accumulated concerning mechanistic understanding of these processes induced by one-electron oxidation of methionine in various structural environments. To analyze the oxidation mechanism of methionine, one has to take into account that methionine can exist in cationic, zwitterionic, or anionic forms in aqueous solutions depending on the pH range with the respective pK_a values: ($COOH/COO^-$) = 2.3 and ($^+NH_3/NH_2$) = 9.2 (Figure 1).

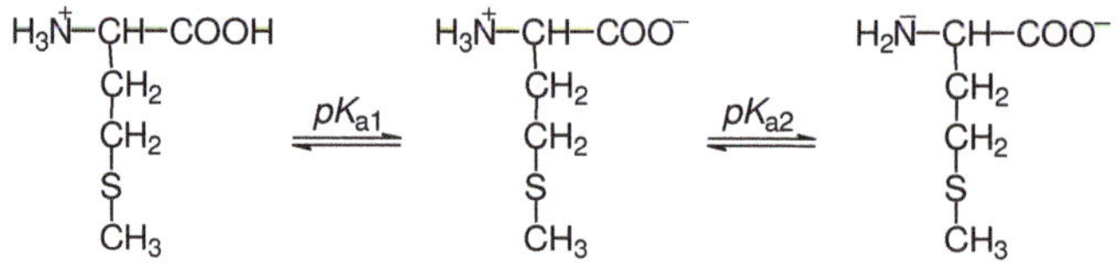

$$H_3\overset{+}{N}-CH(CH_2CH_2SCH_3)-COOH \underset{}{\overset{pK_{a1}}{\rightleftharpoons}} H_3\overset{+}{N}-CH(CH_2CH_2SCH_3)-COO^- \overset{pK_{a2}}{\rightleftharpoons} H_2\bar{N}-CH(CH_2CH_2SCH_3)-COO^-$$

Figure 1. Structures of methionine in aqueous solutions at pH range 1–14.

Neighboring group participation is also an important concept for understanding how the one-electron oxidation of methionine is controlled. This is the case when a neighboring group stabilizes a transition state or intermediate by becoming bonded to the reaction center as a result of "through space" interaction [36]. These neighboring groups can provide a lone pair of electrons which can be shared with the monomeric sulfur cation center, $>S^{\bullet+}$, forming a three-electron-bonded species which effectively stabilizes the radical cation. The same applies to the $^\bullet$OH-induced oxidation of methionine where the $^\bullet$OH radicals directly attack the sulfur, forming three-electron-bonded adduct (>S∴OH), and its subsequent reactions are strongly influenced by the presence of neighboring groups (Figure 2).

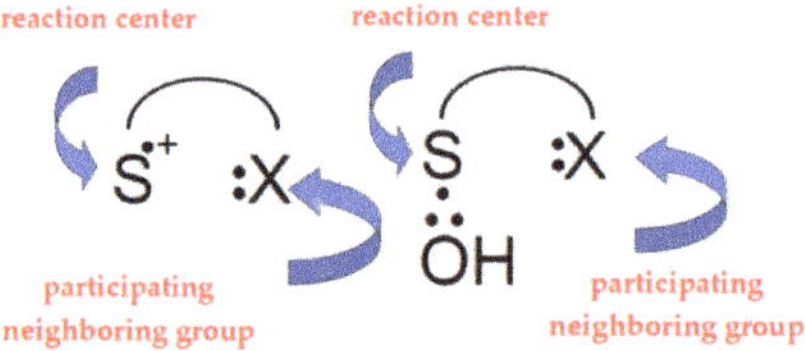

Figure 2. Simplified scheme showing neighboring group participation during oxidation of thioether group of methionine.

Neighboring group participation in the reactions of two intermediates, monomeric sulfur radical cations ($MetS^{\bullet+}$), and OH-adducts to the sulfur atom (MetS∴OH) during Met oxidation is of great importance, particularly within peptides and proteins. This is due to the presence of a manifold of possible participating functionalities such as carboxyl, amine, hydroxyl, amide, and thioether groups [37]. For a long time it was believed that the intramolecular two-centered, three electron (2c-3e) bonds between the oxidized sulfur atom in Met and the lone electron pair on the nitrogen atom in N-terminal amino group and/or the oxygen atoms in the C-terminal carboxyl group are responsible for stabilization of $MetS^{\bullet+}$). Such stabilization leads to five-membered intramolecularly S∴N-bonded species and/or six-membered intramolecularly S∴O-bonded species. Subsequent studies showed that heteroatoms present in the peptide bond can be also involved in the formation of similar 2c-3e bonds with the oxidized sulfur atom. The general description and spectral characterization of two-centered, three electrons (2c-3e) species was summarized and extensively discussed in numerous papers [38–40], book chapters [41,42], and references therein.

The main purpose of this review is to present various factors that might influence the character of intermediates forming during one-electron oxidation of methionine. They can be divided into two parts: those inextricably related to the structure of molecules containing methionine, and those related to the external factors. The former include: (i) protection of terminal amine and carboxyl groups, (ii) location of methionine in the peptide molecule (N/C-terminal and internal), (iii) character of the neighboring amino acid other than methionine, (iv) character of the peptide chain (open vs. cyclic), (v) number of methionine residues in the peptide molecule, and (vi) optical isomerism of methionine residues. In turn, external factors include: (i) type of radiation that induces the oxidation (light vs. high-energy radiation), (ii) type of the oxidant ($^{\bullet}$OH vs one-electron oxidant), and (iii) pH and concentration of Met-containing molecules in the reaction environment.

Application of complementary radiation and photochemical techniques allowed investigators to compare in detail the one-electron oxidation mechanisms initiated either by $^{\bullet}$OH radicals and other one-electron oxidants or the excited triplet state of the sensitizer (CB*). The complete and detailed description of the primary steps of the oxidation mechanisms of Met residues located in the interior of long oligopeptides and proteins is a significant and original contribution in understanding oxidation reactions in real biological systems which might be of great help in imagining new strategies in the struggle against "uncontrolled oxidative stress". The results of these studies and conclusions drawn from them are summarized in this review.

2. Methionine and Methionine Derivatives

2.1. Radiation-Induced Oxidation

2.1.1. Methionine

In the 1980s, the reactions of $^{\bullet}$OH radicals and some selected one-electron oxidants with Met were extensively studied by Asmus' group in order to characterize spectral and kinetic properties of sulfur- and carbon-centered radicals which can be potentially formed [7,43–46]. Mechanistic studies of the radically induced oxidation of Met revealed rather complex reaction schemes which depend mainly on three parameters, the nature of the oxidant, pH of the solution, and concentration of Met. Although $^{\bullet}$OH radicals exhibit strong oxidation properties (the standard reduction potentials of the $HO^{\bullet}/HO^{-}$ and $HO^{\bullet}$, H^{+}/H_2O redox couples are equal to E^0 = +1.90 V and +2.72 V vs. NHE, respectively) [47], their oxidation action is not a straightforward one-electron transfer process. Owing to their high electrophilicity the $HO^{\bullet}$ radicals prefer to add to the sulfur atom which is a reaction center of high electron density.

The underlying mechanism of the $^{\bullet}$OH-induced oxidation of Met is based on an addition of the electrophilic $^{\bullet}$OH radical to the sulfur atom forming MetS∴OH as the initial step. This reaction occurs with an absolute rate constant $k = 2.3 \times 10^{10}$ M^{-1} s^{-1} (vide Table 1), irrespective of pH, and is practically only controlled by diffusion of the reactants. The consecutive steps depend further on pH. In strong acid solutions (pH $\leq$ 3) MetS∴OH will react with external protons yielding $MetS^{\bullet+}$. At a low concentration of Met, $MetS^{\bullet+}$ undergoes deprotonation that leads to α-(alkylthio)alkyl radicals ($k = 2.4 \times 10^5$ s^{-1}) (in the following referred to as α-S1 and α-S2 with estimated $pK_a(MetS^{\bullet}+/Met(-^{\bullet}CHSCH_3) = -6$ for deprotonation in the γ-position and $pK_a(MetS^{\bullet}+/Met(^{\bullet}CH_2SCH_2-) = -2$ for deprotonation in the ε-position [48]. At high concentration of Met, $MetS^{\bullet+}$ can be stabilized via intermolecular equilibration with a second unattacked Met molecule forming a dimeric sulfur radical cation with an intermolecular 2c-3e S∴S bond ($(MetS∴SMet)^{+}$). This dimeric radical cation is characterized by a strong absorption spectrum with λ_{max} = 480 nm. In principle, in acidic conditions Met behaves like an ordinary thioether. At pH $\geq$ 3, where the carboxyl group is deprotonated, the oxidative mechanism is changed and involves an intramolecular process where protons for dihydroxylation of MetS∴OH are provided by the $^{+}NH_3$ group. The latter process leads to formation of an intramolecularly S∴N-bonded intermediate (in the following referred to as S∴N-bonded radical cation, $Met(S∴N)^{+}$) which is assisted by a suitable five membered ring steric arrangement. The $Met(S∴N)^{+}$) is a very-short lived intermediate ($\tau_{1/2}$ = 220 ns) and exhibits a transient absorption spectrum with $\lambda_{max} \approx$ 390–400 nm [46]. The consecutive reactions include opening of the S∴N-bond to the

N-centered radical cation $H_2N^{\bullet+}$-CHR-COO^-, establishment of the mesomeric form, H_2N-CHR-$COO^\bullet$ and CO_2 cleavage of the latter with the formation of 3-methylthiopropylamino radical (in the following referred to as an α-amino radical, α-N). The most important consequences of the mechanism change are decarboxylation of the Met molecule and formation of α-N radicals with strong reductive properties. Formation of the latter radical changes the redox properties of the system from oxidizing to reducing. The general reaction mechanism describing the primary and secondary reactions of radiationally induced $^\bullet OH$ radicals in aqueous solutions is presented in Scheme 1.

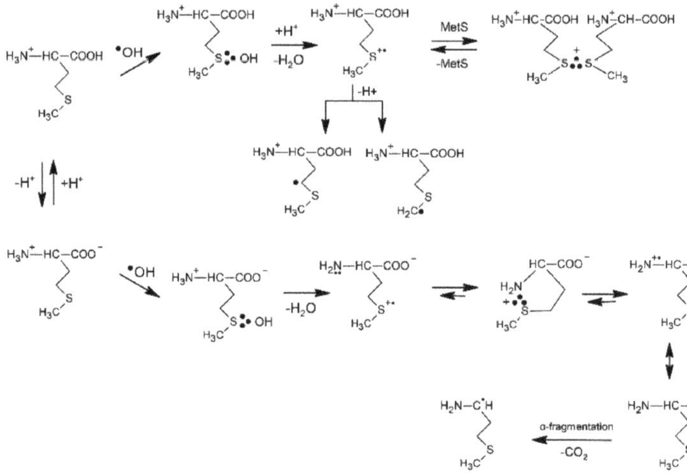

Scheme 1. The reactions mechanism for $^\bullet$OH-induced oxidation of methionine (based on [7,46]).

A significantly different picture with respect to pH is observed in the oxidation of Met by the $CCl_3O_2^\bullet$ radicals (a moderately good one-electron oxidant) [45]. Its reduction potential depends on pH: $E^0(CCl_3OO^\bullet, H^+/CCl_3OOH) = 1.60$ V vs. NHE and $E^0(CCl_3OO^\bullet/CCl_3OO^-) = 1.15$ vs. NHE [49]. The reaction of Met with $CCl_3OO^\bullet$ radicals proceeds with $k = 2.9 \times 10^7\ M^{-1}s^{-1}$ (vide Table 1). The normalized plots for the formation of the (MetS∴SMet)$^+$ and Met(S∴N)$^+$ resulting from the oxidation of Met by $^\bullet$OH and $CCl_3O_2^\bullet$ and based on their absorptions at the respective $\lambda_{max} = 480$ and 390 nm clearly indicate that in contrast to the system containing $^\bullet$OH, the dimeric (MetS∴SMet)$^+$ radical cations are stabilized over the entire acid and neutral pH range before they are replaced by Met(S∴N)$^+$ radical cations. An exclusive formation of the Met(S∴N)$^+$ intermediate occurs at pH > 9 in contrast to the reaction of $^\bullet$OH where the Met(S∴N)$^+$ intermediate is almost exclusively formed at pH > 4. The apparent change in reaction mechanism occurs in slightly more basic solutions, i.e., near the pK_{a2} of the amino group in Met (vide Figure 1). The formation of Met(S∴N)$^+$ in the system containing $CCl_3O_2^\bullet$ requires the availability of the free electron pair of the unprotonated amino function and therefore this species is only formed beyond and near the pK_{a2} [45]. The oxidation mechanism of Met by $CCl_3O_2^\bullet$ and also by other one-electron oxidants such as $CF_3CHClO_2^\bullet$ [45], $CO_3^{\bullet-}$ [50], Tl^{2+} [43], and Ag^{2+} [43] is exemplified in Scheme 2. Their respective rate constants with Met are listed in Table 1. The parameter which essentially controls whether (MetS∴SMet)$^+$ or Met(S∴N)$^+$ is formed is the equilibrium constant K_{a2} (see Figure 1). However, the kinetics of associated processes have to be also taken into account.

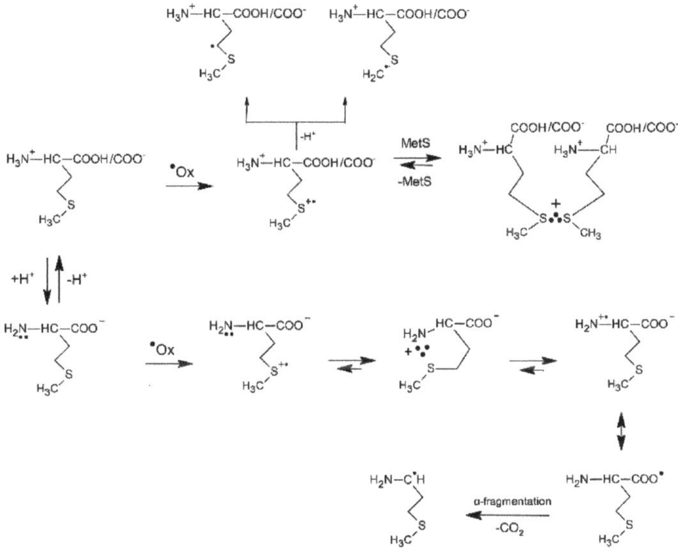

Scheme 2. The general reactions mechanism for $^{\bullet}$Ox—induced oxidation of methionine in aqueous solutions where $^{\bullet}Ox = CCl_3O_2^{\bullet}$, $CF_3CHClO_2^{\bullet}$, $CO_3^{\bullet-}$, Tl^{2+}, Ag^{2+} (based on [43,45]).

Although $Cl_2^{\bullet-}$ radicals exhibit strong oxidation properties (the standard reduction potential of the $Cl_2^{\bullet-}/2Cl^-$ redox couple is equal to E^0 = +2.13 V vs. NHE [47]), their oxidation action is not an adduct mediated one-electron transfer process (an inner-sphere electron transfer) such as in the case of $^{\bullet}OH$ radicals. Moreover, this reaction could be studied only in very acid solutions owing to the complex reaction mechanism leading to $Cl_2^{\bullet-}$ formation in aqueous solutions and involving $^{\bullet}OH$ radicals, $HOCl^{\bullet-}$ radical anions, protons (H^+) and chlorine atoms ($Cl^{\bullet}$) [51]. The primary step of Met oxidation by $Cl_2^{\bullet-}$ radicals was found to be, in principle, similar to that established for the $^{\bullet}OH$ radicals and occurs with absolute rate constant $k = 3.9 \times 10^{10}$ M^{-1} s^{-1} (see Table 1) [52]. It is characterized by a primary attack on the sulfur atom that constitutes a substitution of a chloride anion (Cl^-) in $Cl_2^{\bullet-}$ by Met and leads to the MetS∴Cl three-electron bonded species. Interestingly, from the earlier studies on the oxidation of simple thioethers it is known that MetS∴X species are involved in the following equilibrium with $(MetS∴SMet)^+$ radical cations: MetS∴X + MetS ⇆ $(MetS∴SMet)^+$ + X^- [53].

The reactions mechanism for $Cl_2^{\bullet-}$-induced oxidation of Met is presented in Scheme 3 and clearly shows that a monomeric sulfur radical cation $MetS^{\bullet+}$ can be stabilized not only in a form of the dimeric $(MetS∴SMet)^+$ radical cations but also in a form of MetS∴Cl three-electron bonded species with the chloride anions Cl^- providing a free electron pair. The validity of the mechanism (Scheme 3) was also experimentally demonstrated in the oxidation of Met by $Br_2^{\bullet-}$ where the corresponding MetS∴Br species were also observed at neutral and basic pHs. The observed decrease in rate constant for the reaction of Met with $Br_2^{\bullet-}$ ($k = 2.5 \times 10^{10}$ M^{-1} s^{-1}) correlates well with the standard reduction potential of the $Br_2^{\bullet-}/2Br^-$ redox couple which is equal to E^0 = +1.63 V vs. NHE (see Table 1) [47]. The actual reaction routes are very sensitive to a variety of parameters such as pH, Met and Cl^-/Br^- concentration.

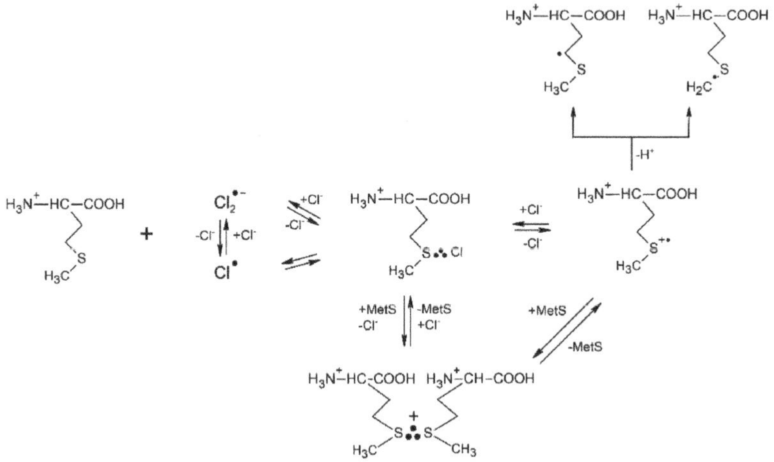

Scheme 3. The reactions mechanism for $Cl_2^{\bullet-}$—induced oxidation of methionine in aqueous solutions (based on [52]).

Table 1. Rate constants of reactions of $^{\bullet}OH$ radical and various one-electron oxidants ($^{\bullet}Ox$) with methionine in aqueous solutions.

Oxidant	Met(NH_3^+, COOH)	Met(NH_3^+, COO^-)	Met(NH_2, COO^-)	E^0 [(a)]	Lit
$^{\bullet}OH$	2.3×10^{10} [(b)]			+2.72	[7]
$CCl_3O_2^{\bullet}$			2.9×10^{7} [(b,c)]	+1.60	[45]
CF_3 $CHClO_2^{\bullet}$			1.4×10^{6} [(b,c)]	+1.15	[45]
$CO_3^{\bullet-}$		3.6×10^{7} [(b)]		+1.57	[50]
$Cl_2^{\bullet-}$	3.9×10^{9} [(b)]			+2.13	[52]
$Br_2^{\bullet-}$	2.5×10^{9} [(b)]	1.7×10^{9} [(b)]	2.0×10^{9} [(b,c)]	+1.63	[52]
Tl^{2+}		2.5×10^{9} [(b)]		+2.23	[43]
Ag^{2+}		3.3×10^{8} [(b)]		+1.98	[43]

[(a)] all reduction potentials in V vs. NHE after reference [47], [(b)] rate constants in $M^{-1}s^{-1}$ units, [(c)] measured by competition method.

2.1.2. Methionine Derivatives

Selective protection of the N-terminal and C-terminal groups in Met enabled investigators to elucidate the role of each individual neighboring group NH_3^+/NH_2 and $COOH/COO^-$ in one-electron oxidation of Met by $^{\bullet}OH$ radicals. The following methionine derivatives were included in the studies: N-acetyl-methionine (N-Ac-Met) [7,54,55], methionine methyl ester (Met-C(=O)CH_3 [54,56], and methionine ethyl ester (Met-C(=O)-CH_2CH_3 [46].

Substitution of one of the hydrogen atoms in the amino group (-NH_2) by an acetyl group (-C=O)-CH_3) does not change the initial reaction of $^{\bullet}OH$ radicals with the sulfur atom in N-Ac-Met. The observed transient absorption bands at pH 7 with λ_{max} = 290, 340, and 480 nm were respectively assigned to α-S1/α-S2 derived from N-Ac-Met, N-Ac-MetS∴OH, and (N-Ac-MetS∴SMet-N-Ac)$^+$ species. However, the following features distinguish N-Ac-Met from Met: (i) prolonged lifetime of N-Ac-MetS∴OH, and (ii) absence of the absorption band which can be assigned to N-Ac-Met(S∴N)$^+$. Based on these observations, it can be concluded that the protonated amino group in Met plays a role of intramolecular proton donor in the disappearance of MetS∴OH, and later being in the deprotonated form (-NH_2) as a donor of free electron pair for stabilization of Met(S∴N)$^+$. Depending on N-Ac-Met concentration, the monomeric sulfur radical cation N-Ac-MetS$^{\bullet+}$, similarly as in Met, can be stabilized via intermolecular equilibration with a second unattacked N-Ac-Met molecule forming a dimeric sulfur radical cation with intermolecular 2c-3e S∴S bond ((N-Ac-MetS∴SMet-N-Ac)$^+$) and further undergo deprotonation to α-S1/α-S2 radicals (see Scheme 1). Detection of CO_2 in the system studied [7,44], clearly indicates that N-Ac-MetS$^{\bullet+}$ can undergo decarboxylation via pseudo-Kolbe mechanism to form N-Ac-substituted α-N radicals [57,58].

The replacement of the carboxyl group with an ester or an amide group in methionine does not significantly change the primary radical reactions related to the oxidation process, except for the fact that the decarboxylation process involving the intramolecular Met(S∴N)$^+$ is inhibited (see Scheme 1). This is reflected in much longer lifetime ($\tau_{1/2}$ ~ 1.1 ms) of this radical, which was directly observed in the example of the methionine ethyl ester [46].

Simultaneous acetylation of the N-terminal amino group and either esterification or amidation of the C-terminal carboxyl group was aimed at elimination of the fast intramolecular proton transfer from the amino group to the MetS∴OH moiety and elimination of decarboxylation of the Met residue via pseudo Kolbe mechanism, respectively. In addition, the insertion of these groups in the methionine molecule allows formation of structural motifs that quite well mimic peptide bonds in peptides and proteins. Two methionine derivatives, N-acetyl methionine amide (N-Ac-Met-NH_2) [10], and methyl ester of N-acetyl-methionine (N-Ac-Met-OCH_3) [11,59], were selected in order to limit stabilization of the monomeric sulfur radical cations (MetS$^{\bullet+}$) only to interactions with N- and O-atoms located in acetyl, ester and amide functional groups. Furthermore, they allow one to establish and to prove the mechanisms for primary and secondary radical reactions following oxidation of the Met residue by $^\bullet$OH radicals with no contribution of free terminal amino and carboxyl groups. Thus, they can be considered to be the simplest models of Met residue incorporated in the interior of oligopeptides and proteins. The general reaction mechanism describing the primary and secondary reactions of radiationally induced $^\bullet$OH radicals with these Met derivatives in aqueous solutions together with the respective intermediates is presented in Section 2.2.2.

2.2. Photo-Induced Oxidation

The mechanism for the quenching of the benzophenone triplet state (^{3}Bz) by sulfur-containing organic compounds, e.g., thioethers, has been studied since the 1970s [60]. It was suggested that quenching leads to the formation of partial charges on sulfur and oxygen atoms (Charge Transfer (CT)-complex) decaying by the proton transfer from the carbon atom bonded to the sulfur atom to form the ketyl radical and α-(alkylthio)alkyl radical (αS) and back electron transfer to regenerate reactants in their ground states (see Scheme 4).

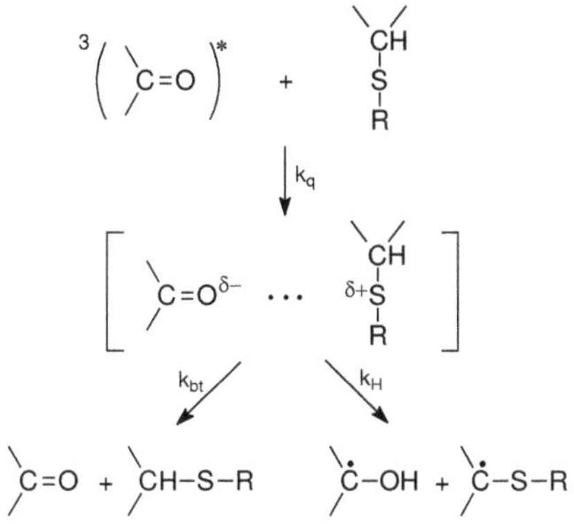

$$^3\left(\,\rangle C{=}O\,\right)^* + \rangle CH{-}S{-}R \xrightarrow{k_q} \left[\,\rangle C{=}O^{\delta-} \cdots {}^{\delta+}S(R){-}CH\langle\,\right]$$

$$\xrightarrow{k_{bt}} \rangle C{=}O + \rangle CH{-}S{-}R \qquad \xrightarrow{k_H} \rangle \dot{C}{-}OH + \rangle \dot{C}{-}S{-}R$$

Scheme 4. The mechanism for the quenching of ^{3}Bz by thioethers (based on [60]).

To analyze the mechanism of quenching of carboxybenzophenone (CB) triplets by methionine, methionine derivatives and peptides (S and N atoms as the expected electron donors), one has to take into account that methionine can exist in zwitterionic (at pH 7) or anionic forms (at pH 11) in aqueous solutions depending on the pH range (see Figure 1). Therefore, the quenching rate constants for quenching the CB triplet in aqueous solutions by methionine, and Met derivatives and for comparison simple Met-containing peptides (Table 2) are presented at two representative pH values 7 and 10/11.

It was proven that the CB triplet quenching by methionine and Met-containing compounds were occurring via an electron transfer mechanism from its S-atom (not its N-atom) based on the following arguments [18,61]:

- as shown in Table 2, the quenching rate constants were found to be in the range of 10^9 M^{-1} s^{-1} (diffusion controlled limit) for methionine and Met-containing compounds and were, respectively, three orders or one order of magnitude lower for amino acids without a sulfur atom (e.g., alanine) at pH 7 and pH 10,
- direct observation of radical-ion products in the transient absorption spectra (electron transfer intermediates, among them various two-center three-electron (2c-3e) bonded species) were products of oxidation of Met and CB radical anions and CB ketyl radicals were products of CB reduction,
- indirectly by the Rehm-Weller correlations of k_q vs. ΔG_{el} (free energy change for electron transfer).

Similar arguments for an electron transfer mechanism of quenching were applied for methionine derivatives with blocked amino and/or carboxylic terminal groups and simple peptides containing methionine residues [16,18,19,22,62,63].

Table 2. Quenching rate constants ($k_q \times 10^{-9}$ M^{-1} s^{-1}) for quenching of CB triplet in aqueous solutions by methionine, Met derivatives and simple Met-containing peptides [15,16,18–20,62].

Amino Acid or Peptide	pH 7	pH 10/11
Methionine	2.5	2.3
N-Ac-Met	1.9	1.6
Met-OCH_3	3.0	2.9
Met-Gly	2.1	2.3
Gly-Met	2.0	2.0
Met-Lys	1.8	1.3
Lys-Met	2.5	1.8
Met-Gly-Gly	2.3	2.2
Gly-Gly-Met	1.8	1.9
Met-Met	2.9	1.8
Met-Enkephalin	1.9	1.8
Alanine	<0.0005	0.18

estimated errors: <10%.

2.2.1. Methionine

The study of the photooxidation mechanism of methionine in aqueous solution started with the pioneering work of Cohen [64]. The suggested mechanism of primary reactions was described by formation of two CT complexes: CT-S [$CB^{\bullet-}$ ... >$S^{\bullet+}$] and CT-N [$CB^{\bullet-}$ ... >$N^{\bullet+}$] but only CT-N complex could undergo decarboxylation leading to the formation of α-amino-alkyl radical (αN). The latter could reduce the ground state CB yielding ketyl radical ($CBH^{\bullet}$) and methional (Scheme 5).

Further studies using nanosecond flash photolysis technique [15,62,65,66] led to a more detailed description of primary and secondary reactions for the photooxidation of methionine in neutral and basic aqueous solutions. These reactions are presented in Schemes 6 and 7.

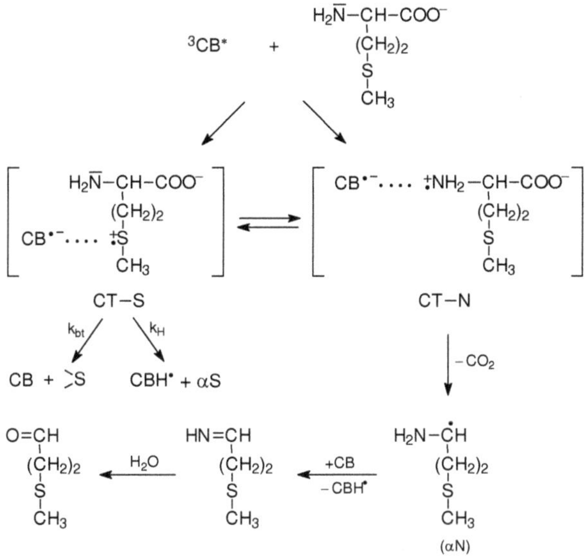

$^3CB^*$ + $H_2\bar{N}-CH-COO^-$ / $(CH_2)_2$ / S / CH_3

[$CB^{\bullet-}\cdots{}^{\bullet+}S$ …] ⇄ [$CB^{\bullet-}\cdots{}^{\bullet+}NH_2-CH-COO^-$ …]

CT−S CT−N

k_{bt}: CB + >S k_H: $CBH^\bullet$ + αS

$-CO_2$

$O{=}CH(CH_2)_2SCH_3$ ← H_2O ← $HN{=}CH(CH_2)_2SCH_3$ ← +CB / −$CBH^\bullet$ ← $H_2N-\dot{C}H(CH_2)_2SCH_3$ (αN)

Scheme 5. The mechanism of CB-sensitized photooxidation of methionine in alkaline aqueous solutions (based on [64]).

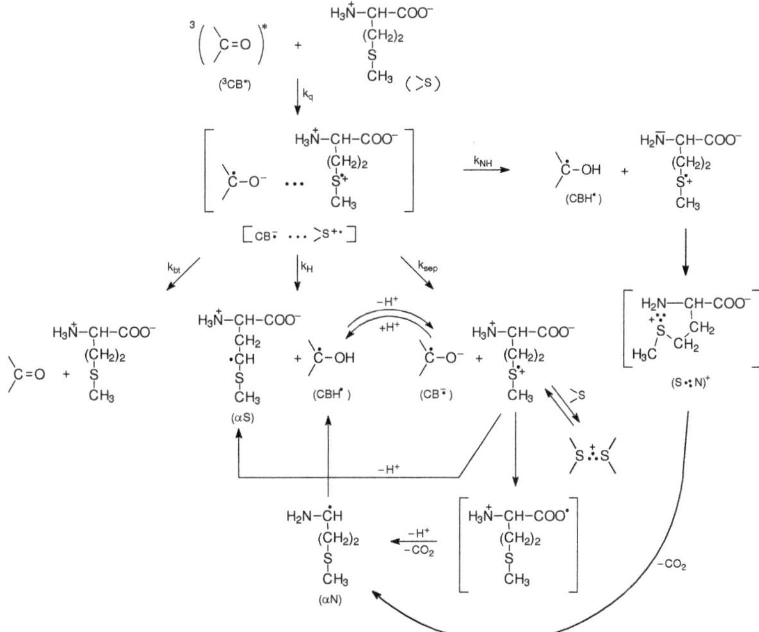

Scheme 6. The mechanism of CB-sensitized photooxidation of methionine in neutral aqueous solutions (pH~7).

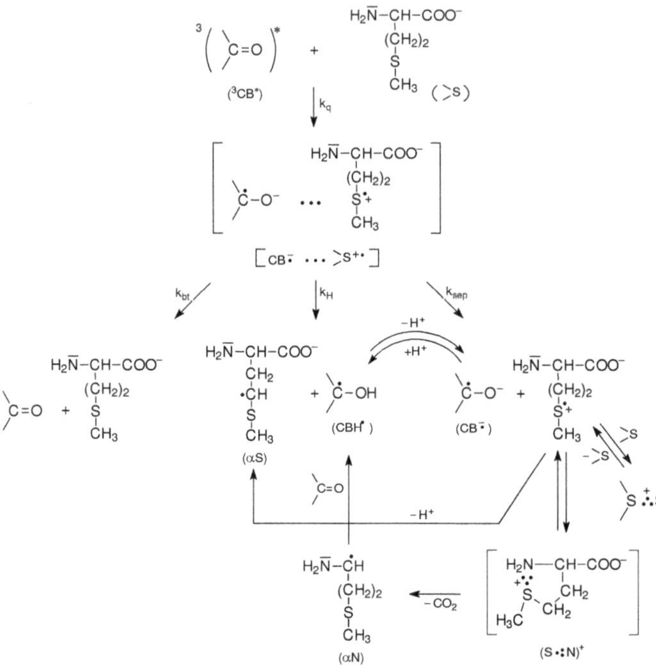

Scheme 7. The mechanism of CB-sensitized photooxidation of methionine in basic aqueous solutions (pH ~ 11).

As presented in Scheme 6 (for methionine with the protonated amino group, pH 7), the electron transfer quenching of the CB triplet led to formation of a CT complex [$CB^{•-}$ ... >$S^{•+}$] that can decay in four main primary reactions: (1) charge separation (k_{sep}) yielding $CB^{•-}$ and >$S^{•+}$ radical ions, (2) proton transfer within the CT complex (k_H) yielding a ketyl radical $CBH^•$ and an α-(alkylthio)alkyl radical (αS), (3) back electron transfer (k_{bt}) leading to regeneration of the reactants in their ground states, and (4) proton transfer from the protonated amino group (-NH_3^+) to the $CB^{•-}$ radical anion within the CT complex (k_{NH}) leading to the ketyl radical $CBH^•$ and sulfur-centered radical cation >$S^{•+}$ with an -NH_2 group (unprotonated) yielding a five-membered cyclic two-center three electron-bonded $(S∴N)^+$ radical cation that can easy decarboxylate to form α-aminoalkyl radicals (αN). These short-lived intermediates can undergo further reactions. The >$S^{•+}$ radical cations can be (i) stabilized by formation of three electron bonds with N, or S atoms (intramolecular $(S∴N)^+$ and/or intermolecular $(S∴S)^+$ in the reaction of >$S^{•+}$ with >S reactant-methionine), (ii) deprotonated to form α-(alkylthio)alkyl radicals (αS), and (iii) decarboxylated to form α-aminoalkyl radicals (αN) via pseudo-Kolbe mechanism [57,58]. The latter radicals (αN), which in general can be formed in two reaction channels, can be involved in further reactions with the CB ground state leading to the formation of ketyl radicals and an imine [15,18,64]. Evidence for this hypothesis was a secondary growth of $CBH^•$ in neutral or $CB^{•-}$ in basic solutions that was observed after the CB triplets had decayed and that was also dependent on the concentration of CB. These observations indicated an additional reaction mechanism for CB reduction. This interpretation was supported by complementary pulse radiolysis experiments, where αN radicals were generated and their reaction with the CB in the ground state was studied [15].

The mechanism of photooxidation of methionine in basic aqueous solutions (methionine with a deprotonated amino group) presented in Scheme 7 is similar to the mechanism for neutral solutions (Scheme 6). However, as one would expect, the k_{NH} reaction is not present and the decarboxylation is due to a single reaction channel occurring via a very short-lived $(S\therefore N)^+$ radical cations (several ns) (Scheme 7) [67].

The primary intermediates presented in Schemes 6 and 7 were observed directly using time-resolved techniques (that included the kinetics of their formations and decays), e.g., nanosecond laser flash photolysis [15,18,67]. The transient absorption spectra obtained in these experiments were resolved at various time delays into the component spectra using a spectral-resolution procedure. This procedure, together with the reference spectra of the expected transients, were described in detail in reference [68]. As a final result of this analysis, the concentrations of all of the transients were determined at various time delays following the laser pulse and were presented as concentration profiles of the transients. Thus, the details of the primary and secondary reactions of CB-sensitized photooxidation of methionine and its derivatives (including formation and decays of intermediates) could be followed by laser flash photolysis techniques. Unfortunately, due to the ground state absorption of the CB sensitizer, the spectral window available for nanosecond laser flash photolysis studies was limited to $\lambda > 360$ nm. As a consequence, some intermediates, including αS–α-(alkylthio)alkyl radical with its maximum absorption at $\lambda = 290$ nm, could not be observed directly.

Using a relative actinometry method with a CB solution as an external actinometer (the absorbances at the excitation wavelength of CB both in the reaction cell and in the actinometer cell, were identical), the initial quantum yields for formation of the intermediates were determined (for experimental details see [68]). The reaction of α-aminoalkyl radicals (αN) with the CB in the ground state led to the slow secondary formation of ketyl radicals ($CBH^{\bullet}$) or ketyl radical anions ($CB^{\bullet-}$) depending on pH and its efficiency can be described as $\Phi''CBH^{\bullet}$ or $\Phi''CB^{\bullet-}$. The quantum yields measured for aqueous solutions at pH ~ 7 and pH ~ 11 for primary and secondary intermediates are presented in Table 3.

Table 3. Quantum yields of primary (Φ) and secondary (Φ″) intermediates and CO_2 in the CB-sensitized photooxidation of Met.

	pH	$\Phi_{CBH^{\cdot}}$	$\Phi_{CB^{\cdot-}}$	$\Phi_{CBH^{\cdot}}$ + $\Phi_{CB^{\cdot-}}$	$\Phi_{(S\therefore N)^+}$	$\Phi_{(S\therefore S)^+}$	$\Phi''_{CBH^{\cdot}}$	Φ_{CO2-}	Lit
Methionine	7	0.11	0.27	0.38	0	0.23	0.25	0.28	[18]
Methionine	11	0.05	0.66	0.71	0.38 [(a)]	0.10	0.66 [(b)]	0.55	[18,67]

error ±20%; [(a)] short-lived transient $\Phi_{(S\therefore N)^+}$, taken at ~20 ns after the laser pulse; [(b)] Φ''_{CB^-}.

Time-resolved and steady-state studies for the CB-sensitized photooxidation of Met (Table 3) led to the following conclusions regarding the primary photochemical reactions of methionine at neutral pH (see Scheme 6): (i) the charge separation reaction is an efficient process yielding the ketyl radical anion ($CB^{\bullet-}$) and the sulfur-centered radical cation $>S^{\bullet+}$ with quantum yield $\Phi_{CB\bullet-} = 0.27$, (ii) $>S^{\bullet+}$ radical cations can react (reversible reaction) with the excess of methionine in its ground state leading to the intermolecular $(S\therefore S)^+$ dimeric sulfur radical cations with $\Phi_{(S\therefore S)+} = 0.23$, approximately equal to the value of the quantum yield of charge separation ($\Phi_{CB\bullet-}$), (iii) k_{NH} proton transfer within the CT complex led to ketyl radicals $CBH^{\bullet}$ and CO_2 as final products with quantum yields ≤ 0.11 ($\Phi_{CBH\bullet}$), (iv) decarboxylation reaction $>S^{\bullet+}$ radicals via pseudo-Kolbe reaction [57,58] cannot be neglected.

In the case of methionine in basic solutions (Scheme 7), the main primary reaction is the charge separation reaction (with $\Phi_{CB\bullet-} = 0.66$ twice that for neutral pH) and the k_H primary reaction is more than ten times less efficient than k_{sep} reaction. Moreover, the intermolecular $(S\therefore S)^+$ intermediate is present along with an efficient decarboxylation ($\Phi_{CO2} = 0.55$) occurring via a very short-lived intramolecular $(S\therefore N)^+$ radical cation (ns time scale).

As shown in Table 3, the quantum yields for secondary reactions at pH 7 and pH 11 ($\Phi''_{CBH\bullet}$ or $\Phi''_{CB\bullet-}$) are equal within the experimental errors with quantum yields of decarboxylation ($\Phi_{CO2} = 0.28$ or $\Phi_{CO2} = 0.55$, respectively). However, the mechanism for the slow formation of the ketyl radical anion $CB^{\bullet-}$ in the secondary reactions was shown to be very complex (e.g., $\Phi_{CB\bullet-}$ was pH dependent and decrease for pH above 11 and at pH 13.2 was approaching zero). This was explained by the involvement of additional acid-base equilibria for sulfur- and nitrogen-centered radical cations, see reference [67].

In addition to the time-resolved kinetic studies using nanosecond laser flash photolysis, the mechanistic conclusions, which are presented in Schemes 6 and 7 for methionine, time-resolved CIDNP, and ESR techniques, have expanded our mechanistic understanding of radical reaction pathways of the Met side chain in various local environments containing differing functional groups. The CIDNP study performed by Goez et al. [23,69] for CB-sensitized photooxidation of methionine confirmed the presence of an electron transfer mechanism involving the sulfur atom (not the nitrogen atom) as the electron donor. The formation of a two-center three electron bonded radical cation $(S\therefore N)^+$ in alkaline solutions (pH = 12) and the presence of the monomeric sulfur-centered radical cation $>S^{\bullet+}$ in neutral solution (pH = 6) were also proved.

A time-resolved ESR spectroscopy study by Yashiro et al. [26] for anthraquinone-sulfonate sensitized photooxidation of methionine in aqueous solution led to the following conclusions: at low pH the intermolecular dimeric sulfur radical cation $(S\therefore S)^+$ was the main intermediate but at high pH, deprotonation of a monomeric sulfur-centered radical cation $>S^{\bullet+}$ to form the aminyl radical (CH_3-S-CH_2-CH_2-CH-(COO-)-$N^{\bullet}H$) was suggested as the main reaction channel. However, the α-(alkylthio)alkyl radicals (αS) were not detected in their ESR experiments.

Some differences in the description of the mechanisms and main intermediates were observed for Met at high pH i the results of ESR by Yashiro et al. [26] and the nanosecond laser flash photolysis studies could be due to differences (i) in the sensitizers used in these two types of experiments (anthraquinone sulfonate (AQS) vs. 4-carboxybenzophenone (CB)) and (ii) in the experimental conditions (e.g., laser pulse energies 90 mJ in comparison with 1–6 mJ, respectively). This may lead to differences in the decay channels of the respective CT complexes [$Sens^{\bullet-}$ … $>S^{\bullet+}$] with AQS and CB as sensitizers. Moreover, the spectral window available in the nanosecond laser flash photolysis did not allow detection of aminyl radical (CH_3-S-CH_2-CH_2-CH-(COO-)-$N^{\bullet}H$) observed in the EPR experiments. Furthermore, as was presented by Bonifačić et al. [70] that type of radicals can undergo further transformation via 1,2-H shift to α-C-centered radicals. On the other hand, identification of radical-radical coupling products such as αS-αS in the steady-state photooxidation of Met derivatives sensitized by CB [22,59] confirmed the presence of αS radicals in the reaction mechanisms.

The mechanism for the photooxidation of methionine was also studied using sensitizers different from CB, namely N-(Methylpurin-6-yl)pyridinium cation (Pyr^+) in aqueous solution at neutral pH [71]. It was shown by nanosecond laser flash photolysis and steady-state photolysis studies that the mechanism of photooxidation is, in general, similar to that for CB-sensitized photooxidation of methionine at pH 7 [15]. The quenching of Pyr^+ excited triplet state led to the decay of the CT complex [$Pyr^{\bullet}$ … $>S^{\bullet+}$] in two primary reactions: (i) charge separation to a $Pyr^{\bullet}$ radical and an $>S^{\bullet+}$ radical cation and (ii) back electron transfer to regenerate reactants in their ground states. The reactions of the resulting sulfur-centered radical cation $>S^{\bullet+}$ are similar to those seen in CB-sensitization (see Scheme 6). These radical cations can undergo decarboxylation yielding an α-aminoalkyl radical (αN) that can reduce Pyr^+ cations to additional $Pyr^{\bullet}$ radicals (slow secondary growth of $Pyr^{\bullet}$) and formation of intermolecular $(S\therefore S)^+$ dimeric sulfur radical cations in reaction to methionine in its ground state. The time-resolved and steady-state studies led to a detailed quantitative description of the primary and secondary reactions for methionine and other sulfur-containing amino acids [71].

2.2.2. Methionine Derivatives

In this subsection, the methionine derivatives, the methyl ester of N-acetyl-methionine (N-Ac-Met-OCH_3) and the N-methylated amide of N-acetyl-methionine (N-Ac-Met-$NHCH_3$)

were selected as simple models of a Met residue incorporated into the interior of oligopeptides and proteins (Scheme 8).

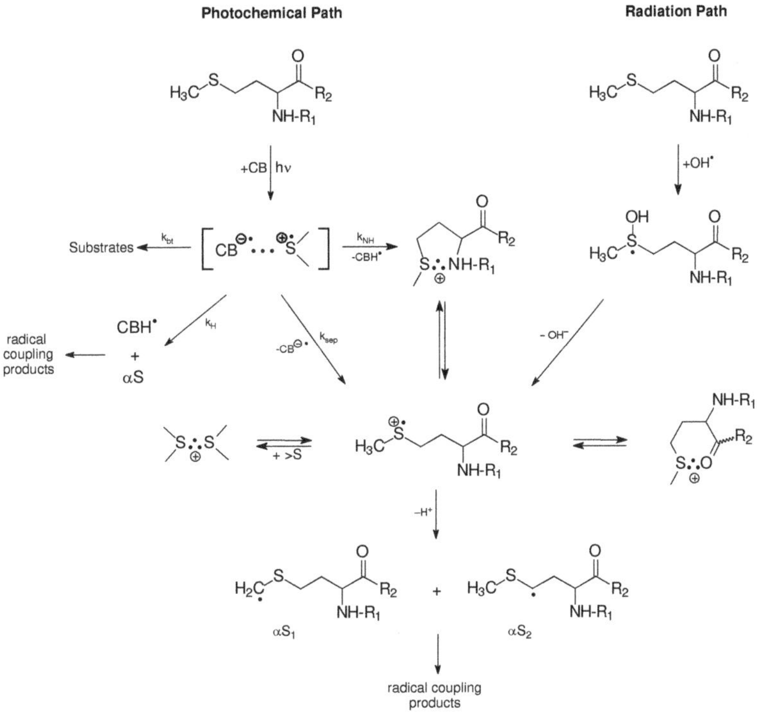

Scheme 8. The mechanism of CB-sensitized photo- and radiation-induced oxidation of methionine derivatives in aqueous solutions (R_1 denotes N-Ac- group or H- and R_2 denotes -OCH_3, -NH_2 and $NHCH_3$ groups).

The purpose for acetylating the N-terminal amino group was to eliminate the fast intramolecular proton transfer from the amino group to the >S∴OH moiety (see Section 2.1.2). A similar purpose for acetylating the N-terminal amino group applies to the CB-sensitized photooxidation of Met. In this case, fast intramolecular proton transfer from the N-terminal amino group of Met to the ketyl radical anion ($CB^{\bullet -}$) within the [$CB^{\bullet -}$ ... >$S^{\bullet +}$] complex [19], (see k_{NH} reaction channel, in Scheme 6) is eliminated. On the other hand, esterification of the C-terminal carboxyl group or amidation by the N-methylated amide group (-$NHCH_3$) eliminates decarboxylation of Met residue via the pseudo-Kolbe mechanism [57,58]. Therefore, the use of N-Ac-Met-OCH_3 or N-Ac-Met-$NHCH_3$ has the following benefit: it allows one to mimic and prove the mechanisms for primary and secondary radical reactions following oxidation of the Met residue with no contribution of N- and C-terminal functional groups.

The results of intensive studies for photo- and radiation-induced oxidation of methionine derivatives (with blocked amino and carboxylic groups) by the complementary laser flash photolysis and pulse radiolysis techniques are summarized in Scheme 8, showing primary photo- and radiation-induced processes.

As presented in Scheme 8, both the photochemical and radiation pathways led to the formation of one common intermediate, namely the sulfur-centered radical cation (>$S^{\bullet +}$). In the case of photochemical pathway the formation of (>$S^{\bullet +}$) is occurring via one of the four competing primary reactions of CT complex, namely the charge separation reaction of [$CB^{\bullet -}$ ... >$S^{\bullet +}$]. For the radiation pathway, a main channel leading to (>$S^{\bullet +}$) is the

HO $^{-}$ release from the MetS∴OH adduct. These pathways exemplify neighboring group participation in the reactions of two intermediates, the sulfur radical cations ($>S^{\bullet+}$) and MetS∴OH during Met oxidation (see Scheme 8) [37]. As was mentioned in the previous sections, two-centered, three electron (2c-3e) bonds between the oxidized sulfur atom and the lone electron pairs located on the nitrogen atom in the N-terminal amino group and the oxygen atoms in the C-terminal carboxyl group are responsible for stabilization of $>S^{\bullet+}$. Such stabilization leads to five-membered intramolecularly S∴N-bonded species and/or six-membered intramolecularly S∴O-bonded species [8,9,16–19]. Subsequent studies showed that heteroatoms present in the peptide bond can be also involved in the formation of similar transient species with 2c-3e bonds with the oxidized sulfur atom [11–13,72–74].

The secondary reactions of intermediates shown in Scheme 8 led to formation of various stable products. According to our knowledge, there were only two attempts to identify stable products after $HO^{\bullet}$ or CB-induced oxidation of methionine derivatives [22,59]. They addressed the full mechanism of Met oxidation from the initial step ($h\nu$ excitation or radiolysis) via identification of short-lived intermediates (by time-resolved techniques) to the analysis of stable products (in steady-state UV irradiation or γ-radiolysis). Knowledge of the structure of stable products was a strong argument confirming the assignments of intermediates detected in the time-resolved experiments.

Quantum yields of primary intermediates in the CB-sensitized photooxidation of Met derivatives are summarized in Table 4. In the case of methionine derivatives with a blocked carboxylic group ($MetOCH_3$ and $MetNH_2$) $(S∴N)^+$ was the main primary intermediate regardless of pH values. At low pH, where amino group is protonated (see Figure 1), the presence of the $(S∴N)^+$ radical cation can be rationalized via the k_{NH} reaction (Scheme 8). For N-Ac-Met: (i) the charge separation reaction (k_{sep}) is the main reaction of the CT complex [$CB^{\cdot-}$ ... $>S^{\cdot+}$] decay, (ii) the decarboxylation reaction can be neglected and (iii) the intermolecular dimeric sulfur radical cation $(S∴S)^+$ is the main intermediate. For methionine derivatives with blocked amino and carboxylic groups (N-Ac-Met-OCH_3, N-Ac-Met-$NHCH_3$ and N-Ac-Met-NH_2), the proton transfer within the CT complex (k_H) yielding ketyl radical $CBH^{\cdot}$ and α-(alkylthio)alkyl radical (αS) was found to be the dominate primary reaction over the charge separation reaction (k_{sep}) that yielded only a small amount of $CB^{\cdot-}$ and $(S∴S)^+$ dimer. The presence of these intermediates was confirmed by the analysis of stable products formed during the steady-state photolysis. The main stable products were found to be the combination products of ketyl radicals with α-(alkylthio)alkyl radical (αS-CBH stable products) [22,59].

Table 4. Quantum yields of primary intermediates and CO_2 in the CB-sensitized photooxidation of Met derivatives.

Met Derivatives	pH	$\Phi_{CBH^{\cdot}}$	$\Phi_{CB^{\cdot-}}$	$\Phi_{CBH^{\cdot}}$ + $\Phi_{CB^{\cdot-}}$	$\Phi_{(S∴N)^+}$	$\Phi_{(S∴S)^+}$	Φ_{CO2-}	Lit
Met-OCH_3	6	0.39	0	0.39	0.20	0.04	0	[19,75]
Met-OCH_3	10	0	0.43	0.43	0.28	0.04	0	[19,75]
Met-NH_2	5	0.39	0	0.39	0.28	0.10	0	[19,75]
Met-NH_2	10	0	0.48	0.48	0.12	0.07	0	[19]
N-Ac-Met	7	0.12	0.29	0.41	0	0.28	0.05	[18,19]
N-Ac-Met	10	0.13	0.29	0.42	0	0.27	–	[19,75]
N-Ac-Met-OCH_3	6	0.22	0.04	0.26	<0.05	0.07	0	[19]
N-Ac-Met-NH_2	6	0.19	0.07	0.26	0	0.09	0	[19]
N-Ac-Met-$NHCH_3$	7	0.32	<0.02	0.34	0	0	0	[22]

errors $\pm$ (10–20)%.

An additional argument for a mechanism of the proton transfer reaction within the CT complex (k_H) for methionine derivatives with substituted amino and carboxylic groups came from the photochemical study of benzophenone-methionine dyads (diketopiperazine-based BP-Met) in acetonitrile solutions [76]. Both the time-resolved laser flash photolysis experiments and the steady-state irradiations of sterically constrained BP-Met dyads identified an electron transfer process from the sulfur atom to the BP triplet followed by proton transfer to form biradical (with quantum yields in the range of 0.6). This biradical was a major precursor of a cyclic stable product (an analogous product to αS-CBH observed in the photolysis of N-Ac-MetOCH_3 and N-Ac-Met$NHCH_3$).

3. Methionine in Linear Peptides

3.1. Radiation-Induced Oxidation

3.1.1. Methionine as the N/C-Terminal Amino Acid Residue

The location of the Met residue in the linear peptide structure was found to affect the type of intermediates formed during $^{\bullet}OH$-induced oxidation. This was demonstrated for simple dipeptides containing Met with the reversed sequence of amino acids: Met-Gly/Gly-Met and Met-Leu/Leu-Met [8,77]. For Met-Gly and Met-Leu dipeptides the observed transient absorption bands in the pH range 5—6 with λ_{max} = 290 and 390 nm were respectively assigned to α-S1/α-S2 radicals derived from Met and Met(S∴N)$^+$ radical cations. Thus, the location of Met residue as the N-terminal amino acid does not significantly change the primary radical reactions related to the oxidation process (see Scheme 1), except for the fact that the decarboxylation process involving the intra-molecular Met(S∴N)$^+$ is inhibited, although the carboxyl group in C-terminal Gly or Leu residues is not protected. This is reflected in a much longer lifetime ($\tau_{1/2}$~200—400 μs) of these radical cations and lack of presence of CO_2 in the system containing Met-Gly [57]. Interestingly, in the presence of superoxide radical anion ($O_2^{\bullet-}$) this species decays much faster with the rate constant k(Met(S∴N)$^+$ + $O_2^{\bullet-}$) = 5.3 × 10^9 M^{-1} s^{-1} [78]. It is important to note that the reaction of $O_2^{\bullet-}$ with superoxide dismutase (SOD) with the respective k = 2.3 × 10^9 M^{-1} s^{-1} proceeds ca. 2.5-fold slower than its reaction with Met(S∴N)$^+$ [79]. From biological point of view it means that reaction of Met(S∴N)$^+$ with $O_2^{\bullet-}$ is a potential source for methionine sulfoxide formation when the system is exposed to high concentration of reactive oxygen species (ROS).

On the other hand, for Gly-Met and Leu-Met dipeptides, the observed transient absorption spectrum in the same pH range is dominated by the absorption band with λ_{max} = 290 nm and a clearly pronounced shoulder in the 360–400 nm range. These absorption bands were assigned to α-S1/α-S2 radicals derived from Met and Met(S∴O)$^+$ radical cation, respectively. Taking into account the character of intermediates and final products detected during $^{\bullet}OH$-induced oxidation of dipeptides containing serine, threonine and γ-glutamic acid as the N-terminal amino acids (see Schemes 9 and 10), it is reasonable to assume that fast intramolecular proton transfer from the free amino group (NH_3^+) of the dipeptide to the MetS∴OH moiety occurs also for dipeptides containing amino acids with alkyl side groups. The sulfur monomeric radical cation (MetS$^{\bullet+}$) resulting from this reaction can decay along three different pathways: (i) forming the very-short-lived multi-membered S∴N-bonded radical cation which can be the precursor of C-centered radicals (on the α-C atom of the Gly residue) formed via analogous consecutive reactions presented in Scheme 11, (ii) forming the Met(S∴O)$^+$ radical cation, (iii) undergoing decarboxylation via the pseudo-Kolbe mechanism to form substituted α-N radicals on the Met residue confirmed by detection of CO_2 [57].

The first convincing evidence for intramolecular proton transfer from the free amino group (NH_3^+) of the dipeptide to the MetS∴OH moiety and formation of the multi-membered S∴N-bonded radical cation was obtained from the studies of $^{\bullet}OH$-induced oxidation of Ser-Met and Thr-Met dipeptides [14]. The general reaction scheme involves an intramolecular proton transfer from the protonated N-terminal amino group to an initially formed MetS∴OH radical on the Met residue. The MetS∴OH undergoes subsequently elimination of water and formation of a Met(S∴N)$^+$ intermediate which was characterized by pulse radiolysis and has a lifetime ($\tau_{1/2)}$ ≈ 300 ns). This intermediate exists in an equilibrium with the open chain N-centered radical cation which further undergoes efficient heterolytic bond cleavage of the $C_\alpha-C_\beta$ bond of the Ser or Thr side chain, leading to an α-aminoalkyl radical of the structure H_2N-C•H-C(=O)-NH-peptide (identified by time-resolved ESR spectroscopy) and formaldehyde or acetaldehyde, respectively (Scheme 9).

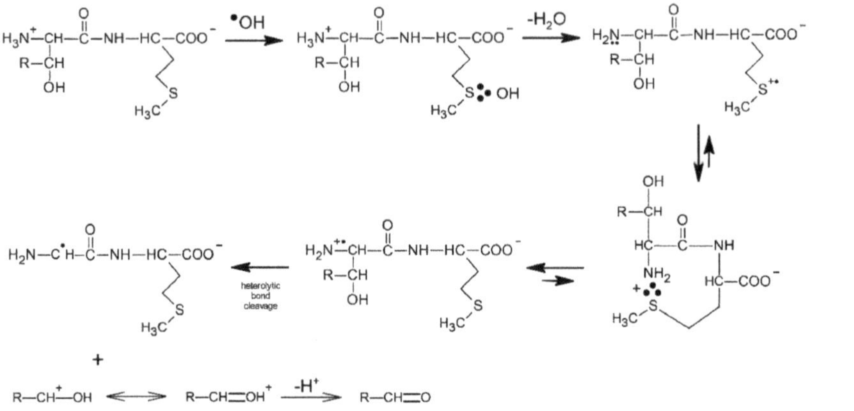

Scheme 9. General reaction scheme for •OH-induced oxidation of methionine in Ser-Met and Thr-Met dipeptides (R = H for serine, R = CH_3 for threonine).

The above mechanism illustrates an interesting feature which has to be considered in any studies on oxidation of peptides containing Met. Although, based on the respective rate constants, Met is the main target of •OH radicals attack, the major final product contains fragments of either Ser or Thr residues. Such transfer of damage constitutes an important mechanistic pathway frequently observed during oxidation of peptides and proteins.

Further evidence of proton transfer from the free amino group of the peptide to the MetS∴OH moiety and formation of the multi-membered S∴N-bonded radical cation was obtained from studies of •OH-induced oxidation of the γ-Glu-Met dipeptide [80]. The mechanism is in principle very similar to that presented for Ser-Met and Thr-Met except for the fact that the open chain N-centered radical cation via establishment of the mesomeric form (H_2N-CH-(COO•))-peptide and subsequent α-fragmentation leads to α-aminoalkyl type radicals and CO_2 (Scheme 10).

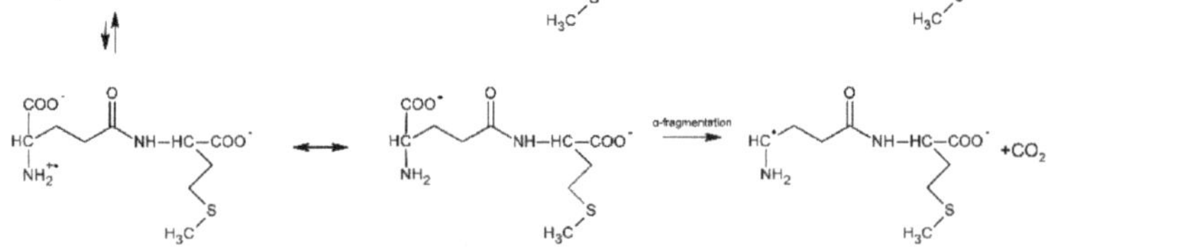

Scheme 10. General reaction scheme for •OH−induced oxidation of methionine in γ-Glu-Met dipeptide.

These radicals obtained from the N-terminal decarboxylation reaction can be easily probed via their reaction with p-nitroacetophenone (PNAP) leading to $PNAP^{\bullet -}$ radical anion [80–82]. The following fact should be emphasized at this point: the radiation chemical yield of CO_2 measured for this dipeptide was not equal to the radiation chemical yield of α-aminoalkyl type radicals [80]. This implies that decarboxylation must proceed via two different routes, namely via the $Met(S\therefore N)^+$ intermediate and via a pseudo-Kolbe reaction. This finding differs from X-Met peptides (X = Gly, Ala, Val, Leu) where decarboxylation was found to occur exclusively via the latter mechanism [57]. Insertion of a Gly residue between the γ-Glu and Met residues in γ-Glu-Gly-Met-Gly peptide increases the separation distance through the bonds as compared with γ-Glu-Met; however, it does not affect the radiation chemical yield of α-aminoalkyl type radicals G(α-N) [81]. This implies that the yields of multi-membered S∴N-bonded radical cations $G(S\therefore N)^+$ have to be similar, which at the same time means that the rate of end-to-end contact formation is similar in both peptides. In other words, the peptide backbone in γ- Glu-Gly-Met-Gly is flexible enough to allow direct interaction between amine and sulfur moieties. The γ-Glu-$(Pro)_n$-Met (n = 0–3) oligopeptides were chosen as oligopeptides with restricted conformational flexibility [82]. The observed continuous decrease of G(αN) with the number of Pro residues (from 0 to 3) indicates that formation of a contact between the S-atom in the C-terminal Met residue and the N-atom of a deprotonated N-terminal amino group of Glu is controlled by the relative diffusion of the $S^{\bullet +}$ and unoxidized N-atom. This study clearly shows that a remote amino group that served as both a proton and a free electron pair donor and that is separated from the Met residue by an oligoproline rigid backbone can still affect the chemistry on Met.

It also worth mentioning that neighboring amino acids (such as tyrosine or tryptophan) can affect the chemistry on Met. Taking into account their respective reduction potentials ($E^0(TrpN^\bullet/TrpN)$ = +1.015 V vs. NHE at pH 7) [83], $E^0(TyrO^\bullet, H^+/TyrOH$ = +0.93 V vs. NHE at pH 7) [84] and $E^0(MetS\therefore OH/Met,HO^-)$ = +1.43 V vs. NHE, $E^0(MetS^{\bullet +}/MetS)$ = 1.66 V vs. NHE [85], they can be easily oxidized by the monomeric sulfur radicals $MetS^{\bullet +}$ or MetS∴OH adducts. This phenomenon was nicely illustrated with the example of Met-enkephalin. Met-enkephalin (Met-enk), is a part of a mixture with Leu-Enk and is present in the central nervous system. Met-enk is a pentapeptide: Tyr-Gly-Gly-Phe-Met and is produced while an organism is under mental and/or physical stress [86]. The degradation induced by $^\bullet$OH radicals in Met-enk is also relevant to the disorders in inflammatory processes.

Nanosecond pulse radiolysis was used to elucidate the oxidation mechanisms of Met-enk by $^\bullet$OH radicals [87]. Based on the respective rate constants of Tyr, Phe, and Met residues with $^\bullet$OH radicals [88], formation of OH adducts on both aromatic (Tyr, Phe) and Met residues is reasonable and expected. Interestingly, fast formation of OH adducts on Tyr and Phe residues was confirmed by pulse radiolysis; however, no transient which can be assigned to MetS∴OH was observed. Moreover, the fast formation of the transient spectrum assigned to $TyrO^\bullet$ radicals was also observed. For a comparison, the absorption spectra of transients derived from Leu-enk (the C-terminal Met is replaced by Leu) were recorded and assigned only to the formation of OH adducts on Tyr and Phe residues. These differences in absorption spectra led to the hypothesis that $TyrO^\bullet$ radicals are formed through a new process not present in Leu-enk, i.e., an intramolecular electron transfer (IET) involving a MetS∴OH intermediate and a Tyr residue. The lower limit for the rate constant of IET was found to be $1.2 \times 10^7\ s^{-1}$. The high value for the rate constant for IET in Met-enk is consistent with the lack of α-(alkylthio)alkyl radicals and with the lack of CO_2 formation [57]. Depending on the pH, the former radicals might be potentially produced by the competitive deprotonation of either MetS∴OH or $MetS^{\bullet +}$. In turn, the formation of CO_2 would have resulted from the competitive oxidation of the carboxylate function in Met via a pseudo Kolbe mechanism. Similar oxidation studies of Met-enk were performed using $Br_2^{\bullet -}$ as a selective one-electron oxidant. They allowed for the determination of the first order rate constant of an IET involving MetS∴Br and Tyr residues ($k = 1.1 \times 10^5\ s^{-1}$) [89]. On the other hand, the rate constant for IET between MetS∴Br and Tyr residues in Tyr-$(Pro)_3$-Met (a peptide with the same number (3) of amino acid residues between Tyr and Met residues as in Met-enk) was found to decrease 10-fold and be equal to $1.1 \times 10^4 s^{-1}$ [90]. In the latter case an IET is likely partitioning along the peptide backbone and direct water mediated contacts between side chains of terminal amino acids residues (Tyr and Met). Therefore, in Met-enk, an IET occurs most probably rather through space (water) pathway with a possible involvement of H-bonds shortcuts. This conclusion is consistent with computational conformational analysis of this pentapeptide [91]. At the end of these considerations, it has to be stressed that the possibility for fast IET between an oxidized Met residue and the easily oxidized Tyr and Trp residues may lead to repairing of Met oxidative damage, along with the formation of damage in other sites of a peptide or protein molecule.

3.1.2. Methionine as the Internal Amino Acid Residue

The Gly-Met-Gly peptide is the simplest model peptide where the Met residue is not a terminal amino acid. Therefore, this peptide can serve as the model of Met residues incorporated in the interior of oligopeptides and proteins. The reaction of $^{\bullet}OH$ radicals with Gly-Met-Gly and its N-acetyl derivative were studied by pulse radiolysis [74]. The transient absorption spectra recorded at short times after the pulse in N_2O-saturated aqueous solutions at pH 5.5 were characterized by a strong UV absorption band with λ_{max} = 270 nm and a very weak and broad shoulder in the range of 400–500 nm. Spectral resolutions of these spectra were necessary in order to resolve them into contributions from various intermediates: the α-C-centered radicals on Met residues (α-C), the α-aminoalkyl radicals (α-N), the α-(alkylthio)alkyl radicals (α-S), the intramolecular $Met(S\therefore N)^+$ radical cations and the intermolecular $Met(S\therefore S)^+$ radical cations. Their radiation chemical yields and percentage contribution (in parenthesis) to the total yield of radicals formed in the reaction of OH is listed in Table 5.

Table 5. The radiation chemical yield (G) [(a)] of radicals and radical cations and their percentage contribution (in parenthesis) to the total yield of radicals formed in the reaction of $^{\bullet}OH$ radicals with Gly-Met-Gly [(b)] [74].

Time	α-N	α-S	α-C	$Met(S\therefore N)^+$	$Met(S\therefore S)^+$
1.4 μs	0.34 (57.1%)	0.14 (23.6%)	0.052 (8.6%)	0.035 (5.9%)	0.029 (4.8%)
3.5 μs	0.35 (58.7%)	0.13 (21.8%)	0.061 (10.2%)	0.029 (4.8%)	0.027 (4.5%)

[(a)] G-values are in $\mu mol\ J^{-1}$ units; [(b)] [Gly-Met-Gly] = 0.2 mM.

It is worthwhile to note that the contribution of $MetS\therefore OH$ is neglected for short times after the pulse, which means that its lifetime is very short. This observation can be rationalized by analogy with peptides containing C-terminal Met (see Section 3.1.1), in terms of a concerted process which involves a fast proton transfer from the N-terminal $-NH_3^+$ group to the $MetS\therefore OH$ moiety, which eliminates HO^- (in the form of water) occurs and leads by sequential steps to α-N radicals (Scheme 11). It has to be stressed at this point that α-N and α-S radicals are the most abundant radicals resulting from the $^{\bullet}OH$-induced oxidation of Gly-Met-Gly and constitute more than 80% of all of the radicals. Despite the fact that the carboxyl group in Gly residue is free, the pseudo Kolbe reaction does not operate in this case which was confirmed by the lack of CO_2 [57]. Acetylation of the N-terminal amino group in Gly-Met-Gly eliminates any fast proton transfer to the $MetS\therefore OH$ moiety which is reflected in the longer lifetime of $MetS\therefore OH$ radicals, and as a consequence, their substantial contribution at short times after the electron pulse (Table 6).

Table 6. The radiation chemical yield (G) [(a)] of radicals and radical cations and their percentage contribution (in parenthesis) to the total yield of radicals formed in the reaction of $^{\bullet}$OH radicals with N-Ac -Gly-Met-Gly [(b)] [74].

Time	MetS.·.OH	α-S	α-C	Met(S.·.N)$^+$	Met(S.·.S)$^+$
1.4 μs	0.18 (30.6%)	0.17 (29.9%)	0.05 (8.1%)	0.10 (17.7%)	0.08 (13.6%)
3.5 μs	0.07 (11.8%)	0.29 (49.0%)	0.02 (3.8%)	0.11 (19.2%)	0.10 (16.3%)

[(a)] G-values are in μmol J^{-1} units; [(b)] [N-Ac-Gly- Met-Gly] = 0.2 mM.

Moreover, formation of the multi-membered S.·.N-bonded radical cation (precursors of α-N radicals) is not possible and the sequential reaction steps leading to α-N radicals are switched off (Scheme 11). This is confirmed by the fact that α-N radicals were not necessary for inclusion in the spectral resolutions.

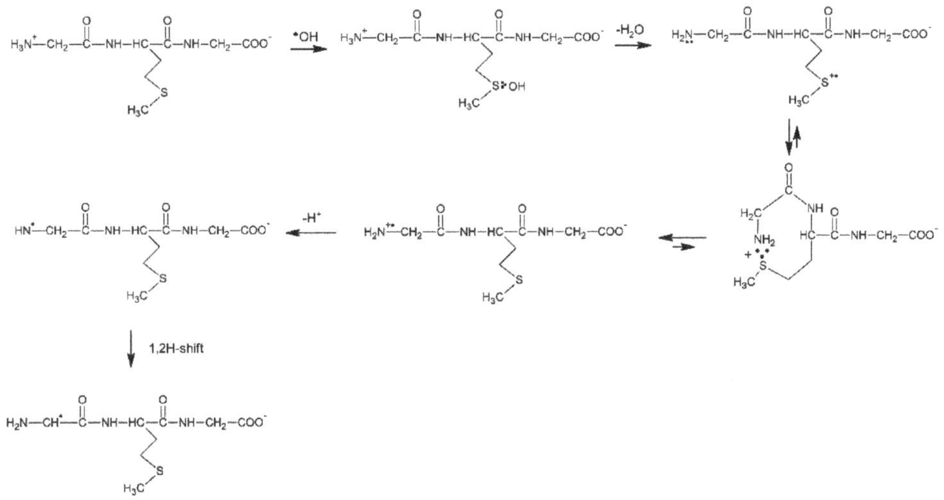

Scheme 11. General reaction scheme for $^{\bullet}$OH−induced oxidation of methionine in Gly-Met-Gly tripeptide.

The first experimental proof that Met can be oxidized by radicals with the reduction potentials lower than +1.4–1.6 V vs. NHE (which is equal to $E^0(MetS^{\bullet+}/Met)$) came from one-electron oxidation of β-amyloid peptide (β-AP1-40) using $N_3^{\bullet}$ radicals with $E^0(N_3^{\bullet}/N_3^-)$ = 1.33 V vs. NHE [92]. Thermodynamic considerations indicate that $N_3^{\bullet}$ should not oxidize Met residues unless the one-electron reduction potential of Met is lowered because of favorable environment. It was shown that Met^{35} is the target in β-AP1-40 oxidation. Molecular modeling studies showed that β-AP26-40 (a representative fragment of the native β-amyloid peptide (β-AP1-42) has a large probability of forming S−O bonds between Ile^{31} and Met^{35} due to the specific structural properties of the polypeptide fragment (an α-helical motif) (Figure 3) [93,94]. These results corroborate the reported role of Met^{35} in the toxicity and protein oxidation capacity of β-AP(1–40) [95–97].

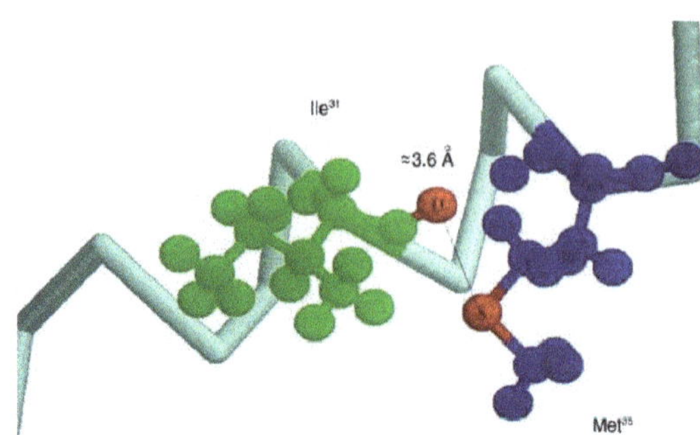

Figure 3. Schematic drawing of the α-helical portion of the C-terminal region that encompasses Met^{35} (reprinted with permission from [94]).

3.1.3. Linear Peptides with Two Methionine Residues

Contrary to cyclic dipeptides (see Section 4) extended conformers dominate in the linear (open chain) Met-Met dipeptides due to expressed tendency toward trans amides. Therefore, the close contacts between side chains of Met residues should be enhanced in D,L- and, L,D-stereoisomers in comparison to L,L- and D,D-stereoisomers (see Section 4.1.2). However, one should expect weaker steric constraints in linear peptides in comparison to cyclic dipeptides (see Section 4).

The reaction of $^{\bullet}OH$ radicals with four stereoisomers was investigated by pulse radiolysis with time-resolved UV-vis spectrophotometry in Ar or N_2O saturated aqueous solutions at pH 1 and pH 5.2, respectively [9]. The absorption bands observed at pH 1 for D,L-, L,D-, L,L-,and D,D-stereoisomers were characterized by a maximum with λ_{max} = 490 nm and similar molar absorption coefficients ε_{490} = 5900, 5550, 5500, and 5800 $M^{-1}cm^{-1}$, and were unequivocally assigned to intramolecular $Met(S\therefore S)^+$ radical cations. The similarity between the maximum positions and the molar absorption coefficients suggests similar overlap of p-orbitals of sulfur atoms and thus a similarity in the strength of the S∴S-bond in the radical cations derived from all of the optical isomers [98,99] and thus confirming weaker steric constrains in linear peptides than in cyclic dipeptides (see Section 4.1.2), whereas the influence of the stereoisomerism was manifested in the decay rates of $Met(S\therefore S)^+$ radical cations: first order decay rate constants were measured at pH 1: $6.2 \times 10^3\ s^{-1}$ (L,L-isomer), $8.3 \times 10^3\ s^{-1}$ (D,D-isomer), $3.9 \times 10^3\ s^{-1}$ (L,D-isomer), and $3.7 \times 10^3\ s^{-1}$ (D,L-isomer). These $Met(S\therefore S)^+$ radical cations are in an equilibrium with the respective monomeric sulfur radical cations ($MetS^{\bullet+}$) The observed differences can be understood in terms of different equilibrium constants (K_{eq}) due to different distances between the NH_3^+ group and the positively charged S-atom. At pH = 5 the transient absorption observed for L-Met-L-Met was characterized by a strong absorption band with λ_{max} = 390 nm which was assigned inter alia to $Met(S\therefore N)^+$ radical cation which are formed via dehydroxylation of the primary MetS∴OH radical at the N-terminal Met assisted by an intramolecular proton transfer from the NH_3^+ group. A weaker absorption band with λ_{max} = 290 nm was assigned to α-(alkylthio)alkyl radicals (α-S). Because of neutral character of $^{\bullet}OH$ radicals it is reasonable to assume that both sulfur atoms located in the N-terminal and C-terminal Met residues are attacked by $^{\bullet}OH$ radicals with the same efficiency. Dehydroxylation of MetS∴OH at the C-terminal Met can be also assisted by an intramolecular proton transfer from the NH_3^+ in the process analogous to Gly-Met-Gly tripeptide (see Section 3.1.2, Scheme 11). In this case, the monomeric sulfur cations ($MetS^{\bullet+}$) can form an intramolecular $Met(S\therefore O)^+$ radical cation, deprotonate to α-S radicals, and undergo a pseudo Kolbe reaction that leads to decarboxylation. The last reaction can be excluded based on the photochemical experiments where L-Met-L-Met was oxidized by the triplet of 4-carboxybenzophenone (see Section 3.2.2) and no CO_2 was detected.

Similarly, as for the γ-Glu-(Pro)$_n$-Met (n = 0–3) oligopeptides (see Section 3.1.1), the Met-(Pro)$_n$-Met (n = 0–4) oligopeptides were chosen as oligopeptides with restricted conformational flexibility. The radiation chemical yields of intramolecular Met(S∴S)$^+$ radical cations depend on the number of Pro-residues; however, they do not depend in a simple way on the average distance between the sulfur atoms in the Met residues. The analysis showed that formation of a contact between terminal Met residues in the peptides with 0–2 Pro residues is controlled by the activated formation of Met(S∴S)$^+$ whereas in the peptides with 3–4 Pro residues, by the relative diffusion of the MetS$^{\bullet+}$ and the unoxidized sulfur atom. A decrease in the yields of Met(S∴S)$^+$ species with an increase in the number of Pro residues occurs at the expense of an increase in the yields of intramolecular Met(S∴O)$^+$ radical cations, and α-(alkylthio)alkyl radicals (α-S) [100]. These findings were similar to those obtained for the photo-induced oxidation of Met-(Pro)$_n$-L-Met peptides by the triplet state of CB (see Section 3.2.2, Scheme 12).

3.2. Photo-Induced Oxidation

3.2.1. Methionine as the N/C-Terminal and the Internal Amino Acid Residue

The quenching of CB triplets by methionine-containing peptides occurs with rate constants of the order of 2–3 $\times$ 10^9 $M^{-1}s^{-1}$ (Table 2). Small differences in the quenching rate constants (indicating the existence of multiple sulfur targets in the quenching process) were interpreted and discussed in reference [16]. The formation of the electron transfer products such as CB$^{\cdot-}$ radical anion, CBH$^{\cdot}$ ketyl radical and (S∴N)$^+$ and (S∴S)$^+$ transients observed in the nanosecond laser flash photolysis indicated that the quenching is electron transfer in nature. As in the case of methionine derivatives, the mechanism of the primary steps in the photooxidation can be described by three primary reactions of the CT complex [CB$^{\cdot-}$ ... >S$^{\cdot+}$]: (i) charge separation (k_{sep}) yielding CB$^{\cdot-}$ and >S$^{\cdot+}$ radical ions, (ii) proton transfer within the CT complex (k_H) yielding a ketyl radical CBH$^{\cdot}$ and an α-(alkylthio)alkyl radical (αS), and (iii) back electron transfer (k_{bt}) (Scheme 8). In the case of peptides with an N-terminal methionine residue at low pH, an additional, fourth primary reaction (namely proton transfer from the protonated amino group(-NH_3^+) to the CB$^{\cdot-}$ radical anion within the CT complex (k_{NH} reaction–see Scheme 8) was present. The quantum yields of primary intermediates in the CB-sensitized photooxidation of Met peptides are summarized in Table 7.

An analysis of these results led to conclusions similar to those for methionine derivatives. For peptides with the N-terminal methionine residue (Met-Gly, Met-Gly-Gly, Met-Lys, Met-Met) at low pH, the k_{NH} reaction was responsible for the large values of the quantum yields of CBH$^\bullet$ and the (S∴N)$^+$ intermediate. At high pH, the charge separation reaction (k_{sep}) was the main primary reaction of the CT complex [CB$^{\bullet-}$ ... >S$^{\bullet+}$] and large values of $\Phi_{CB\bullet-}$ and $\Phi_{(S∴N)+}$ were observed. Secondary reactions of the (S∴N)$^+$ radical cation in alkaline solutions for Met-Gly peptide (pH 9–11) studied by laser flash photolysis were described in detail in reference [17]. In the case of peptides with C-terminal methionine regardless of pH, the intermolecular dimeric sulfur radical cations (S∴S)$^+$ were the main intermediates. The decarboxylation reaction was not observed (or could be neglected in comparison to the other efficient reactions) for Met-containing peptides. This indicates that for peptides with a C-terminal methionine the main irreversible reaction of the sulfur-centered radical cation >S$^{\bullet+}$ was deprotonation leading to the α-(alkylthio)alkyl radical (αS) but a competing decarboxylation reaction could be neglected. This interpretation was additionally confirmed in the steady-state irradiation experiments, where radical coupling stable products CBH-αS were detected [20].

Table 7. Quantum yields of primary intermediates and CO_2 in the CB-sensitized photooxidation of Met-containing linear peptides.

Met-Containing Peptides [(a)]	pH	$\Phi_{CBH^{\cdot}}$ [(b)]	$\Phi_{CB^{\cdot -}}$ [(b)]	$\Phi_{CBH^{\cdot}}$ + $\Phi_{CB^{\cdot -}}$ [(b)]	$\Phi_{(S \therefore N)^+}$ [(b)]	$\Phi_{(S \therefore S)^+}$ [(b)]	$\Phi_{CO_2^-}$ [(c)]	Lit
Met-Gly	6	0.25	0.13	0.38	0.18	~0.04	0.01	[16,18]
Met-Gly	11	0.06	0.55	0.61	0.51	~0.04	0.01	[16,18,75]
Gly-Met	6	0.15	0.22	0.37	0	0.22	0.02	[18]
Gly Met	11	0.10	0.32	0.42	0	0.24	0.02	[18]
Met-Gly-Gly	6	~0.29	~0.06	~0.35	0.14	~0.08	0.01	[16]
Met-Gly-Gly	11	0.09	0.49	0.58	0.30	<0.03	–	[16]
Gly-Gly-Met	6	0.19	0.25	0.44	0	0.24	0.02	[16]
Gly-Gly-Met	11	0.13	0.30	0.43	0	0.29	–	[16]
Gly-Met-Gly	6	0.24	0.05	0.29	0	0.08	0.02	[16]
Gly-Met-Gly	11	0.10	0.21	0.31	0	0.07	–	[16]
Met-Lys	6	0.55	~0.03	0.58	0.33	0	0	[20]
Lys-Met	6	0.63	0	0.63	0	0.19	0	[20]
Met-Enkephalin	7	–	–	0.82	0	0	0	[61]

[(a)] [CB] = 2–4 mM, [Met peptides] = 5–20 mM; [(b)] error ± 20%; [(c)] error ± 50%.

In the case of methionine as the internal amino acid residue (Gly-Met-Gly), the results from the CB-sensitized photooxidation were discussed based on the general mechanism presented in Scheme 8. Four primary reactions of CT complex [$CB^{\cdot -}$ ... $>S^{\cdot +}$] could occur:

(i) charge separation (k_{sep}) yielding $CB^{\cdot -}$ and $>S^{\cdot +}$ radical ions, (ii) proton transfer within the CT complex (k_H) yielding a ketyl radical $CBH^{\cdot}$ and an α-(alkylthio)alkyl radical (αS), (iii) back electron transfer (k_{bt}), and (iv) proton transfer from the protonated amino group(-NH_3^+) of Gly to the $CB^{\cdot -}$ radical anion within the CT complex (k_{NH}). At low pH the k_{NH} and k_H reactions were the main reaction channels ($\Phi_{CBH\cdot}$ = 0.24) in comparison with the less efficient reaction of charge separation ($\Phi_{CB\bullet -}$ = 0.05) (Table 5). It is noteworthy that in this case, the k_{NH} reaction does not lead to the $(S \therefore N)^+$ transient as for an N-terminal methionine residue in peptides. (For expected products from a decay of $>S^{\cdot +}$ radical ions containing unprotonated amino group in the Gly residue see Scheme 11). In basic solutions, the charge separation reaction (k_{sep}) was the main reaction channel yielding $CB^{\cdot -}$ and $>S^{\cdot +}$ radical ions ($\Phi_{CB\bullet -}$ = 0.21). $(S \therefore N)^+$ cyclic radical cations involving a nitrogen atom from a peptide bond was not observed.

The mechanism of CB-sensitized oxidation of Met-Gly and Gly-Met peptides was also studied by the time-resolved CIDNP technique by Morozova et al. [24] at various pH values. It was shown that at low pH for both peptides a sulfur-centered radical cation $>S^{\bullet +}$ was formed. For Met-Gly, the $>S^{\bullet +}$ radical cation was converted (after deprotonation) to the (S∴N)+ radical cation with five-membered cyclic structure. In basic solutions the main transients for Met-Gly were $(S \therefore N)^+$ radical cations and aminyl type radicals; however, for Gly-Met, the $>S^{\bullet +}$ radical cations and aminyl type radicals were the main transients.

The mechanism proposed on the basis of the CIDNP experiments correlates with that obtained on the basis of laser flash photolysis experiments with an exception of aminyl radicals. (These radicals were not identified in the laser flash photolysis experiments, see Section 2.2.1).

3.2.2. Linear Peptides with Two Methionine Residues

In the case of Met-Met peptides (Table 8), as was presented in reference [63], the CB-sensitized photooxidation led to formation of intramolecular two-centered three-electron bonded $(S \therefore S)^+$ radical cations that can compete with the formation of intramolecular

$(S\therefore N)^+$ radical cations at low pH via the k_{NH} reaction (see Scheme 6) and at high pH via cyclization of $>S^{\bullet+}$ radical cations.

Table 8. Quantum yields of primary intermediates and CO_2 in the CB-sensitized photooxidation of peptides containing two methionine residues.

Met-Containing Peptides [(a)]	pH	$\Phi_{CBH^{\cdot}}$ [(b)]	$\Phi_{CB^{\cdot-}}$ [(b)]	$\Phi_{CBH^{\cdot}}$ + $\Phi_{CB^{\cdot-}}$ [(b)]	$\Phi_{(S\therefore N)^+}$ [(b)]	$\Phi_{(S\therefore S)^+}$ [(b)]	Φ_{CO2^-} [(c)]	Lit
L-Met-L-Met	6	0.23	0.15	0.38	0.19	0.07 [(d)]	<0.01	[63]
L-Met-L-Met	10	0.13	0.41	0.54	0.43	0.01 [(d)]	<0.01	[63]
D-Met-D-Met	6	0.24	0.18	0.42	0.27	0.06 9 [(d)]	-	[63]
D-Met-D-Met	10	0.09	0.40	0.49	0.41	(~0.02 [d)]	-	[63]
L-Met-D-Met	6	0.17	0.23	0.40	0.18	0.15 [(d)]	<0.01	[63]
L-Met-D-Met	10	0.08	0.37	0.45	0.34	0.06 [(d)]	<0.01	[63]
D-Met-L-Met	6	0.15	0.25	0.40	0.15	0.16 [(d)]	<0.01	[63]
D-Met-L-Met	10	0.10	0.38	0.48	0.36	0.06 [(d)]	<0.01	[63]
c-(L-Met-L-Met)	7	0.17	0.19	0.36	0	0.14 [(d)]	-	[63]

[(a)] [CB] = 2 mM, [Met-peptides] = 0.5 mM; [(b)] errors ± 10–20%; [(c)] error ± 50%; [(d)] intramolecular $(S\therefore S)^+$ (adapted after [3]).

At high pH, the charge separation reaction (k_{sep}) yielding $CB^{\cdot-}$ and $>S^{\cdot+}$ radical ions was shown to be the main primary process involved in the decay of the $[CB^{\cdot-} \ldots >S^{\cdot+}]$ complex and the quantum yield of $CB^{\cdot-}$ was equal to the sum of quantum yields of the sulfur radical cations $(S\therefore N)^+$ and $(S\therefore S)^+$. At low pH, in addition to the charge separation reaction (k_{sep}) reaction the proton transfer rection from the protonated amino group(-NH_3^+) to the $CB^{\cdot-}$ radical anion within the CT complex (k_{NH} reaction–see Scheme 8) was present. Small differences in the values of quantum yields for $(S\therefore S)^+$ and $(S\therefore N)^+$ radical cations were found for mixed L,D or D,L stereoisomers of Met-Met and L,L or D-D stereoisomers. However, for mixed stereoisomers the quantum yields of $(S\therefore N)^+$ radical cations went down at the expense of higher values for $(S\therefore S)^+$ radical cations in both acidic and basic solutions. In addition, for mixed stereoisomers the quantum yields of $(S\therefore S)^+$ radical cations were more than twice higher of those for L,L and D,D stereoisomers. These observations were rationalized by the higher propensity of $((S\therefore S)^+$ formation for mixed stereoisomers.

As was mentioned previously (see Section 3.1.3), the L-Met-$(Pro)_n$-L-Met peptides with Met residues located on N- and C- termini were shown to be excellent models of oligopeptides for studying the intramolecular interaction between two sulfur atoms in oligopeptides since this series of Met-$(Pro)_n$-Met peptides provides the investigator with the possibility for controlling the S ... S distance (Scheme 12).

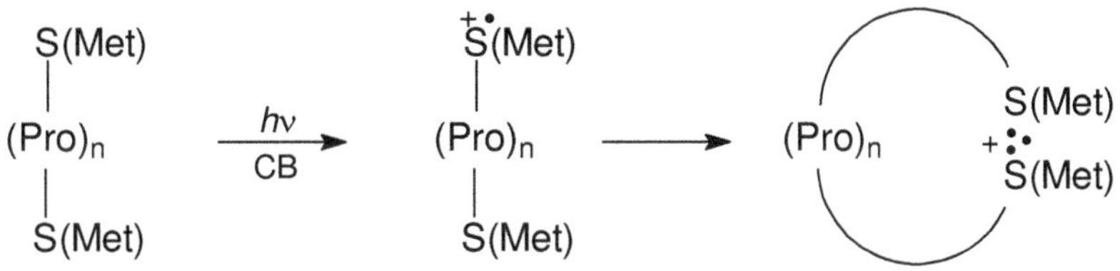

Scheme 12. Formation of intramolecular dimeric sulfur radical cations $(S\therefore S)^+$ in photo-induced oxidation of Met-$(Pro)_n$-Met oligopeptides.

This is important because amino acid chains can serve as relay stations for electron transfer processes. As shown in Scheme 8, in the CB-sensitized photooxidation of methionine containing compounds (photochemical path), the sulfur-centered monomeric radical cations ($>S^{\cdot+}$) were formed. The $>S^{\cdot+}$ radical cations (regardless of location on N or C termini) can be stabilized by formation of the intramolecular dimeric sulfur radical cations $(S\therefore S)^+$, and $(S\therefore O)^+$ or $(S\therefore N)^+$ intermediates or decay by deprotonation leading to α-(alkylthio)alkyl radicals (αS). The nanosecond laser flash photolysis experiments and the resolution of transient absorption spectra allowed the investigators to monitor kinetics and quantum yields of the intermediates at various time delays after the excitation laser pulse [68]. It was found that a decrease in the quantum yield (Φ) of the $(S{\therefore}S)^+$ intermediate with the number of proline residues occurred at the expense of quantum yields (Φ) of the other intermediates: $(S{\therefore}O)^+/(S{\therefore}N)^+$ and αS radicals. However, the dependence of $\Phi(S{\therefore}S)^+$ on the average distance between sulfur atoms was not linear (the largest change in $\Phi(S{\therefore}S)^+$ was observed when the number of proline residues was changed from two to three). These observations were reproduced by Langevin dynamics and statistical mechanical theory showing that for peptides with zero to two Pro residues contact between sulfur atoms was controlled by the activated formation of $(S{\therefore}S)^+$ but for the peptides with three and four proline residues was controlled by relative diffusion of the $>S^{\cdot+}$ radical cation and an unoxidized S atom [68]. These findings were similar to those obtained for the radiation-induced oxidation of Met-$(Pro)_n$-L-Met peptides by $^{\bullet}OH$ radicals (see Section 3.1.3).

4. Methionine in Cyclic Peptides

Non-extended conformers dominate in the cyclic structures of Met-Met dipeptides due to the expressed tendency toward cis amides. Therefore, the side chains of Met residues in L, L-configured cyclic dipeptides are on the same side of a diketopiperazine ring and, as a consequence, close contacts between them are enhanced (Figure 4a). On the other hand, the side chains of Met residues in L,D-configured cyclic dipeptides are on opposite sides of a diketopiperazine ring and their close contacts are difficult if not impossible, due to the strong steric constraints in the diketopiperazine ring (Figure 4b). Moreover, the presence of free terminal amino and carboxyl groups is eliminated and, thus, stabilization of monomeric sulfur radical cations ($MetS^{\bullet+}$) can come solely from interactions with lone electron pairs on the sulfur atoms in the side chains and/or with lone pairs on the heteroatoms (N- and O-atoms) associated with the peptide bonds.

Figure 4. Schematic drawing of structures of (**a**) the cyclic L-Met-L-Met dipeptide and (**b**) the cyclic L-Met-D-Met dipeptide.

4.1. Radiation-Induced Oxidation

4.1.1. Cyclic Dipeptides with Single Met Residue

Oxidation of c-(Gly-L-Met) dipeptide by $^{\bullet}OH$ radicals was mainly aimed at determining whether in the case of c-(D-Met-L-Met) intramolecular $Met(S{\therefore}S)^+$ radical cations or intermolecular dimeric radical cations $(MetS{\therefore}SMet)^+$ are formed (see Section 4.1.2) [13]. The oxidation pattern observed in c-(Gly-L-Met) dipeptide in which there is no chance of forming an intramolecular $Met(S{\therefore}S)^+$ radical cation is exactly the same as observed in c-(D-Met-L-Met). Similarities in the spectral and kinetic features observed in c-(Gly-L-Met) and c-(D-Met-L-Met) clearly indicate that the mechanisms of OH-induced oxidation of both peptides and the nature of the intermediates are the same for both dipeptides (see Section 4.1.2, Scheme 14).

4.1.2. Cyclic Dipeptides with Two Met Residues

For the above conformational reasons, cyclic Met-Met dipeptides are suitable models compounds to study reaction of MetS∴OH and $MetS^{\bullet+}$ in oligopeptides and proteins containing multiple and adjacent Met residues (e.g., calmodulin CaM-Ca_4 (see Section 5.1), human prion proteins hPrP [101], α-synuclein [102]). A small model cyclic dipeptide c-(L-Met-L-Met) (see Figure 3a) was oxidized by $^{\bullet}OH$ radicals generated by pulse radiolysis and the ensuing reactive intermediates were monitored by time-resolved UV-vis spectroscopy and conductometry [12]. Nearly similar radiation chemical yields of intramolecular Met(S∴N) radicals and intramolecular $Met(S∴S)^+$ radical cations are formed (see Table 9) in the competitive processes from the primary formed MetS∴OH adduct (Scheme 13).

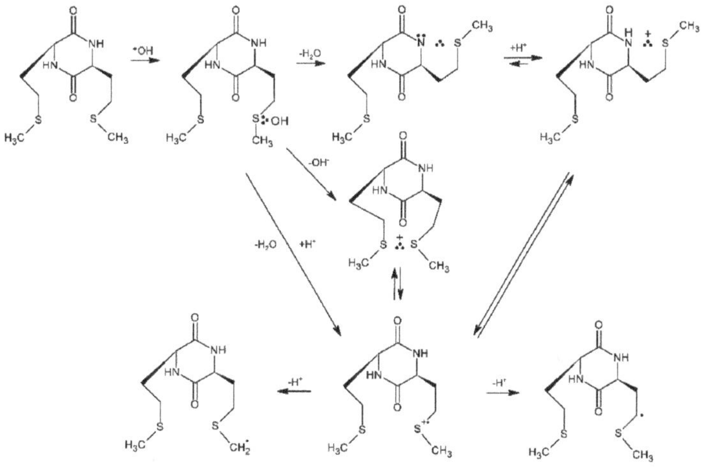

Scheme 13. General reaction scheme for $^{\bullet}OH$-induced oxidation of methionine in c-(L-Met-L-Met).

One feature that requires a note is that the neutral Met(S∴N) radical seems to arise directly from an intermediary MetS∴OH adduct rather than going through the monomeric $MetS^{\bullet+}$ radical cation. Mechanistically, this involves a concerted deprotonation and HO^- elimination from MetS∴OH. which frees up an electron pair on the N-atom of the peptide bond. Ultimately, the Met(S∴N) radicals decayed via two-different pH-dependent reaction pathways: (i) at low pH by conversion into additional $Met(S∴S)^+$ radical cations via the $Met(S∴NH)^+$ intermediate with the rate constant $k_{Met(S∴N) + H^+} = (2.0 \pm 0.1) \times 10^9$ $M^{-1}s^{-1}$, and (ii) at pH close to neutral by a series of consecutive reactions involving, protonation, hydrolysis, and electron transfer followed by decarboxylation of the N-centered radical on Met (not shown in Scheme 13) [12]. The observed decrease of formation of $Met(S∴S)^+$ radical cations via pathway (i) with pH is clearly corroborated by the respective $G(Met(S∴S)^+$ values recorded at various pH values before and after decay of Met(S∴N) (Table 9). Interestingly, an absorption band assigned to $Met(S∴S)^+$ in c-(L-Met-L-Met) is characterized by λ_{max} = 520 nm which is red shifted in comparison to λ_{max} = 480 nm of the absorption band of intermolecular dimeric radical cations $(MetS∴SMet)^+$ in Met [7] or λ_{max} = 490 nm of the absorption band of intramolecular $Met(S∴S)^+$ radical cations in linear L-Met-L-Met dipeptides [9]. This phenomenon can be rationalized by the less favorable overlap of the p orbitals of sulfur atoms in c-(L-Met-L-Met) due to stronger steric constraints of diketopiperazine ring [98,99].

Table 9. Radiation chemical yields of intramolecular Met(S∴N) and Met(S∴S)$^+$ formed during $^{\bullet}$OH-induced oxidation of the cyclic L-Met-L-Met dipeptide in N_2O-saturated aqueous solutions.

Type of the Radical	pH 4	pH 4.3	pH 4.9	pH 5.3
Intramolecular Met(S∴N)	0.26 [(a)] 0	0.28 [(a)] 0	0.25 [(a)] 0	0.27 [(a)] 0
Intramolecular Met(S∴S)$^+$	0.22 [(a)] 0.45 [(a)]	0.23 [(a)] 0.45 [(a)]	0.22 [(a)] 0.32 [(a)]	0.23 [(a)] 0.34 [(a)]

[(a)] in μM J^{-1}; [L-Met-L-Met] = 0.2 mM.

Oxidation of the cyclic dipeptide c-(D-Met-L-Met) by $^{\bullet}$OH radicals induced by pulse radiolysis was also studied with combined time-resolved UV-vis spectroscopy and conductometry. This isomer has geometric restrictions which should impose limitations on stabilization of intramolecular Met(S∴S)$^+$ radical cations (see Figure 3b). In contrast to the previously observed intramolecular stabilization of MetS$^{\bullet+}$ by the unoxidized sulfur atom in the neighboring Met, in the isomer c-(L-Met-L-Met) (Scheme 13 and Table 9), no similar intramolecular stabilization of MetS$^{\bullet+}$ was observed in c-(D-Met-L-Met). One feature that requires a note at this point is that the species absorbing in the 480–490 nm range has to be assigned to intermolecular dimeric radical cations (MetS∴SMet)$^+$ based on the spectral and kinetic features observed in c-(Gly-L-Met) (see Section 4.1.1).

However, formation of Met(S∴N) occurs via an analogous concerted deprotonation and HO^- elimination mechanism from MetS∴OH as in c-(L-Met-L-Met) (Scheme 14).

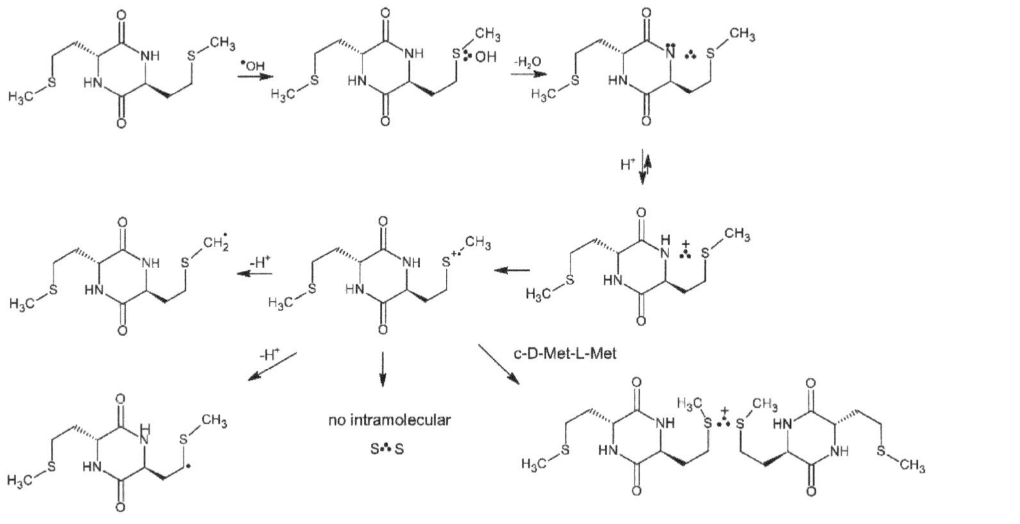

Scheme 14. General reaction scheme for $^{\bullet}$OH-induced oxidation of methionine in c-(D-Met-L-Met).

4.2. Photo-Induced Oxidation

4.2.1. Cyclic Dipeptides with Single Met Residue

One of the simplest model compounds for sensitized photooxidation are diketopiperazine-based benzophenone-methionine dyads: c-(L-BP-L-Met) and c-(L-BP-D-Met) with their structures presented in Figure 5 [76]. They allowed the investigators to study the intramolecular quenching of BP triplets by Met in conformationally controlled donor-acceptor pairs and to study the dynamics of the intermediates formed in the oxidation process.

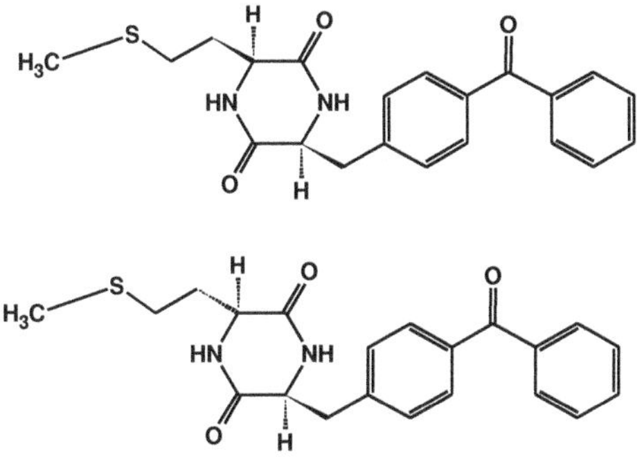

Figure 5. Structures of diketopiperazine-based benzophenone-methionine dyads: c-(L-BP-L-Met) (top) and c-(L-BP-D-Met) (bottom).

The combined results from time-resolved and steady-state experiments showed that for c-(L-BP-L-Met), the quenching occurred via electron transfer from the sulfur atom to the benzophenone triplet (via the [$BP^{\cdot-}$ ... $>S^{\cdot+}$] complex) was followed by the proton transfer reaction that led to a biradical intermediate with the quantum yield $\Phi_{ketyl} = 0.6$. These biradicals recombined to form a cyclic stable product via C-C coupling reaction. For the c-(L-BP-D-Met) dyad, with its two reactive moieties on the opposite side of diketopiperazine ring, the quenching process and formation of biradical were shown to be less effective ($\Phi_{ketyl} < 0.18$). It was shown that differences in the reaction rates for both isomeric BP-Met dyads were attributed to the differences between the distribution of the carbonyl/sulfur atom distances in the two stereoisomers.

4.2.2. Cyclic Dipeptides with Two Met Residue

Cyclic Met-Met dipeptides are also suitable model compounds for proteins to study the photo-induced oxidation. They have the unique features of having no terminal groups and therefore they are good model to study interactions between a side chain and peptide bonds.

The mechanism of the CB-sensitized photooxidation of c-(L-Met-L-Met) with Met residues in the cis position was studied using nanosecond flash photolysis techniques [63]. The photochemical path of c-(L-Met-L-Met) oxidation led to the $>S^{\cdot+}$ radical cation (as presented in Scheme 8). This radical cation can decay in two competing reactions: the formation of intramolecular dimeric sulfur radical cations $(S\therefore S)^+$ and a deprotonation yielding αS radicals. Similar to the results presented above for the radiation-induced oxidation, the efficiency of intramolecular $(S\therefore S)^+$ formation was relatively large ($\Phi_{(S\therefore S)+} = 0.14$ (see Table 8) showing that this reaction was a main path of $>S^{\cdot+}$ radical cation decay. The mechanism of $(S\therefore S)^+$ formation was shown to be similar to that presented in Scheme 13 for the radiation-induced oxidation of c-(L-Met-L-Met).

The CB-sensitized photooxidation of cyclic dipeptides containing methionine residues was also studied using time-resolved CIDNP [25]. For c-(D-Met-L-Met) with Met residues on the opposite side of the ring and for c-(Gly-Met) the sulfur-centered radical cations $>S^{\cdot+}$ were detected. In the case of c-(L-Met-L-Met the intramolecular cyclic $(S\therefore S)^+$ transients with three-electron bond between two sulfur atoms were detected.

5. Methionine in Proteins

5.1. *Radiation-Induced Oxidation*

Pulse radiolysis is a very useful method for studies of free radical processes in biological systems, including proteins. However, the vast majority of studies was focused on the generation of unstable intermediates in proteins through one-electron reduction [103]. There were also some attempts aimed at studying the oxidation processes in proteins. However, they only concerned the oxidation of amino acids located in proteins which are very prone to oxidation, such as Trp or Tyr [5,104–107]. To the best of our knowledge, only two papers addressed radiation-induced oxidation of Met in proteins and the subsequent steps following oxidation [108,109].

Thioredoxin Ch1 (Trx) from Chlamydomonas reinhardtii contains two Met residues: Met^{41} and Met^{79}. The one-electron oxidation of the W35A mutant of Trx by pulse radiolytically generated $N_3^{\bullet}$ radicals led to a transient absorption spectrum with three absorption bands characterized by λ_{max} = 390, 420, and 480 nm. The absorption band with λ_{max} = 420 nm indicated formation of $TyrO^{\bullet}$ radicals via direct reaction of Tyr with $N_3^{\bullet}$ radicals. On the other hand, the absorption bands with λ_{max} = 390 and 480 nm were assigned to $Met(S\therefore O)^+$ which results from the interaction of the monomeric sulfur radical cation ($MetS^{\bullet+}$) with the O-atom of the carbonyl function of Phe^{31} (which is less than 4 Å away from S-atom) and $Met(S\therefore S)^+$ which results from the interaction of $MetS^{\bullet+}$ with the S-atom of Cys^{49}, respectively. With time evolution, no oxidized products derived from Met were detected but there was a further increase of the absorption band assigned to $TyrO^{\bullet}$ radicals. An intramolecular electron transfer from the Tyr residue to Met-derived radicals seems to be responsible for the latter observation [108] (see Scheme 15 for an analogous reaction involving $Met(S\therefore N)^+$ radical cation).

In turn, calmodulin (CaM-Ca_4) (a regulatory "Ca-sensor protein") contains nine Met residues: Met^{36}, Met^{51}, Met^{71}, Met^{72}, Met^{76}, Met^{109}, Met^{124}, Met^{144}, and Met^{145} which are located in a variety of local environments. Therefore, the stabilization of the monomeric sulfur radical cation ($MetS^{\bullet+}$) can be realized in various ways depending upon local structures. The one-electron oxidation of CaM-Ca_4 using pulse radiolytically generated $^{\bullet}OH$ radicals led to a transient absorption with λ_{max} = 390 nm which shows a close similarity to that characteristic for S∴N-bonded radical cations observed in cyclic Met-Met dipeptides [12,13]. The lack of the characteristic spectrum of intramolecular S∴S-bonded radical cation, despite the proximity of methionine residues in two pairs Met^{71}/Met^{72} and Met^{144}/Met^{145} can be rationalized by local steric constraints which prevent close contact of their side chains [13]. With time evolution, the transient absorption is shifted to λ_{max} = 410 nm which is characteristic for $TyrO^{\bullet}$ radicals. This observation was rationalized in terms of intramolecular electron transfer from Tyr to $Met(S\therefore N)^+$ radical cations (see Scheme 15) [109].

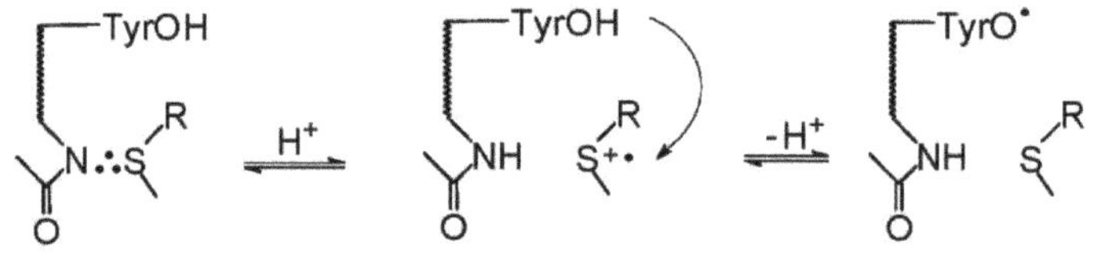

Scheme 15. electron transfer between $Met(S\therefore N)^+$ and Tyr in CaM-Ca_4 (adapted from [109]).

5.2. *Photo-Induced Oxidation*

Despite the significant role of protein oxidation reactions in biology and medicine the sensitized photooxidation of Met within proteins was the subject of only a few papers. This may be due to complexity of the reactants, the diversity of possible target positions in the oxidation of proteins, and the experimental difficulties in analyzing the oxidation products. The mechanism of CB-sensitized photooxidation of plant cytokinin-specific binding proteins (VrPhBP) was recently studied using nanosecond flash photolysis and various methods for analyzing the stable photoproducts (chemical analysis, silver-staining gel electrophoresis, chromatography, and mass spectrometry including peptide mapping by proteolysis and coupling with chromatography) [110]. The results indicated oxidation of methionine and tyrosine residues within the protein.

Based on the crystallographic structure of the VrPhBP protein and taking into account knowledge from similar studies for model amino acids and peptides, methionine (Met^{141}) and tyrosine (Tyr^{142}) residues were suggested as being the most prone to oxidation. It was shown that the CB triplets were mainly quenched by methionine and partially by tyrosine residues via the electron transfer mechanism leading to the formation of sulfur-centered monomeric radical cations $>S^{\cdot+}$ (stabilized by $(S\therefore N)^+$ formation), tyrosyl radicals $TyrO^{\cdot}$ and ketyl radicals $CBH^{\cdot}$. As was already shown for the model peptide (Met-enkephalin) [61], the intramolecular electron transfer from the Tyr residue to the oxidized sufur atom of the Met residue can take place leading to regeneration of the Met residue and formation of additional $TyrO^{\cdot}$ radicals. The presence of the transient species observed in the time-resolved experiments were additionally confirmed by the detection of such stable products as methionine sulfoxide, the Met-CBH adduct (radical coupling products (αS-CBH) and dityrosine cross links (see Scheme 16). The results from the study of sensitized photooxidation of model compounds including methionine derivatives and peptides were very helpful to understand the mechanism of sensitized photoxidation of proteins containing methionine residues.

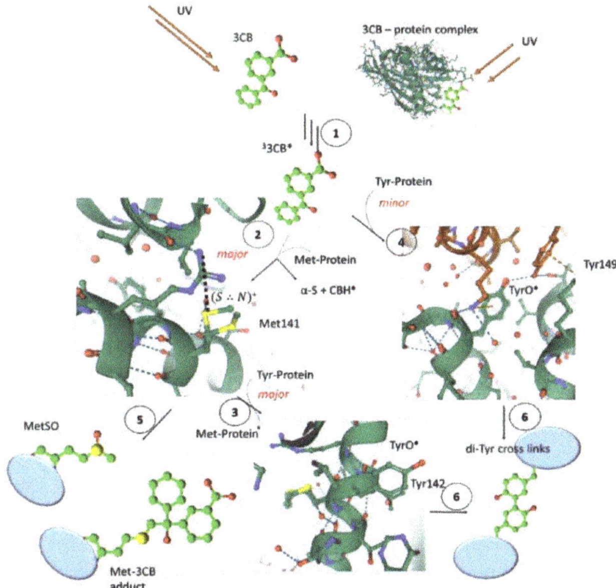

Scheme 16. Graphical scheme reaction of the mechanism of VrPhBp protein oxidation sensitized by CB in aqueous solutions methionine in Gly-Met-Gly tripeptide (adapted from [110]).

6. Conclusions

A wealth of knowledge has been accumulated over the years concerning the mechanistic understanding of one-electron oxidation of methionine in various structural environments using two-complementary radiation and photochemical time-resolved techniques such as pulse radiolysis and laser flash photolysis coupled with UV-vis spectrophotometry, conductometry, ESR spectroscopy and CIDNP. Creation of a variety of very reactive transients such as •OH-adducts to methionine (MetS∴OH), monomeric sulfur radical cations ($MetS^{\bullet+}$), intramolecularly (S∴N)-bonded radicals (Met(S∴N)) or radical cations ($Met(S\therefore N)^+$), α-alkylthio)alkyl radicals (αS1, αS2), α-aminoalkyl radicals (αN) intermolecularly S∴S-bonded dimeric radical cations ($(MetS\therefore SMet)^+$), intramolecularly S∴S-bonded radical cations ($Met(S\therefore S)^+$), and intramolecularly S∴O-bonded radical cations ($Met(S\therefore O)^+$), which are precursors for the subsequent final products responsible for protein damage, requires a knowledge of their spectral parameters as well as the mechanisms of their formation and decay along with appropriate kinetic parameters. Neighboring group participation seems to be an essential parameter which controls one-electron oxidation of methionine by various one-electron oxidants in various structural and pH environments. Provided that photo- and radiation-induced reaction pathways lead to the formation of a common intermediate, namely the monomeric sulfur-centered radical cation ($MetS^{\bullet+}$), its secondary reactions in peptides and proteins will be the same. This review was also aimed at demonstrating the complementarity of photo- and radiation-chemical studies using the example of methionine oxidation processes.

Future studies on photo- and radiation-induced oxidation of Met containing compounds aimed at analysis of stable products formed during the steady state photo- and γ-irradiations (similar to those reported in the references [20,22,59,76] are expected to help in understanding the overall mechanism of oxidation from short-lived intermediates to the final products.

Author Contributions: Conceptualization, B.M. and K.B.; writing—original draft preparation, B.M. and K.B.; writing—review and editing B.M. and K.B. All authors have read and agreed to the published version of the manuscript.

Funding: This research was funded by the Polish National Science Centre (NCN) under grant no. UMO 2017/27/B/ST4/00375 (B.M.) and by statutory funds of the Institute of Nuclear Chemistry and Technology (INCT) (K.B.)

Institutional Review Board Statement: Not applicable.

Informed Consent Statement: Not applicable.

Data Availability Statement: Not applicable.

Acknowledgments: We acknowledge the co-authors of our papers cited in the references for the fruitful exchange of ideas and collaborations. We are very grateful to Gordon L. Hug from the Notre Dame Radiation Laboratory (USA) for his insightful comments and helpful suggestions. Thanks are also due to Konrad Skotnicki (K.B.) and Jarogniew Bartoszewicz (B.M.) for their invaluable help in editing schemes presented in the manuscript.

Conflicts of Interest: The authors declare no conflict of interest. The funders had no role in the design of the study; in the collection, analyses, or interpretation of data; in the writing of the manuscript, or in the decision to publish the results.

References

1. Levine, R.L.; Mosoni, L.; Berlett, B.S.; Stadtman, E.R. Methionine residues as endogenous antioxidants in proteins. *Proc. Natl. Acad. Sci. USA* **1996**, *93*, 15036–15040. [CrossRef] [PubMed]
2. Vogt, W. Oxidation of Methionyl Residues in Proteins: Tools, Targets, and Reversal. *Free Radic. Biol. Med.* **1995**, *18*, 93–105. [CrossRef]
3. Davies, M. The oxidative environment and protein damage. *Biochim. Biophys. Acta* **2005**, *1703*, 93–109. [CrossRef] [PubMed]
4. Stadtman, E.R.; Van Remmen, H.; Richardson, A.; Wehr, N.B.; Levine, R.L. Methionine oxidation and aging. *Biochim. Biophys. Acta* **2005**, *1703*, 135–140. [CrossRef] [PubMed]
5. Houée-Levin, C.; Bobrowski, K. The use of methods of radiolysis to explore the mechanisms of free radical modifications in proteins. *J. Proteom.* **2013**, *92*, 51–62. [CrossRef] [PubMed]
6. Ignasiak, M.T.; Marciniak, B.; Houée-Levin, C. A Long Story of Sensitized One-Electron Photo-oxidation of Methionine. *Isr. J. Chem.* **2014**, *54*, 248–253. [CrossRef]
7. Hiller, K.-O.; Masloch, B.; Göbl, M.; Asmus, K.-D. Mechanism of the $OH^{\cdot}$ radical induced oxidation of methionine in aqueous solution. *J. Am. Chem. Soc.* **1981**, *103*, 2734–2743. [CrossRef]
8. Bobrowski, K.; Holcman, J. Formation of three-electron bonds in one-electron oxidized methionine dipeptides: A pulse radiolytic study. *Int. J. Radiat. Biol. Relat. Stud. Phys. Chem. Med.* **1987**, *52*, 139–144. [CrossRef]
9. Bobrowski, K.; Holcman, J. Formation and stability of intramolecular three-electron S∴N, S∴S, and S∴O bonds in one-electron-oxidized simple methionine peptides. Pulse radiolysis study. *J. Phys. Chem.* **1989**, *93*, 6381–6387. [CrossRef]
10. Schöneich, C.; Pogocki, D.; Wisniowski, P.; Hug, G.L.; Bobrowski, K. Intramolecular Sulfur-Oxygen Bond Formation in Radical Cations of N-Acetylmethionine Amide. *J. Am. Chem. Soc.* **2000**, *122*, 10224–10225. [CrossRef]
11. Schöneich, C.; Pogocki, D.; Hug, G.L.; Bobrowski, K. Free radical reactions of methionine in peptides: Mechanisms relevant to β-amyloid oxidation and Alzheimer's disease. *J. Am. Chem. Soc.* **2003**, *125*, 13700–13713. [CrossRef]
12. Bobrowski, K.; Hug, G.L.; Pogocki, D.; Marciniak, B.; Schöneich, C. Stabilization of sulfide radical cations through complexation with the peptide bond: Mechanisms relevant to oxidation of proteins containing multiple methionine residues. *J. Phys. Chem. B* **2007**, *111*, 9608–9620. [CrossRef]

13. Hug, G.L.; Bobrowski, K.; Pogocki, D.; Hörner, G.; Marciniak, B. Conformational influence on the type of stabilization of sulfur radical cations in cyclic peptides. *ChemPhysChem* **2007**, *8*, 2202–2210. [CrossRef]
14. Schöneich, C.; Zhao, F.; Madden, K.P.; Bobrowski, K. Side chain fragmentation of N-terminal threonine or serine residue induced through intramolecular proton transfer to hydroxy sulfuranyl radical formed at neighboring methionine in dipeptides. *J. Am. Chem. Soc.* **1994**, *116*, 4641–4652. [CrossRef]
15. Bobrowski, K.; Marciniak, B.; Hug, G.L. 4-Carboxybenzophenone-sensitized photooxidation of sulfur-containing amino acids. Nanosecond laser flash photolysis and pulse radiolysis studies. *J. Am. Chem. Soc.* **1992**, *114*, 10279–10288. [CrossRef]
16. Marciniak, B.; Hug, G.L.; Bobrowski, K.; Kozubek, H. Mechanism of 4-carboxybenzophenone-sensitized photooxidation of methionine-containing dipeptides and tripeptides in aqueous solution. *J. Phys. Chem.* **1995**, *99*, 13560–13568. [CrossRef]
17. Hug, G.L.; Marciniak, B.; Bobrowski, K. Acid-base equilibria involved in secondary reactions following the 4-carboxybenzophenone sensitized photooxidation of methionyl-glycine in aqueous solution. Spectral and time resolution of the decaying $(S\therefore N)^+$ radical cation. *J. Phys. Chem.* **1996**, *100*, 14914–14921. [CrossRef]
18. Hug, G.L.; Marciniak, B.; Bobrowski, K. Sensitized photo-oxidation of sulfur-containing amino acids and peptides in aqueous solution. *J. Photochem. Photobiol. A* **1996**, *95*, 81–88. [CrossRef]
19. Hug, G.L.; Bobrowski, K.; Kozubek, H.; Marciniak, B. Photooxidation of Methionine Derivatives by the 4-Carboxybenzophenone Triplet State in Aqueous Solution. Intramolecular Proton Transfer involving the Amino Group. *Photochem. Photobiol.* **1998**, *68*, 785–796. [CrossRef]
20. Ignasiak, M.T.; Pedzinski, T.; Rusconi, F.; Filipiak, P.; Bobrowski, K.; Houée-Levin, C.; Marciniak, B. Photosensitized Oxidation of Methionine-Containing Dipeptides. From the Transients to the Final Products. *J. Phys. Chem. B* **2014**, *118*, 8549–8558. [CrossRef]
21. Pedzinski, T.; Markiewicz, A.; Marciniak, B. Photosensitized oxidation of methionine derivatives. Laser flash photolysis studies. *Res. Chem. Intermed.* **2009**, *35*, 497–506. [CrossRef]
22. Pedzinski, T.; Grzyb, K.; Kaźmierczak, F.; Frański, R.; Filipiak, P.; Marciniak, B. Early Events of Photosensitized Oxidation of Sulfur-Containing Amino Acids Studied by Laser Flash Photolysis and Mass Spectrometry. *J. Phys. Chem. B* **2020**, *124*, 7564–7573. [CrossRef] [PubMed]
23. Goez, M.; Rozwadowski, J.; Marciniak, B. CIDNP spectroscopic observation by magnetic resonance of SN+ radical cations with two-center-three-electron-bond during the photooxidation of methionine. *Angew. Chem. Int. Ed.* **1998**, *37*, 785–796. [CrossRef]
24. Morozova, O.B.; Korchak, S.E.; Vieth, H.-M.; Yurkovskaya, A.V. Photo-CIDNP Study of Transient Radicals of Met-Gly and Gly-Met Peptides in Aqueous Solution at Variable pH. *J. Phys. Chem. B* **2009**, *113*, 7398–7406. [CrossRef]
25. Köchling, T.; Morozova, O.B.; Yurkovskaya, A.V.; Vieth, H.-M. Magnetic Resonance Characterization of One-Electron Oxidized Cyclic Dipeptides with Thioether Groups. *J. Phys. Chem. B* **2016**, *120*, 9277–9286. [CrossRef] [PubMed]
26. Yashiro, H.; White, R.C.; Yurkovskaya, A.V.; Forbes, M.D.E. Methionine Radical Cation: Structural Studies as a Function of pH Using X- and Q-Band Time-Resolved Electron Paramagnetic Resonance Spectroscopy. *J. Phys. Chem. A* **2005**, *109*, 5855–5864. [CrossRef] [PubMed]
27. Bobrowski, K.; Houée-Levin, C.; Marciniak, B. Stabilization and Reactions of Sulfur Radical Cations: Relevance to One-Electron Oxidation of Methionine in Peptides and Proteins. *Chimia* **2008**, *62*, 728–734. [CrossRef]
28. Schöneich, C. Redox processes of methionine relevant to β-amyloid oxidation and Alzheimer's disease. *Arch. Biochem. Biophys.* **2002**, *397*, 370–376. [CrossRef]
29. Schöneich, C. Methionine oxidation by reactive oxygen species: Reaction mechanisms and relevance to Alzheimer's disease. *Biochim. Biophys. Acta* **2005**, *1703*, 111–119. [CrossRef]
30. Glass, R.S. Sulfur Radicals and Their Application. *Top. Curr. Chem.* **2018**, *376*, 22. [CrossRef]
31. Schöneich, C.; Zhao, F.; Yang, J.; Miller, B. Mechanisms of Methionine Oxidation in Peptides. In *Therapeutic Protein and Peptide Formulation and Delivery*; Shahrokh, Z., Sluzky, V., Cleland, J.L., Shire, S.J., Randolph, T.W., Eds.; ACS Symposium Series; American Chemical Society: Washington, DC, USA, 1997; Volume 675, pp. 79–89.
32. Bobrowski, K. Chemistry of Sulfur-centered Radicals. In *Recent Trends in Radiation Chemistry*; Wishart, J.F., Rao, B.S.M., Eds.; World Scientific: Singapore, 2010.
33. Bobrowski, K. Radiation-induced radicals and radical ions in amino acids and peptides. In *Selectivity, Control, and Fine Tuning in High Energy Chemistry*; Staas, D.V., Feldman, V.I., Eds.; Research Signpost: Trivandrum, India, 2011.
34. Houée-Levin, C. One-electron redox processes in proteins. In *Selectivity, Control, and Fine Tuning in High-Energy Chemistry*; Stass, D.V., Feldman, V.I., Eds.; Research Signpost: Trivandrum, India, 2011; p. 59.
35. Schöneich, C. Radical-Based Damage of Sulfur-Containing Amino Acid Residues. In *Encyclopedia of Radical in Chemistry, Biology and Materials*; Chatgilialoglu, C., Studer, A., Eds.; Chemical Biology; John Wiley & Sons, Ltd.: Chichester, UK, 2012; Volume 3, pp. 1459–1474.
36. Capon, B.; Mc Manus, S.P. *Neighboring Group Participation*; Plenum Press: New York, NY, USA, 1976; Volume 1.
37. Glass, R.S.; Hug, G.L.; Schöneich, C.; Wilson, G.S.; Kuznetsova, L.; Lee, T.; Ammam, M.; Lorance, E.; Nauser, T.; Nichol, G.S.; et al. Neighboring Amide Participation in Thioether Oxidation: Relevance to Biological Oxidation. *J. Am. Chem. Soc.* **2009**, *131*, 13791–13805. [CrossRef]
38. Fourre, I.; Silvi, B. What Can We Learn from Two-Center Three-Electron Bonding with the Topological Analysis of ELF? *Heteroat. Chem.* **2007**, *18*, 135–160. [CrossRef]

39. Fourre, I.; Bergès, J.; Houée-Levin, C. Structural and Topological Studies of Methionine Radical Cations in Dipeptides: Electron Sharing in Two-Center Three-Electron Bonds. *J. Phys. Chem. A* **2010**, *114*, 7359–7368. [CrossRef]
40. Brunelle, P.; Rauk, A. One-electron oxidation of methionine in peptide environments: The effect of three-electron bonding on the reduction potential of the radical cation. *J. Phys. Chem. A* **2004**, *108*, 11032–11041. [CrossRef]
41. Asmus, K.-D. Sulfur-centered three-electron bonded radical species. In *Sulfur-Centered Reactive Intermediates in Chemistry and Biology*; Chatgilialoglu, C., Asmus, K.-D., Eds.; NATO ASI Series, Series A: Life Sciences; Plenum Press: New York, NY, USA, 1990; Volume 197, pp. 155–172.
42. Asmus, K.-D. Sulfur-centered Free Radicals. In *Methods in Enzymology*; Packer, L., Ed.; Academic Press: Orlando, FL, USA, 1990; Volume 186, pp. 168–180.
43. Hiller, K.-O.; Asmus, K.-D. Tl^{2+} and Ag^{2+} metal-ion-induced oxidation of methionine in aqueous solution. A pulse radiolysis study. *Int. J. Radiat. Biol. Relat. Stud. Phys. Chem. Med.* **1981**, *40*, 597–604. [CrossRef]
44. Hiller, K.-O.; Asmus, K.-D. Formation and reduction reactions of α-amino radicals derived from methionine and its derivatives in aqueous solutions. *J. Phys. Chem.* **1983**, *87*, 3682–3688. [CrossRef]
45. Mönig, J.; Göbl, M.; Asmus, K.-D. Free Radical One-electron versus Hydrogen Radical-induced Oxidation. Reaction of Trichloromethyl Peroxyl Radicals with Simple and Substituted Aliphatic Sulphides in Aqueous Solution. *J. Chem. Soc. Perkin Trans.* **1985**, 647. [CrossRef]
46. Asmus, K.-D.; Göbl, M.; Hiller, K.-O.; Mahling, S.; Mönig, J. S∴N and S∴O three-electron-bonded radicals and radical cations in aqueous solutions. *J. Chem. Soc. Perkin Trans. 2* **1985**, 641–646. [CrossRef]
47. Armstrong, D.A.; Huie, R.E.; Koppenol, W.H.; Lymar, S.V.; Merenyi, G.; Neta, P.; Ruscic, B.; Stanbury, D.M.; Steenken, S.; Wardman, P. Standard electrode potentials involving radicals in aqueois solutions: Inorganic radicals (IUPAC Technical Report). *Pure Appl. Chem.* **2015**, *87*, 1139–1150. [CrossRef]
48. Rauk, A.; Armstrong, D.A.; Fairlie, D.P. Is oxidative damage by β-amyloid and prion peptides mediated by hydrogen atom transfer from glycyne α-carbon to methionine sulfur within β-sheets? *J. Am. Chem. Soc.* **2000**, *122*, 9761–9767. [CrossRef]
49. Merényi, G.; Lind, J.; Engman, L. One- and Two-electron reduction Potentials of Peroxyl Radicals and Related Species. *J. Chem. Soc. Perkin Trans. 2* **1994**, 2551–2553. [CrossRef]
50. Chen, S.N.; Hoffman, M.Z. Rate constants for the reaction of the carbonate radical with compounds of biochemical interest in neutral aqueous solution. *Radiat. Res.* **1973**, *56*, 40–47. [CrossRef]
51. Wojnárovits, L.; Takács, E. Rate constants of dichloride radical anion reactions with molecules of environmental interest in aqueous solution: A review. *Environ. Sci. Pollut. Res.* **2021**, *28*, 41552–41575. [CrossRef]
52. Hiller, K.-O.; Asmus, K.-D. Oxidation of methionine by $X_2^{\cdot-}$ in aqueous solution and characterization of some >SX three-electron bonded intermediates. A pulse radiolysis study. *Int. J. Radiat. Biol.* **1981**, *40*, 583–595.
53. Bonifačić, M.; Asmus, K.-D. Stabilization of oxidized sulfur centres by halide ions. Formation and properties of R_2SX radicals in aqueous solutions. *J. Chem. Soc. Perkin Trans. 2* **1980**, 758–762. [CrossRef]
54. Pogocki, D.M. Investigation of Radical Processes Induced by Hydroxyl Radical in Amino Acids and Peptides Containing Thioether Group. Ph.D. Thesis, Institute of Nuclear Chemistry and Technology, Warsaw, Poland, 1996.
55. Mishra, B.; Priyadarsini, K.I.; Mohan, H. Pulse radiolysis studies on reaction of OH radical with N-acetyl methionine in aqueous solutions. *Res. Chem. Interm.* **2005**, *31*, 625–632. [CrossRef]
56. Shirdhonkar, M.; Maity, D.K.; Mohan, H.; Rao, B.S.M. Oxidation of methionine methyl ester in aqueous solution: A combined pulse radiolysis and quantum chemical study. *Chem. Phys. Lett.* **2006**, *417*, 116–123. [CrossRef]
57. Bobrowski, K.; Schöneich, C.; Holcman, J.; Asmus, K.-D. •OH radical induced decarboxylation of methionine-containing peptides. Influence of peptide sequence and net charge. *J. Chem. Soc., Perkin Trans. 2* **1991**, 353–362. [CrossRef]
58. Steffen, L.K.; Glass, R.S.; Sabahi, M.; Wilson, G.S.; Schöneich, C.; Mahling, S.; Asmus, K.-D. ˙OH radical induced decarboxylation of amino acids. Decarboxylation vs bond formation in radical intermediates. *J. Am. Chem. Soc.* **1991**, *113*, 2141–2145. [CrossRef]
59. Pedzinski, T.; Grzyb, K.; Skotnicki, K.; Filipiak, P.; Bobrowski, K.; Chatgilialoglu, C.; Marciniak, B. Radiation- and Photo-Induced Oxidation Pathways of Methionine in Model Peptide Backbone under Anoxic Conditions. *Int. J. Mol. Sci.* **2021**, *22*, 4773. [CrossRef]
60. Guttenplan, J.B.; Cohen, S.G. Quenching and reduction of photoexcited benzophenone by thioethers and mercaptans. *J. Org. Chem.* **1973**, *38*, 2001–2007. [CrossRef]
61. Wojcik, A.; Lukaszewicz, A.; Brede, O.; Marciniak, B. Competitive photosensitized oxidation of tyrosine and methionine residues in enkephalins and their model peptides. *J. Photochem. Photobiol. A Chem.* **2008**, *198*, 111–118. [CrossRef]
62. Marciniak, B.; Bobrowski, K.; Hug, G.L. Quenching of triplet states of aromatic ketones by sulfur-containing amino acids in solution. Evidence for electron transfer. *J. Phys. Chem.* **1993**, *97*, 11937–11943. [CrossRef]
63. Hug, G.L.; Bobrowski, K.; Kozubek, H.; Marciniak, B. Photo-oxidation of Methionine-containing Peptides by the 4-Carboxybenzophenone Triplet State in Aqueous Solution. Competition between Intramolecular Two-centered Three-electron Bonded $(S∴S)^+$ and $(S∴N)^+$ Formation. *Photochem. Photobiol.* **2000**, *72*, 1–9. [CrossRef]
64. Cohen, S.G.; Ojanpera, S. Photooxidation of methionine and related compounds. *J. Am. Chem. Soc.* **1975**, *97*, 5633–5634. [CrossRef]
65. Encinas, M.V.; Lissi, E.A.; Olea, A.F. Quenching of triplet benzophenone by vitamins E and C and by sulfur containing aminoacids and peptides. *Photochem. Photobiol.* **1985**, *42*, 347–352. [CrossRef]

66. Encinas, M.V.; Lissi, E.A.; Vasquez, M.; Olea, A.F.; Silva, E. Photointeraction of benzophenone triplet with lysozyme. *Photochem. Photobiol.* **1989**, *49*, 557–563. [CrossRef] [PubMed]
67. Bobrowski, K.; Hug, G.L.; Marciniak, B.; Kozubek, H. 4-Carboxybenzophenone-sensitized photooxidation of sulfur-containing amino acids in alkaline aqueous solutions. Secondary photoreactions kinetics. *J. Phys. Chem.* **1994**, *98*, 537–544. [CrossRef]
68. Filipiak, P.; Bobrowski, K.; Hug, G.L.; Pogocki, D.; Schöneich, C.; Marciniak, B. New Insights into the Reaction Paths of 4-Carboxybenzophenone Triplet with Oligopeptides Containing N- and C-Terminal Methionine Residues. *J. Phys. Chem. B* **2017**, *121*, 5247–5258. [CrossRef]
69. Goez, M.; Rozwadowski, J. Reversible Pair Substitution in CIDNP: The Radical Cation of Methionine. *J. Phys. Chem. A* **1998**, *102*, 7945–7953. [CrossRef]
70. Bonifačić, M.; Stefanic, I.; Hug, G.L.; Armstrong, D.A.; Asmus, K.-D. Glycine decarboxylation: The free radical mechanism. *J. Am. Chem. Soc.* **1998**, *120*, 9930–9940. [CrossRef]
71. Marciniak, B.; Hug, G.L.; Rozwadowski, J.; Bobrowski, K. Excited Triplet State of N-(9-methylpurin-6-yl)pyridinium Cation as as Efficient Photosensitizer in the Oxidation of Sulfur-containing Amino Acids. Laser Flash and Steady-state Photolysis Studies. *J. Am. Chem. Soc.* **1995**, *117*, 127–134. [CrossRef]
72. Bobrowski, K.; Hug, G.L.; Pogocki, D.; Marciniak, B.; Schöneich, C. Sulfur radical cation peptide bond complex in the one-electron oxidation of S-methylglutathione. *J. Am. Chem. Soc.* **2007**, *129*, 9236–9245. [CrossRef] [PubMed]
73. Filipiak, P.; Hug, G.L.; Bobrowski, K.; Pedzinski, T.; Kozubek, H.; Marciniak, B. Sensitized Photooxidation of S-Methylglutathione in Aqueous Solution: Intramolecular (S∴O) and (S∴N) Bonded Species. *J. Phys. Chem. B* **2013**, *117*, 2359–2368. [CrossRef]
74. Barata-Vallejo, S.; Ferreri, C.; Zhang, T.; Permentier, H.; Bischoff, R.; Bobrowski, K.; Chatgilialoglu, C. Radiation chemical studies of Gly-Met-Gly in aqueous solution. *Free Rad. Res.* **2016**, *50*, S24–S39. [CrossRef]
75. Hug, G.L.; Bobrowski, K.; Kozubek, H.; Marciniak, B. pH effects on the photooxidation of methionine derivatives by the 4-carboxybenzophenone triplet state. *Nukleonika* **2000**, *45*, 63–71.
76. Lewandowska-Andralojć, A.; Kazmierczak, F.; Hug, G.L.; Hörner, G.; Marciniak, B. Photoinduced CC-coupling Reactions of Rigid Diastereomeric Benzophenone-Methionine Dyads. *Photochem. Photobiol.* **2013**, *89*, 14–23. [CrossRef]
77. Ignasiak, M.T. Study of the Mechanism of Radiation- and Photo-Induced Oxidation of Methionine Containing Peptides. Ph.D. Thesis, Adam Mickiewicz University, Poznań, Poland, 2014.
78. Bonifačić, M.; Hug, G.L.; Schöneich, C. Kinetics of the reactions between sulfide radical cation complexes, $[S∴S]^+$ and $[S∴N]^+$, and superoxide or carbon dioxide radical anions. *J. Phys. Chem. A* **2000**, *104*, 1240–1245. [CrossRef]
79. Klug, D.; Rabani, J.; Fridovich, I. A direct demonstration of the catalytic action of superoxide dismutase through the use of pulse radiolysis. *J. Biol. Chem.* **1972**, *247*, 4839–4842. [CrossRef]
80. Bobrowski, K.; Schöneich, C.; Holcman, J.; Asmus, K.-D. •OH radical induced decarboxylation of γ-glutamylmethionine and S-alkylglutathione derivatives: Evidence for two different pathways involving C- and N-terminal decarboxylation. *J. Chem. Soc. Perkin Trans. 2* **1991**, 975–980. [CrossRef]
81. Bobrowski, K.; Schöneich, C. Decarboxylation mechanism of the N-terminal glutamyl moiety in γ-glutamic acid and methionine containing peptides. *Radiat. Phys. Chem.* **1996**, *47*, 507–510. [CrossRef]
82. Filipiak, P.; Bobrowski, K.; Hug, G.L.; Schöneich, C.; Marciniak, B. N-Terminal Decarboxylation as a Probe for Intramolecular Contact Formation in γ-Glu-(Pro)n-Met Peptides. *J. Phys. Chem. B* **2020**, *124*, 8082–8098. [CrossRef]
83. Harriman, A. Further comment on the redox potentials of tryptophan and tyrosine. *J. Phys. Chem.* **1987**, *91*, 6102–6104. [CrossRef]
84. Folkes, L.K.; Trujillo, M.; Bartesaghi, S.; Radi, R.; Wardman, P. Kinetics of reduction of tyrosine phenoxyl radicals by glutathione. *Arch. Biochem. Biophys.* **2011**, *506*, 242–249. [CrossRef]
85. Merenyi, G.; Lind, J.; Engman, L. The Dimethylhydroxysulfuranyl Radical. *J. Phys. Chem.* **1996**, *100*, 8875–8881. [CrossRef]
86. Stryer, L. *Biochemistry*; W. H. Freeman and Company: New York, NY, USA, 1988.
87. Mozziconacci, O.; Mirkowski, J.; Rusconi, F.; Kciuk, G.; Wisniowski, P.; Bobrowski, K.; Houée-Levin, C. Methionine residue acts as a prooxidant in the OH-induced oxidation of enkephalins. *J. Phys. Chem B* **2012**, *116*, 9352–9362. [CrossRef]
88. Buxton, G.V.; Greenstock, C.L.; Helman, W.P.; Ross, A.B. Critical review of rate constants for reactions of hydrated electrons, hydrogen atoms and hydroxyl radicals ($^{\cdot}OH/^{\cdot}O^-$) in aqueous solution. *J. Phys. Chem. Ref. Data* **1988**, *17*, 513–886. [CrossRef]
89. Bobrowski, K.; Wierzchowski, K.L.; Holcman, J.; Ciurak, M. Intramolecular electron transfer in peptides containing methionine, tryptophan, and tyrosine: A pulse radiolysis study. *Int. J. Radiat. Biol.* **1990**, *57*, 919–932. [CrossRef]
90. Bobrowski, K.; Wierzchowski, K.L.; Holcman, J.; Ciurak, M. Pulse radiolysis, of intramolecular electron transfer in model peptides. IV. Met/S∴Br -> Tyr/O˙ radical transformation in aqueous solution of H-Tyr-$(Pro)_n$-Met-OH peptides. *Int. J. Radiat. Biol.* **1992**, *62*, 507–516. [CrossRef]
91. Bergès, J.; Trouillas, P.; Houée-Levin, C. Oxidation of protein tyrosine, or methionine residues: From the amino acid to the peptide. *J. Phys. Conf. Ser.* **2011**, *261*, 012003. [CrossRef]
92. Kadlčik, V.; Sicard-Roselli, C.; Mattioli, T.A.; Kodiček, M.; Houée-Levin, C. One-electron oxidation of β−amyloid peptide: Sequence modulation of reactivity. *Free Radic. Biol. Med.* **2004**, *37*, 881–891. [CrossRef] [PubMed]
93. Pogocki, D.; Schöneich, C. Redox properties of Met^{35} in neurotoxic β-amyloid peptide. A molecular modeling study. *Chem. Res. Toxicol.* **2002**, *15*, 408–418. [CrossRef] [PubMed]
94. Butterfield, D.A.; Bush, A.I. Alzheimer's amyloid β-peptide (1-42): Involvement of methionine residue 35 in the oxidative stress and neurotoxicity properties of this peptide. *Neurobiol. Aging* **2004**, *25*, 563–568. [CrossRef] [PubMed]

95. Kanski, J.; Varadarajan, S.; Aksenova, M.; Butterfield, D.A. Role of glycine-33 and methionine-35 in Alzheimer's amyloid beta peptide 1–42-associated oxidative stress and neurotoxicity. *Biochim. Biophys. Acta* **2001**, *1586*, 190–198. [CrossRef]
96. Varadarajan, S.; Yatin, S.; Kanski, J.; Jahanshahi, F.; Butterfield, D.A. Methionine residue 35 is important in amyloid beta peptide-associated free radical oxidative stress. *Brain Res. Bull.* **1999**, *50*, 133–141. [CrossRef]
97. Butterfield, D.A.; Kimball-Boyd, D. The critical role of methionine 35 in Alzheimer's amyloid β-peptide (1-42)-induced oxidative stress and neurotoxicity. *Biochim. Biophys. Acta* **2005**, *1703*, 149–156. [CrossRef] [PubMed]
98. Asmus, K.-D. Stabilization of oxidized sulfur centers in organic sulfides. Radical cations and odd-electron sulfur-sulfur bonds. *Acc. Chem. Res.* **1979**, *12*, 436–442. [CrossRef]
99. Asmus, K.-D.; Bahnemann, D.; Fischer, C.-H.; Veltwisch, D. Structure and stability of radical cations from cyclic and open-chain dithia compounds in aqueous solutions. *J. Am. Chem. Soc.* **1979**, *101*, 5322–5329. [CrossRef]
100. Filipiak, P.; Bobrowski, K.; Hug, G.L.; Pogocki, D.; Schöneich, C.; Marciniak, B. Formation of a Three-Electron Sulfur−Sulfur Bond as a Probe for Interaction between Side Chains of Methionine Residues. *J. Phys. Chem. B* **2016**, *120*, 9732–9744. [CrossRef]
101. Zahn, R.; Liu, A.; Luhrs, T.; Riek, R.; Von Schroetter, C.; Lopez Garcia, F.; Billeter, M.; Calzolai, L.; Wider, G.; Wüthrich, K. NMR solution structure of the human prion protein. *Proc. Nat. Acad. Sci. USA* **2000**, *97*, 145–150. [CrossRef]
102. Glasser, C.B.; Yamin, G.; Uversky, V.N.; Fink, A.L. Methionine oxidation, α−synuclein and Parkinson's disease. *Biochim. Biophys. Acta* **2005**, *1703*, 157–169. [CrossRef]
103. Kobayashi, K. Pulse Radiolysis Studies for Mechanism in Biochemical Redox Reactions. *Chem. Rev.* **2019**, *119*, 4413–4462. [CrossRef]
104. Bobrowski, K.; Holcman, J.; Poznanski, J.; Wierzchowski, K.L. Pulse radiolysis of intramolecular electron transfer in model peptides and proteins. 7. Trp → TyrO radical transformation in hen-egg white lysozyme. Effects of pH, temperature, Trp62 oxidation and inhibitor binding. *Biophys. Chem.* **1997**, *63*, 153–166. [CrossRef]
105. Houée-Levin, C.; Bobrowski, K. Pulse radiolysis studies of free radical processes in peptides and proteins. In *Radiation Chemistry: From Basics to Applications in Material and Life Sciences*; Spotheim-Maurizot, M., Mostafavi, M., Douki, T., Belloni, J., Eds.; EDP Sciences: Bonchamp-Les Laval, France, 2008; pp. 233–247.
106. Butler, J.; Land, E.J.; Prütz, W.A.; Swallow, A.J. Charge transfer between tryptophan and tyrosine in proteins. *Biochim. Biophys. Acta* **1982**, *705*, 150–162. [CrossRef]
107. Audette-Stuart, M.; Blouquit, Y.; Faraggi, M.; Sicard-Roselli, C.; Houee-Levin, C.; Jolles, P. Re-evaluation of intramolecular long-range electron transfer between tyrosine and tryptophan in lysozymes. *Eur. J. Biochem.* **2003**, *270*, 3565–3571. [CrossRef]
108. Sicard-Roselli, C.; Lemaire, S.; Jacquot, J.-P.; Favaudon, V.; Marchand, C.; Houee-Levin, C. Thioredoxin Ch1 of Chlamydomonas reinhardtii displays an unusual resistance toward one-electron oxidation. *Eur. J. Biochem.* **2004**, *271*, 3481–3487. [CrossRef]
109. Nauser, T.; Jacoby, M.; Koppenol, W.H.; Squier, T.C.; Schöneich, C. Calmodulin methionine residues are targets for one-electron oxidation by hydroxyl radicals: Formation of S∴N three-electron bonded radical complexes. *Chem. Commun.* **2005**, 587–589. [CrossRef]
110. Ignasiak, M.; Nowicka-Bauer, K.; Grzechowiak, M.; Sikorski, M.; Shashikadze, B.; Jaskolski, M.; Marciniak, B. Sensitized photo-oxidation of plant cytokinin-specific binding protein—Does the environment of the thioether group influence the oxidation reaction? From primary intermediates to stable products. *Free Radic. Biol. Med.* **2021**, *165*, 411–420. [CrossRef]

Review

Direct and Indirect Chemiluminescence: Reactions, Mechanisms and Challenges

Marina A. Tzani [1], Dimitra K. Gioftsidou [1], Michael G. Kallitsakis [1], Nikolaos V. Pliatsios [1], Natasa P. Kalogiouri [1], Panagiotis A. Angaridis [1], Ioannis N. Lykakis [1,*] and Michael A. Terzidis [2,*]

[1] Department of Chemistry, Aristotle University of Thessaloniki, University Campus, 54124 Thessaloniki, Greece; marina_tzani@hotmail.com (M.A.Tz.); dimgpan10@yahoo.gr (D.K.G.); kallitsos29@gmail.com (M.G.K.); npliatsios@gmail.com (N.V.P.); kalogiourin@gmail.com (N.P.K.); panosangaridis@chem.auth.gr (P.A.A.)

[2] Department of Nutritional Sciences and Dietetics, International Hellenic University, Sindos Campus, 57400 Thessaloniki, Greece

* Correspondence: lykakis@chem.auth.gr (I.N.L.); mterzidis@ihu.gr or mterzidi@gmail.com (M.A.T.)

Abstract: Emission of light by matter can occur through a variety of mechanisms. When it results from an electronically excited state of a species produced by a chemical reaction, it is called chemiluminescence (CL). The phenomenon can take place both in natural and artificial chemical systems and it has been utilized in a variety of applications. In this review, we aim to revisit some of the latest CL applications based on direct and indirect production modes. The characteristics of the chemical reactions and the underpinning CL mechanisms are thoroughly discussed in view of studies from the very recent bibliography. Different methodologies aiming at higher CL efficiencies are summarized and presented in detail, including CL type and scaffolds used in each study. The CL role in the development of efficient therapeutic platforms is also discussed in relation to the Reactive Oxygen Species (ROS) and singlet oxygen ($^1O_2$) produced, as final products. Moreover, recent research results from our team are included regarding the behavior of commonly used photosensitizers upon chemical activation under CL conditions. The CL prospects in imaging, biomimetic organic and radical chemistry, and therapeutics are critically presented in respect to the persisting challenges and limitations of the existing strategies to date.

Keywords: chemiluminescence; reaction mechanisms; singlet oxygen; reactive oxygen species; light emission

Citation: Tzani, M.A.; Gioftsidou, D.K.; Kallitsakis, M.G.; Pliatsios, N.V.; Kalogiouri, N.P.; Angaridis, P.A.; Lykakis, I.N.; Terzidis, M.A. Direct and Indirect Chemiluminescence: Reactions, Mechanisms and Challenges. *Molecules* **2021**, *26*, 7664. https://doi.org/10.3390/molecules26247664

Academic Editor: Chryssostomos Chatgilialoglu

Received: 24 November 2021
Accepted: 14 December 2021
Published: 17 December 2021

1. Introduction

Chemiluminescence (CL) is the spontaneous emission of light from an electronically excited state of a species produced by a chemical reaction [1]. Nature uses CL (called bioluminescence) in many living organisms such as fireflies, mushrooms, shells, jellyfish, [2] worms and bacteria, mainly for communication or defense roles [3]. In biological systems, bioluminescence occurs from *in situ* enzyme-catalyzed chemical transformations, for example luciferin [4] reacts with oxygen in the presence of luciferase enzyme [5], magnesium [6–9] or calcium ions and adenosine triphosphate (ATP), leading to luminescence [10–12]. The reaction involves chemical activation of specific molecules (A) via oxidation, resulting in a chemiexcited intermediate (CEI) (C* or D*; excited species indicated by *) that releases its energy either via light emission (direct) or by transferring it, through a chemiluminescence resonance energy transfer (CRET) process, to an adjacent fluorophore (E) that becomes excited (E*); this fluorophore subsequently releases part of its energy by emitting light. These two distinct mechanisms dictate the two types of CL, direct and indirect, respectively, as shown in Figure 1.

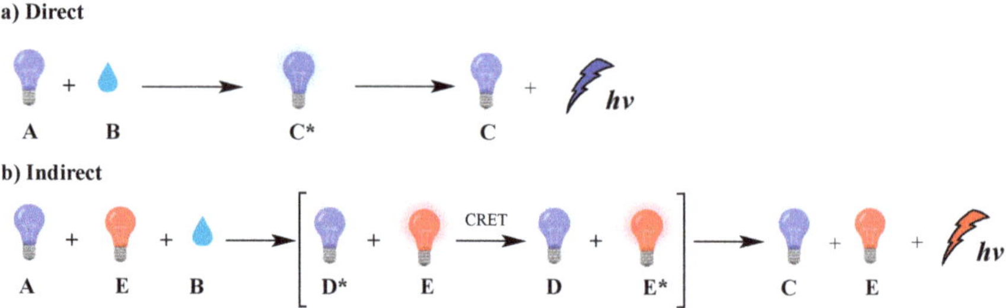

Figure 1. Direct (**a**) and indirect (**b**) chemiluminescence.

Typical examples of molecules that emit light via direct CL are luminol and luciferin, two of the most well-studied luminogenic substances which are susceptible to oxidation by hydrogen peroxide (H_2O_2) and superoxide radical anion ($O_2^{\bullet-}$). The emitted light is the result of changes in the chemical structure of the produced chemiexcited intermediates. Under certain conditions, singlet oxygen ($^1O_2$) can also be produced at high amounts, as chemiexcitation product, by peroxides via direct CL. $^1O_2$ upon radiative relaxation emits infrared light at 1270 nm if not previously trapped, affording oxidation products [13]. The production of $^1O_2$ under CL conditions is discussed later in the text (*vide infra*). In addition, oxalate esters and 1,2-dioxetane derivates serve as efficient precursors both for direct and indirect CL providing tunable emission systems depending on the photophysical properties of the energy acceptor fluorophore, with a wide range of applications in sensing and diagnosis, molecular imaging, and cancer treatment [14,15].

In indirect CL, energy transfer takes place, resulting in excitation of a fluorophore or a photosensitizer that can further act independently. In the presence of molecular oxygen (O_2), the photosensitizer can generate reactive oxygen species (ROS) such as $O_2^{\bullet-}$, hydroxyl radicals ($HO^\bullet$) or $^1O_2$. These species have a broad spectrum of reactivity with biomolecules [16–19] and can be used in photodynamic therapy (PDT) to induce selectively cancer cell death, as well as other chemical transformations [20–23].

Most of the studied CL systems exhibit flash-type single-color light emissions that usually limit their applications. Long-lasting multicolor CL in aqueous solutions is highly desirable, especially in biological applications, however a controllable chemical procedure still remains a challenge. CL is a powerful tool in analytical chemistry that is used for the detection and quantification of reactive oxygen and nitrogen species [24,25], and numerous biological materials such as DNA [26], RNA [27], proteins [28], microorganisms, cells, metals [14,29], etc. [30,31]. Lately, immunoassays have been also used for the detection of SARS-CoV-2 [32].

Moreover, CL has been considered as an alternative photo-uncaging option in drug-delivery technologies for meeting the increasing demand for efficient strategies [15]. However, in biological systems, several issues have to be addressed, such as: (i) the tissue penetration of the incident light used for photoactivation of the emitted photons represents one of major barriers limiting efficiency in imaging or other applications, (ii) light can be scattered or absorbed by other biomolecules inside the cell, and (iii) light intensity is inversely proportional to the square of the distance from the light source, in this way causing weak interaction between an incident photon and a molecule (see Jablonski energy diagram, Figure 2) which leads to poor photochemical interactions. CL has increasingly attracted the interest of the research community, providing solutions to persisting problems of typical photochemical procedures. Recent successful examples from imaging and therapeutics fields highlight the great potential of CL-induced applications. To date, the reported reviews discuss CL mainly focused on chemiluminescent molecules such as luminol, cypridina luciferin and peroxyoxalates or analogues that can be used in sensing, imagine and therapeutic applications [33–36].

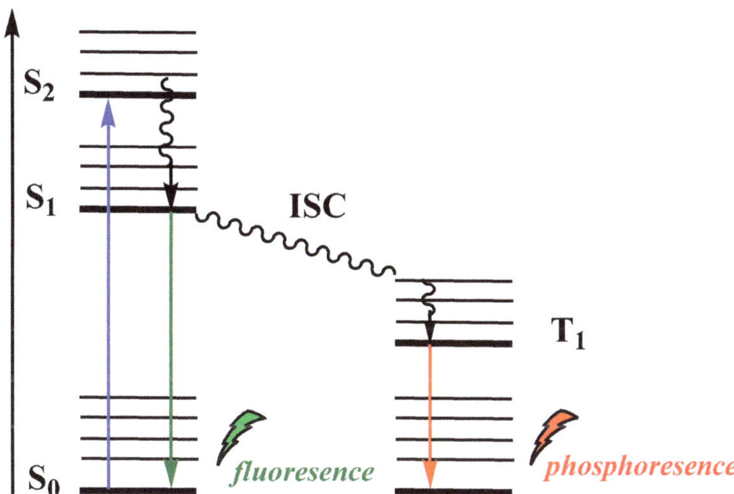

Figure 2. Jablonski diagram showing typical photoexcitation and radiative relaxation.

The aim of this review is to provide a summary of the recent developments in this field, based on direct and indirect CL. An overview of the strategies leading to better analytical sensitivity is included along with recent indicative examples. The role of $^1O_2$ as a chemiexcitation product is also described in relation to the application and further development of more efficient CL-induced therapeutic platforms. Finally, unpublished results of our research regarding the behavior of commonly used photosensitizers as energy acceptors under CL conditions are also included together with discussion on the prospects and the challenges of CL applications in imaging, biomimetic organic and radical chemistry, and therapeutics.

2. Mechanistic Aspects of the Types of Chemiluminescence: Direct and Indirect

Among the different types of luminescence, the most typical ones are photoluminescence (PL), chemiluminescence (CL), bioluminescence (BL), and electrochemiluminescence (ECL). A list of recent references for each luminescence type is provided in Table 1.

Table 1. Types of luminescence based on different activation mode.

Type of Luminescence	Activation Mode	References
Photoluminescence (PL)	Photoexcitation	[37–39]
Chemiluminescence (CL) and Bioluminescence (BL)	Chemiexcitation	[40,41]
Electrochemiluminescence (ECL)	Electrochemical Excitation	[42]

In PL, incident photons are the driving force for the emission of light by a molecule. Upon absorption of a photon of a particular energy, the molecule becomes photoexcited to an electronically excited state (Figure 2). Among the various paths that can be followed to dissipate the excess energy, radiative relaxation can occur via phosphorescence ($T_1 \rightarrow S_0$) or fluorescence ($S_1 \rightarrow S_0$).

Electrochemiluminescence (ECL) is the production of light by an excited species that has been generated electrochemically, i.e., an electron transfer reaction at the surface of an electrode (Figure 3). Numerous review articles and monographs are found in the literature to date on the electrogenerated CL, describing the fundamentals, mechanisms and applications. Thus, in this review we will not further discuss this type of CL [42].

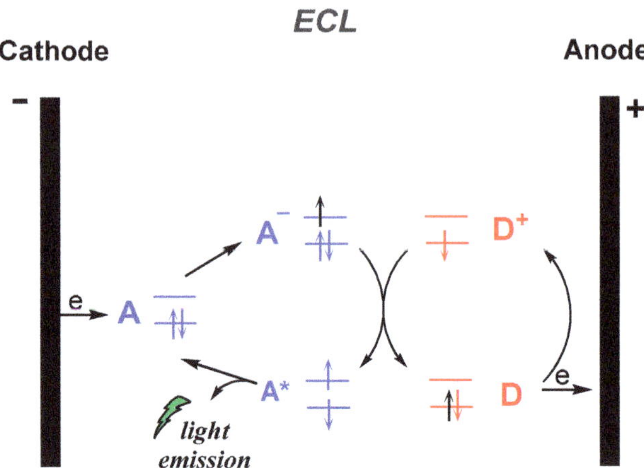

Figure 3. Electrogenerated chemiluminescence (ECL) mechanism.

In bioluminescence (BL), which is naturally occurring CL, light is emitted by live organisms. Various luminophore molecules, enzymes and cofactors are involved in the more than 30 different bioluminescent systems known to date [40]. As was mentioned before, CL is a process in which chemical reactions can generate an excited intermediate who can emit light directly—direct emission—or transfer its energy to an acceptor fluorophore—indirect emission—(Figure 1). The FRET (Förster Resonance Energy Transfer or Fluorescence Resonance Energy Transfer), which refers to the nonradiative transfer of energy from an excited donor to an acceptor mediated by electronic dipole–dipole coupling, between the oxidized CL reagent (excited state donor) and a fluorophore (acceptor) is referred as chemiluminescence resonance energy transfer (CRET). Apart from bioimaging and sensing applications, recent advances in the PDT field revealed that CL could act as an alternative to conventional light sources. The proximity of the light source to the "target" molecule, as is the case for CL, assists the photochemical transformations to progress faster and photon flux requirements are lower in respect to the procedures based on conventional type of light sources [41].

3. Direct Chemiluminescence

To the best of our knowledge, one of the very first examples of direct CL is dated back in 1928 [43], and it is based on the oxidation of luminol by H_2O_2 (Scheme 1) [44,45]. After luminol's oxidation, a strong blue emission with λ_{max} = 425 nm is recorded through a reverse intersystem crossing (rISC) [46–48] that can last from a second up to few hours, depending on the quantity of the reacting species, the presence of specific additives and the "feeding" process. The light emission of luminol and its derivatives can be assisted by catalysts such as peroxidase and heme, which are usually used as additives. For example, horseradish peroxidase (HRP) is involved in the production of a luminol free radical by two molecules of luminol anion [49].

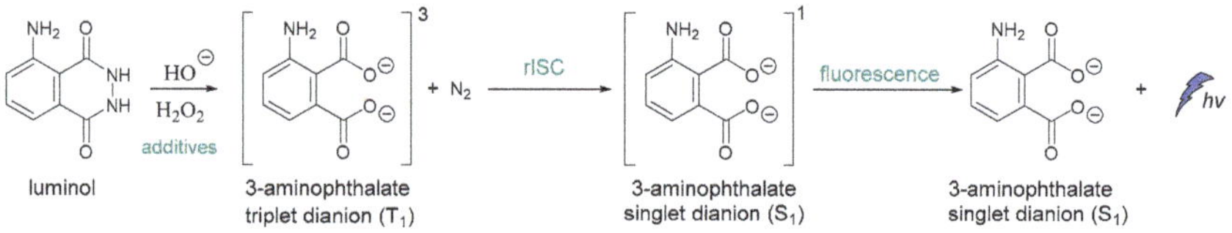

Scheme 1. Luminol chemiluminescence mechanism.

Similarly to luminol, cypridina luciferin emits blue light after oxidation by molecular oxygen catalyzed by cypridina luciferase (Scheme 2) [50–52]. Moreover, emission was also recorded after oxidation of 2-methyl-6-phenyl-3,7-dihydroimidazo[1,2-a]pyrazin-3-one, a cypridina luciferin analogue, by $O_2^{\bullet-}$. At the time of its discovery, this selective oxidation provided a unique opportunity for direct CL detection of $O_2^{\bullet-}$ and further development of other similar chemiluminescent probes [6].

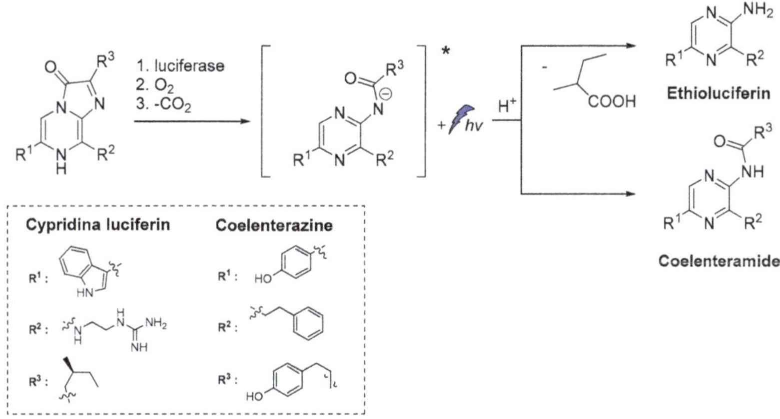

Scheme 2. CL of cypridina luciferin, coelenterazine and their derivatives after oxidation [6–9].

A D-luciferin hydrazine analogue has been synthesized for the selective imaging of Cu^{2+}. The CL reaction relies on the catalytic transformation of hydrazide analogue to D-luciferin that is further converted to oxyluciferin by luciferase in the presence of ATP and Mg^{2+} (Scheme 3) [53].

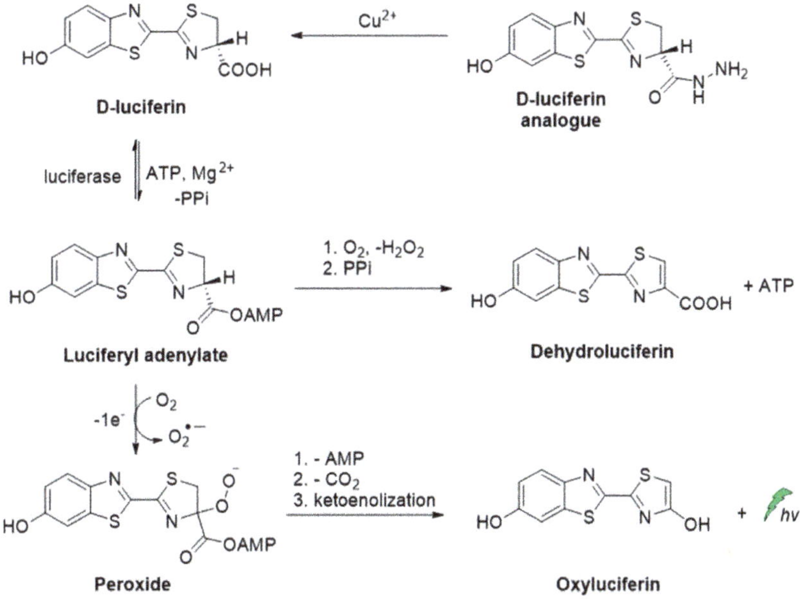

Scheme 3. CL of D-luciferin hydrazine analogue catalyzed by Cu^{2+}.

Besides the commonly used luminol and cypridina luciferin scaffolds, other compounds such as peroxyoxalates, dioxetane, acridinium and chromone derivatives were also developed and have been used lately as efficient, direct CL reagents in various applications. A summary of recent direct CL applications is provided in Table 2 along with details on the system and the scaffold used.

Table 2. Summary of recent scaffolds, systems and applications based on direct chemiluminescence [54].

Scaffold	System	Applications	References
Luminol	Without System, Cobalt-Doped Hydroxyapatite Nanoparticles, Cu-Metal-Organic Frameworks	Monitoring of Advanced Oxidation Processes, Sensing, H_2O_2 and $O_2^{\bullet-}$ Detection	[55–59]
ADLumin-1		Amyloid Beta (Aβ) Species Probe	[60]
Dioxetane	Precursor of Dioxetane Containing the Dicyanomethylchromone Moiety, Functional Self-immolative Molecular Scaffolds	$^1O_2$ Detection in vitro and in vivo, Monitoring of β-Galactosidase Activity—Drug Uncaging, Carbapenemase Activity, Enzymatic Activity of Prostate Specific Antigen (PSA)	[54,61–64]

3.1. Direct Chemiluminescence by Luminol and Its Derivatives

In aqueous luminol solutions, H_2O_2 usually acts as an oxidant causing CL. However, luminol can also be oxidized catalytically by ozone, $^1O_2$, and hypochlorites in the presence of various transition metal ions. In contrast, in aprotic solvents (i.e., DMSO) luminol's CL depends on the pH (basic conditions are required) and the presence of O_2. Recently, it was proposed that the decomposition of the dianionic cyclic peroxide intermediate (CP^{2-}), i.e., 1,2-dioxane-3,6-dione dianion, causes the chemiexcitation of luminol, as shown in Scheme 4. Computational studies have supported a stepwise single-electron transfer (SET) from the amino-phthaloyl to the O−O bond, initiating the decomposition of CP^{2-} [65].

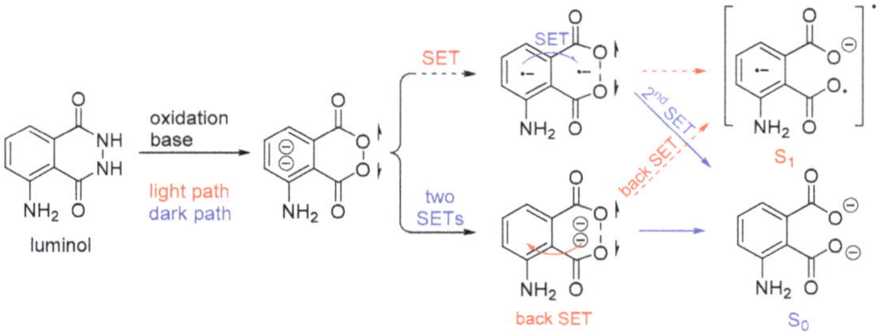

Scheme 4. Proposed luminol chemiexcitation mechanism.

It is also reported that luminol CL could be significantly enhanced in the presence of hydroxylated intermediates generated by persulfate-based advanced oxidation processes (AOPs) of 1,2-divinylbenzene (DVB). Since the emission of light found to be dependent on the oxidation efficiency of the recalcitrant pollutants, luminol CL was proposed (Figure 4) for tracking a wider range of hydroxylated intermediates generated during persulfate-based AOPs under different degradation conditions, and other relevant organic compounds [55].

In most cases, where transition metals are involved, metal-peroxide complexes and metal-luminol or metal-luminol-O_2 intermediates (M-LHOOH) are formed, which are believed to be the key intermediates for the CL of luminol-based systems (Scheme 5) [65].

Cobalt-doped hydroxyapatite nanoparticles (CoHA-NPs) have been reported to enhance luminol CL intensity by more than 75 times compared to the conventional luminol-H_2O_2 system. During investigations on the impact of CoHA-NPs sizes, composition, and crystallinity on the CL intensity, it was found that the maximum CL intensity is achieved by polycrystalline NPs, owing to their larger surface area. The proposed mechanism involves the ionization of H_2O_2 under basic conditions, leading to the formation of hydroperoxide-anions, which subsequently interact with the Co atoms on the NPs surface, forming the metal-hydroperoxide intermediate. The peroxyl radicals, after decomposition of the intermediate, are degraded into $HO^{\bullet}$, which ultimately oxidizes luminol leading to the chemiexcited intermediate (Figure 5) [56].

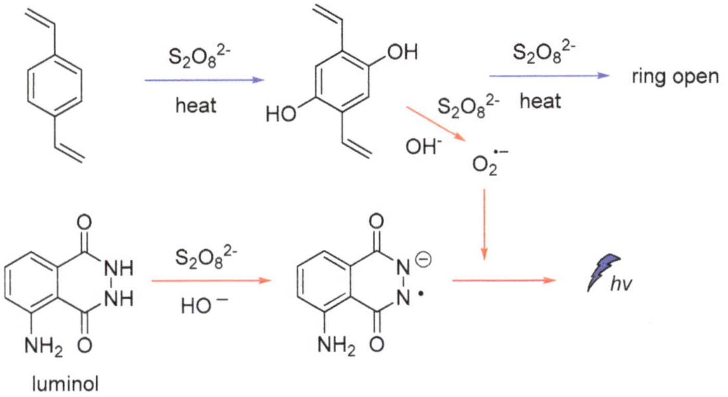

Figure 4. CL mechanism for monitoring hydroxylated intermediates generated on persulfate-based AOPs.

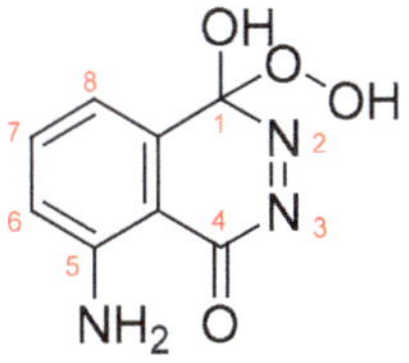

Scheme 5. LHOOH intermediate proposed as the key compound for luminol's CL.

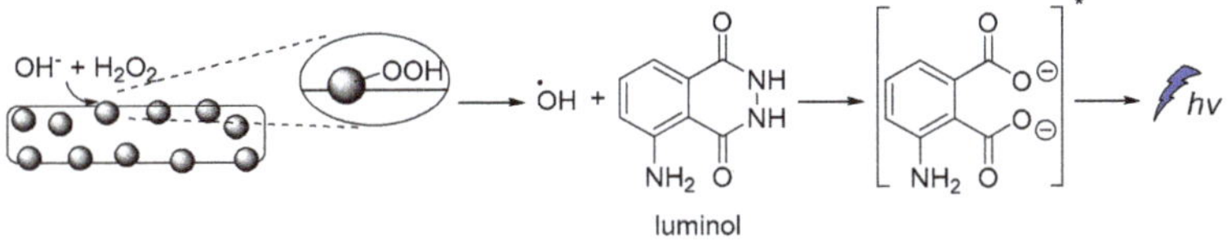

Figure 5. Free radical CL mechanism for the luminol-H_2O_2-Co-HAp NPs.

A new artificial heme-enzyme, so-called FeMC6*a [57], was also found to enhance luminol CL emission signal in comparison to the horseradish peroxidase (HRP)-based sensors (Figure 6), when used in a H_2O_2-detecting bioassay [49,66–71].

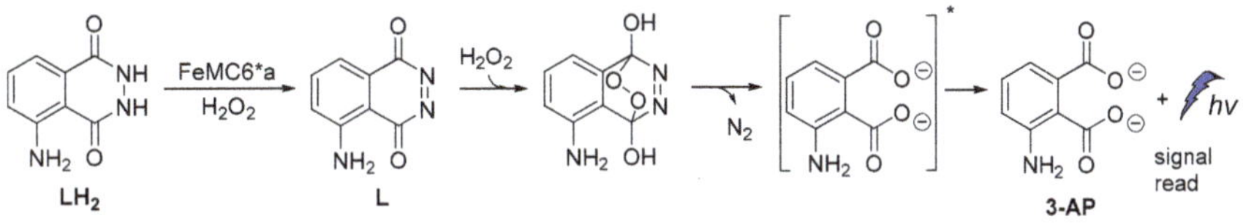

Figure 6. Proposed reaction mechanism of luminol oxidation catalyzed by FeMC6*a.

The catalytic activity of Cu-metal-organic frameworks (Cu-MOFs) with flower morphology, obtained from Cu-based metal-organic gels, was reported to promote persistent

CL in luminol–H_2O_2 systems. According to the proposed mechanism, luminol's CL was attributed to the gradual generation of $HO^{\bullet}$, $O_2^{\bullet -}$ and $^1O_2$ species (Figure 7) and recombination of $HO^{\bullet}$ and $O_2^{\bullet -}$ into $^1O_2$ on the surface of Cu-MOFs, which also resulted in prolonged CL [58].

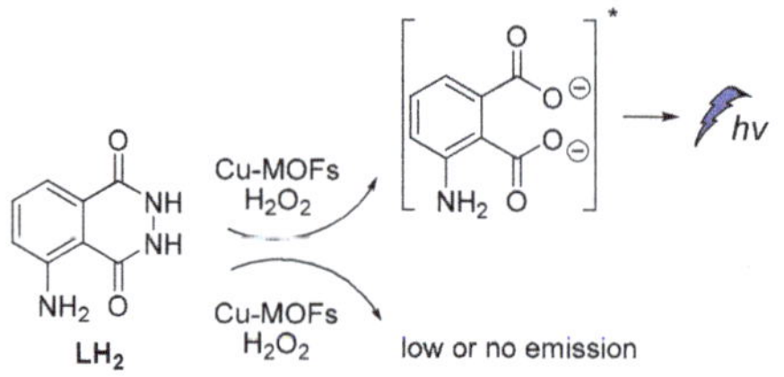

Figure 7. Cu-MOFs catalyzed luminol CL.

Recently, it was also reported that luminol's CL efficiency can be tuned by manipulating the electronic properties of the aromatic ring via substitution, with the 8-methyl substituent providing the highest Φ_{CL} enhancement (Figure 8). *In silico* studies revealed that the 8-substitution assists the cleavage of the O-O bond via the negative charge donation to the *in situ* formed peroxide [72]. Previous studies on the factors that could possibly influence luminol's CL indicated that both steric and electronic effects play a significant role [73].

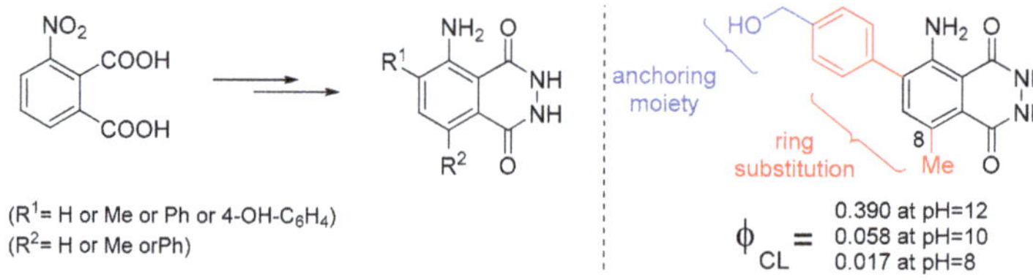

Figure 8. Substitute luminol derivatives with enhanced CL at pH 8, 10 and 12.

3.2. *Direct Chemiluminescence from Probes Other Than Luminol*

Apart from luminol derivatives, isoluminol [73], pyridopyridazines and phthalhydrazides [59] that were studied and used as CL probes [74], other molecules have been also designed and used in various CL applications which are not based on luminol's chemical structure. J. Yang and coworkers reported the synthesis and validation of a new CL molecule, so-called ADLumin-1, as a turn-on probe for amyloid beta (Aβ) species with maximum emission at 590 nm. This probe was then found that can be used in indirect CL via CRET to a curcumin-based NIR fluorescent imaging probe, so-called CRANAD-3, thus more information will be provided in the section on indirect CL (*vide infra*) [60].

Peroxyoxalate chemiluminescence (PO-CL) is one of three most common CL reactions [75]. It is a base-catalyzed reaction of an aromatic oxalic ester with H_2O_2 in the presence of a fluorophore, also commonly referred to as activator (ACT) [76]. When ACT interacts with the *in situ*-generated high-energy intermediate (HEI), usually a 1,2-dioxetanedione, light emission occurs. The process relies on the reaction between the diphenyloxalate and H_2O_2 that generates 1,2-dioxetanedione [77] which spontaneously decomposes to carbon dioxide via a paramagnetic oxalate biradical intermediate [78,79]. Several catalysts have been proposed so far, with imidazole being the most commonly used one [80]. Imidazole acts as a nucleophilic and base catalyst at different reaction steps [81]. Apart from imidazole, only few catalysts appear to work efficiently, and among them are: sodium salicylate, 4-*N*,*N*-dimethylaminopyridine and its derivatives, and a mixture of 1,2,4-triazole and 1,2,2,6,6-pentamethylpiperidine [82,83]. Cabello et al., working on the kinetics of the PO-CL, used three oxalic esters of diverse reactivity (e.g., bis(2,4,6-trichlorophenyl) oxalate (TCPO), bis(4-methylphenyl)oxalate (BMePO), bis[2-(methoxycarbonyl)-phenyl]oxalate (DMO)). The mechanistic studies performed in the presence of sodium salicylate in partially aqueous medium showed that the reaction rate constants were determined by H_2O_2 and sodium salicylate amounts (Scheme 6). Further studies showed that by increasing the salicylate concentration there is a decrease in the CL efficiency, probably due to the interaction between salicylate and the in situ-produced HEI. From experiments using different oxalic esters, a variation in the CL emission was reported, with DMO showing the highest quantum yields in the aqueous medium used (1:1 *v*/*v* of DME/H_2O) [84].

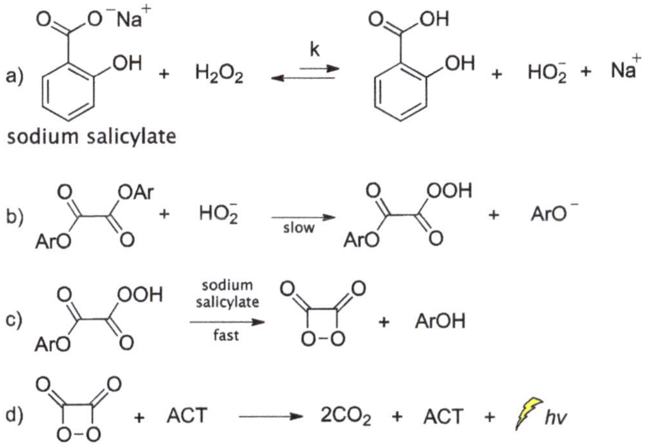

Scheme 6. Mechanism of PO-CL between an oxalic ester and H_2O_2, occurring via specific base catalysis (SBS) by sodium salicylate, through the formation of (**a**) hydroperoxy anion, (**b**) monoperoxalic ester and (**c**) 1,2-dioxetanedione, which subsequently leads to (**d**) carbon dioxide and light emission.

When 2,6-lutidine was used as a non-nucleophilic base catalyst on the TCPO CL reaction, it was found that the concentrations of 2,6-lutidine and H_2O_2 play a crucial role in the rate-determining step. The reaction was found to follow a general base catalysis mechanism (Scheme 7) [85].

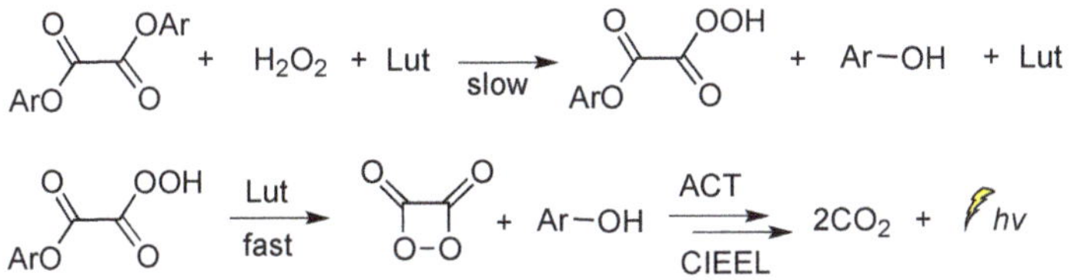

Scheme 7. Reaction mechanism of TCPO with H_2O_2 in DME under general base catalysis by 2,6-lutidine (Lut) towards CL emission in the presence of a fluorophore.

Among other CL molecules of interest are the 1,10-phenanthroline dioxetanes, generated after treatment of 1,10-phenanthroline with IO_4^- in the presence of H_2O_2. It has been reported that 1,10-phenanthroline dioxetanes enhance light emission in respect to the IO_4^--H_2O_2 system alone. The proposed radical-based mechanism involves the C^5=C^6 bond of 1,10-phenanthroline that is initially attacked by $HO^\bullet$, leading to a dioxetane intermediate responsible for the CL emission (Scheme 8) [86]. The CL of the IO_4^-/peroxide system and the factors influencing the emission have been recently reviewed elsewhere [87].

Investigation into the CL efficiency of four bicyclic dioxetanes bearing 2-hydroxybiphenyl-4-yl (**a**), 2-hydroxy-*p*-terphenyl-4-yl (**b**), 2-hydroxy-*p*-quaterphenyl-4-yl (**c**), or 2-hydroxy-*p*-quinquephenyl-4-yl (**d**) (Figure 9) moieties, showed that **a-c** emit light from 466 nm to 547 nm during their decomposition while **d** was found to emit light of much weaker intensity from 490 nm up to 607 nm, depending on the solvent (water, acetonitrile or

DMSO). The CL efficiencies of the latter dioxetanes in aqueous media were found to be enhanced in the presence of additives such as *β*-MCD (*β*-methylated cyclodextrin) or TBHP (tributylhexadecylphosphonium bromide) [88].

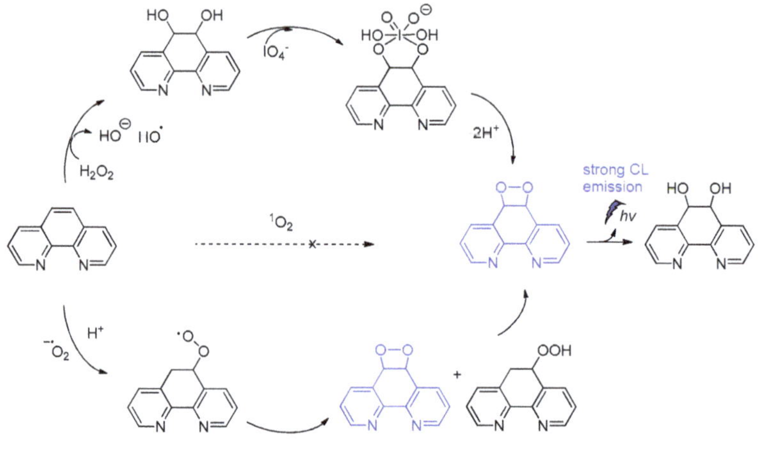

Scheme 8. CL of 1,10-phenanthroline dioxetanes.

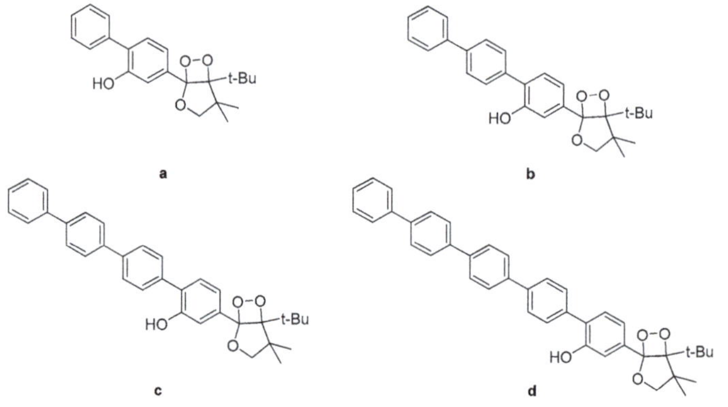

Figure 9. Bicyclic dioxetanes bearing 3-hydroxyphenyl substituted with a p-oligophenylene moiety. (**a**) 2-hydroxybiphenyl-4-yl; (**b**) 2-hydroxy-*p*-terphenyl-4-yl; (**c**) 2-hydroxy-p-quaterphenyl-4-yl; (**d**) 2-hydroxy-*p*-quinquephenyl-4-yl.

M. Yang et al. constructed a NIR chemiluminescent probe by caging the precursor of Schaap's dioxetane scaffold and a dicyanomethylchromone acceptor for selective $^1O_2$ detection. In the presence of $^1O_2$, the probe is oxidized, affording the phenol-dioxetane moiety which undergoes spontaneous bond cleavage leading to the formation of the chemiexcited product responsible for the CL emission (Figure 10). With this probe, they were able to detect $^1O_2$ in vitro with a turn-on bathochromic CL signal in the NIR region at 700 nm and image intracellular $^1O_2$ produced by a photosensitizer under simulated photodynamic therapy conditions [54]. Similarly, hyperconjugated push-pull dicyanomethylchromone systems were used for the *in vivo* detection of H_2O_2 and *β*-galactosidase activity [61].

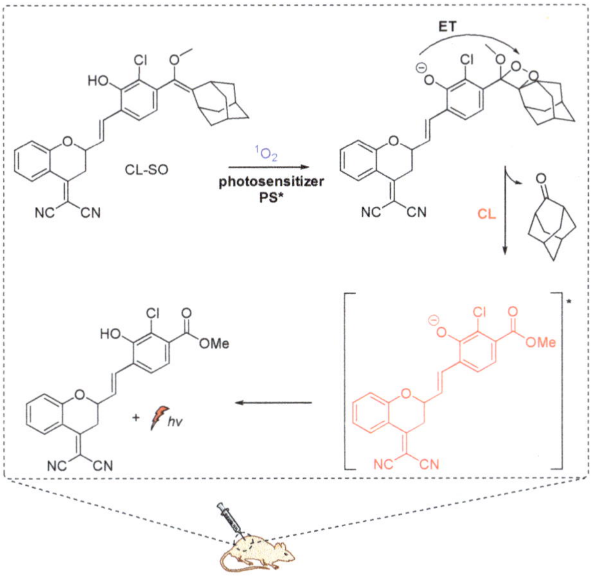

Figure 10. Chemiluminescent probe activated by $^1O_2$ with maximum emission at 700 nm.

The CL emission mechanism proposed for the Schaap's dioxetane probes [89–91], involves Chemically Initiated Electron Exchange Luminescence (CIEEL) [1,92] with a solvent-cage back electron-transfer step, which affords the 2-adamantanone and the chemiexcited phenolate that is responsible for the light emission (Scheme 9) [92,93].

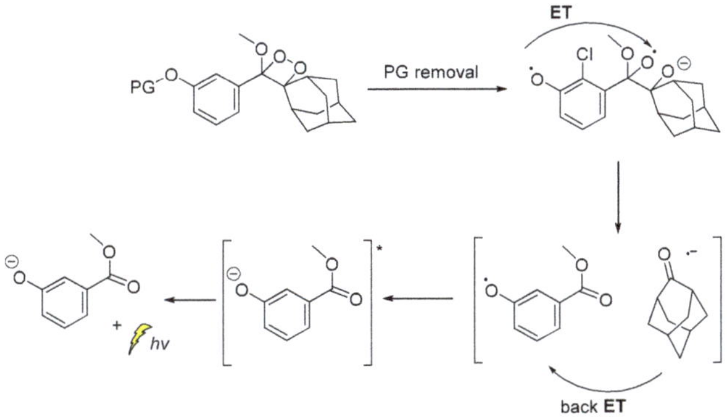

Scheme 9. Proposed CIEEL mechanism for triggered CL of Schaap's dioxetane probes.

Recently, modified ortho-nitrobenzyl and spiropyran 1,2-dioxetane probes have been synthesized and studied towards photoactivation by 254 nm or 365 nm photons. Irradiation of ortho-nitrobenzyl 1,2-dioxetanes with 254 nm light resulted in the formation of the chemiexcited phenolate after deprotection, while the spiropyran 1,2-dioxetane irradiation at $\lambda = 254$ nm afforded the open merocyanine chromophore intermediate that was finally converted to the corresponding excited phenolate with maximum emission at 500 nm. This open merocyanine intermediate was able to act as a photoswitch, thus it could return to its spiropyran form under white light irradiation subsequently after UV light irradiation (Scheme 10) [94].

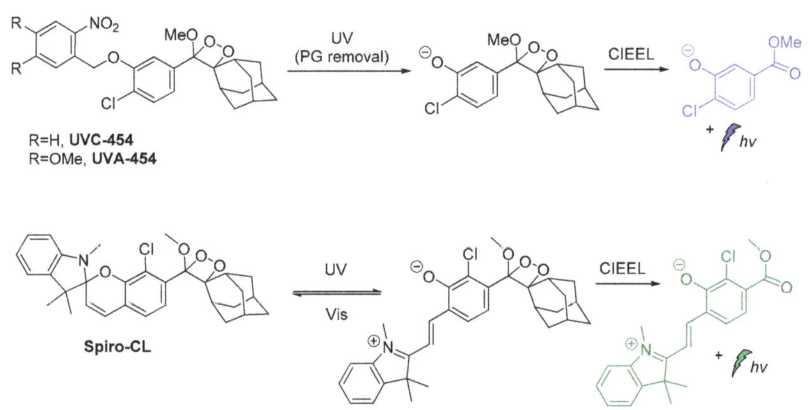

Scheme 10. Modified 1,2-dioxetane probes activated by UV light.

Shabat and coworkers described the development of functional self-immolative molecular scaffolds based on Schaap's dioxetane probes, and described their use in signal amplification and real-time monitoring of enzymatic activity or drug release. The general mode of action involves a triggering event for releasing a chemiexcited product, the CL reporter unit, and the payload which can be a reagent molecule, a drug or a prodrug. This strategy was used successfully in monitored release of the cytotoxic agent monomethyl auristatin E (MMAE) from prodrug **Gal-A-Pept** (Scheme 11). The prodrug is initially recognized by a β-galactosidase, releasing intermediate **A-Pept** which undergoes 1,6-elimination to afford the MMAE drug and quinone-methide **A**. Then, intermediate **B** is produced by the reaction of **A** with H_2O, subsequently releasing 2-adamantanone and the chemiexcited phenolate **C***. Green emission at 550 nm is observed due to radiative decay of the phenolate to the ground state [62].

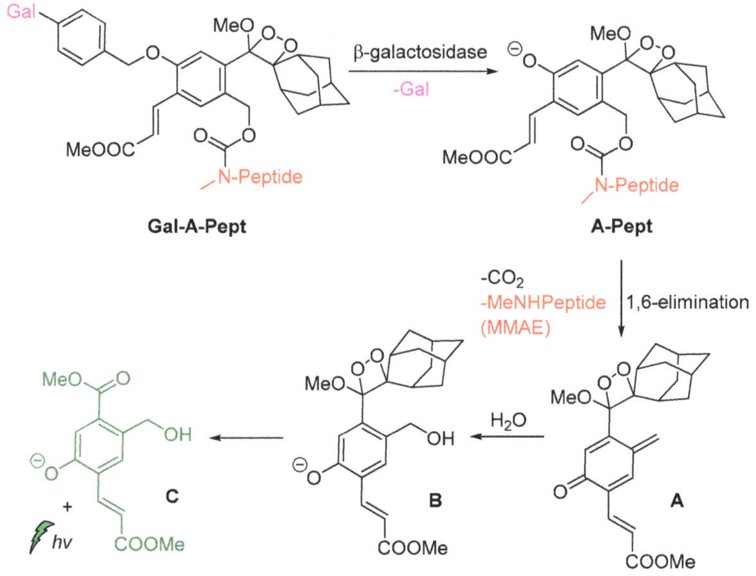

Scheme 11. Activation mode of the self-immolative chemiluminescence prodrug **Gal-A-Pept** by β-galactosidase.

Carbapenemase-producing organisms (CPOs) pose a severe threat to antibacterial treatments due to the acquisition of antibiotic resistance. Recently, Das et al. demonstrated the first carbapenemase CL probe (CPCL) for the detection of carbapenemase and CPOs. The unique structural design of CPCL enables the probe to rapidly disassemble via an efficient 1,8-elimination process upon hydrolysis of the β-lactam ring by a carbapenemase (Figure 11). This process results in the release of a chemiexcited phenolate that undergoes radiative relaxation, producing photons at 540 nm that enable a rapid and selective detection of carbapenemase activity [63].

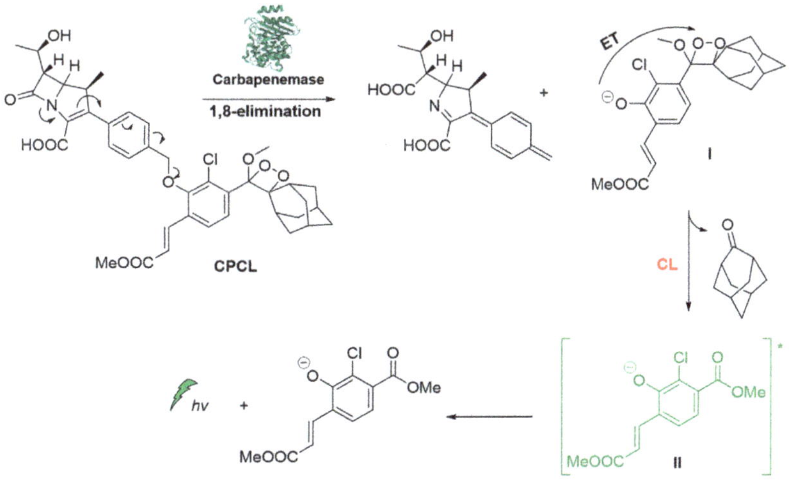

Figure 11. Carbapenemase activity monitoring based on chemiluminescent probe.

The same as above reporter for monitoring carbapenemase activity CL was also used for detection of prostate-specific antigen (PSA) enzymatic activity by using artificial metalloenzymes (ArMs). The probe activation mechanism is based on a catalytic cleavage of a specific peptidyl substrate, followed by a release of a phenoxy-dioxetane intermediate affording a chemiexcited benzoate that emits a green photon. The benefit and simplicity of a CL assay to detect seminal fluid was effectively demonstrated by on-site measurements using a small portable luminometer [64].

Other efficient CL molecules are based on acridinium ester dyes analogues. The acridan dyes are quaternary acridine derivatives that become luminescent directly after their activation [95–100]. The mechanism of CL involving a dioxetane intermediate was first introduced by McCapra (Scheme 12) [101,102]. The simplicity of their synthesis and the high quantum yield (up to ca. 7%) [102], together with their ability to become oxidized under alkaline conditions, making acridinium derivatives highly suitable for sensing and chemiluminometric analysis [95,103]. Research on the design of new acridinium derivatives with even higher activity is an ongoing process [104].

Acridine-9-carboxylate derivatives [e.g., (4-(methoxycarbonyl)phenyl-10-methyl-10λ^4-2,7-dimethyl-acridine-9-carboxylates)] were used for studying the CL efficiency in the 7–10 pH range. Reactions of acridine-9-carboxylates with H_2O_2 result in the formation of a relatively unstable dioxetanone intermediates. The chemiexcited methyl-acridone products formed after the decomposition of the dioxetanone intermediates are responsible for light emission. Among the compounds studied, 4-(methoxycarbonyl)phenyl- derivative gave the highest CL intensities at pH 7−10, indicating that the introduction of electron-donating groups at the 2,7-positions on the acridine moiety plays a crucial role in the whole process (Scheme 13) [105].

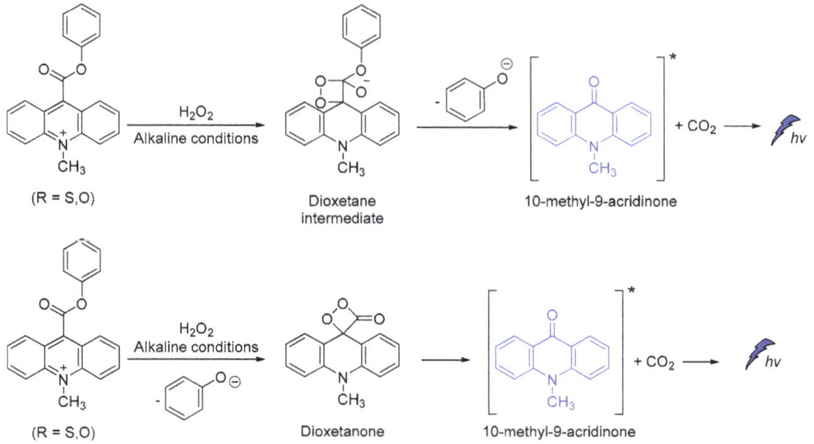

Scheme 12. Proposed CL mechanism of acridinium derivatives.

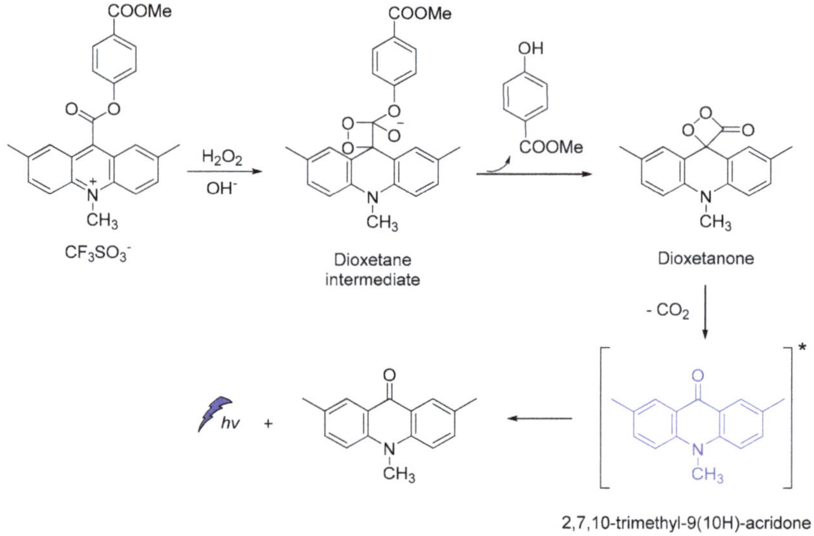

Scheme 13. Proposed CL mechanism of 4-(methoxycarbonyl)phenyl-10-methyl-10λ^4-2,7-dimethyl-acridine-9-carboxylates.

4. Indirect Chemiluminescence

In the indirect CL process, energy is transferred by the chemiexcited molecule (energy donor, D) to a second molecule (acceptor, E), which is further excited during the process. In the resonance excitation transfer mechanism, D and E are not in contact and may be separated by as much as 10 nm, although 1 nm or smaller transfer distances are more convenient. Except for distance, other factors dictating the energy transfer include the lifetime of the excited state of D and the degree of the overlap between the emission spectrum of D and the absorption spectrum of E. Usually, the higher overlapping the more efficient the energy transfer from D to E. The solvent is another crucial factor for the energy transfer between D and E molecules, however this effect includes a lot of specific characteristics properties of each system, such as the solvent cage factor. The short-range energy transfer by the wavefunction overlap between the donor and the acceptor, meaning overlap between the electronic clouds, is also a crucial factor. Particularly, both the Förster Resonance Energy Transfer or Fluorescence Resonance Energy Transfer (FRET) and Dexter Energy Transfer, with the second occurring when the D–E wavefunctions overlap, contribute to the energy transfer of the system, promoting its scope. FRET, which refers to the nonradiative energy transfer from an excited donor to an acceptor mediated by electronic dipole–dipole coupling, between a CL reagent as the excited donor and a fluorophore acceptor, is referred to as chemiluminescence resonance energy transfer (CRET) [106].

Two different modes of indirect CL emission can be distinguished, intermolecular and intramolecular CL, as shown in a simplified manner in Scheme 14A,B, respectively. After the triggering event, chemiexcited D is formed and then its energy is transferred to E, e.g., a photosensitizer (PS). When E is not directly linked with the chemiexcited species (Scheme 14A), the intermolecular distance between D and E (photosensitizer) plays crucial role in the energy transfer, with a direct emission ("escaping" photons) taking place together with the secondary emission. Sometimes this mode of action is crucial for distinguishing the two different events and their efficacy rate: one of the initial CL emission (direct emission) and the second is processed from the excited acceptor properties (Scheme 14A, secondary emission). On the other hand, in the case of an intramolecular system, the chemiexcited donor (D) is linked with the energy acceptor molecule (E) and a secondary emission pathway is favored (Scheme 14B).

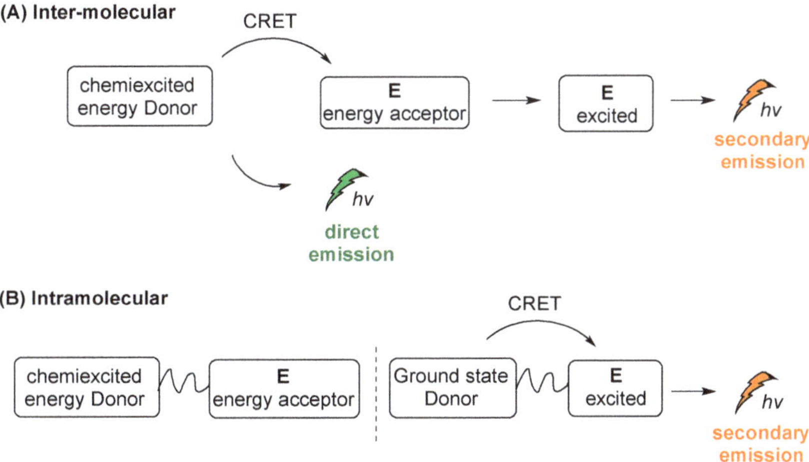

Scheme 14. Intermolecular (**A**) and intramolecular (**B**) indirect CL emission modes.

The absorbance band of conventional photosensitizers such as Rose Bengal and TetraPhenylPorphyrin (TPP) (Type II photosensitizers) [107–110], as well as $W_{10}O_{32}{}^{4-}$ (Type I′ photocatalyst) [111–113], fit well into the requirements for efficient CRET and high CL emissions [114]. Quantum Dots (QDs) have been also used successfully in various CRET studies. QDs are luminescent semiconductor nanocrystals with very distinct, size-dependent physicochemical behavior [106]. It should be noted that the excited photosensitizer or QDs can trigger various reaction cascades. For example, in the presence of molecular oxygen they can result in the production of ROS (e.g., $^1O_2$, $HO^{\bullet}$, $O_2{}^{\bullet-}$). This function is used in CL-induced photodynamic therapies and is described later in the text in more details (*vide infra*) [115].

A summary of recent direct CL scaffolds, systems and their applications are provided in Tables 3 and 4, based on intra- and inter-molecular modes of reaction.

Table 3. Summary of recent scaffolds, systems and applications based on intermolecular indirect chemiluminescence.

Scaffold	System	Applications	References
ADLumin-1	CRANAD-3	Aβ Species Probe	[60]
Luminol	Chlorine6 Containing QDs Polymeric NPs, Fe(III) Deuteroporphyrin IX Chloride NPs, CdSeTe Core QDs	H_2O_2 Imaging and Therapy	[116–118]
Peroxyoxalate	9,10-Diphenylanthracene	Cationic Polymerization	[119]
Dioxetane	Emerald^{TM} and Emerald II^{TM} Enhancers, Carboxy-SNARF-1 Dye	Imaging of β Galactosidase and Nitroreductase Activities, H_2S, Ratiometric pH	[116,120–122]

Table 4. Summary of imaging, sensing and therapy approaches based on intramolecular indirect chemiluminescence.

Scaffold	System	Applications	References
Dioxetane	Molecular pH-Sensitive Carbofluorescein Probe Precursor of Dioxetane	Ratiometric pH Imaging in Live Animals	[123]
Cypridina Luciferin	Sulforhodamine 101	$O_2^{\bullet-}$ Detection	[124]
Phenoxy-Dioxetane 7-Hydroxycoumarin	FRET Quencher	Monitoring of Matrix Metalloproteinase Activity	[125]

4.1. Intermolecular Chemiluminescence Emission Mode

J. Yang and coworkers, as mentioned above (*vide supra*), reported on the ADLumin-1 probe for Aβ species (Figure 12). To overcome problems related to short emission of ADLumin-1 they used the CRET strategy for transferring the energy to **CRANAD-3**, a curcumin-based, NIR fluorescent imaging probe emitting in the near-infrared region, at 730 nm (*vide supra*) [60].

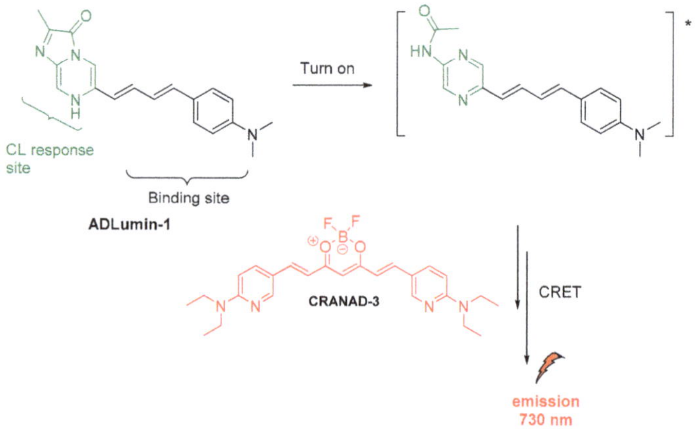

Figure 12. Indirect CL from **ADLumin−1/CRANAD-3** system.

Spiroadamantane 1,2-dioxetanes have been used in CL assay for *in vivo* visualization of β-galactosidase enzymatic activity in transgenic mice. For achieving this, an accelerant (diethanolamine in buffer containing EmeraldTM enhancer, T2081) was administered along with the β-D-galactopyranoside trigger (3-chloro-5-(5′-chloro-4-methoxyspiro[1,2-dioxetane-3,2′-tricyclo[3.3.1.13,7]decan]-4-yl)phenyl β-D-galacto pyranoside Galacton Plus). Thus, the energy was transferred by the produced chemiexcited intermediate to the Emerald^{TM} accelerant, causing emission at 540 nm (Figure 13) [120].

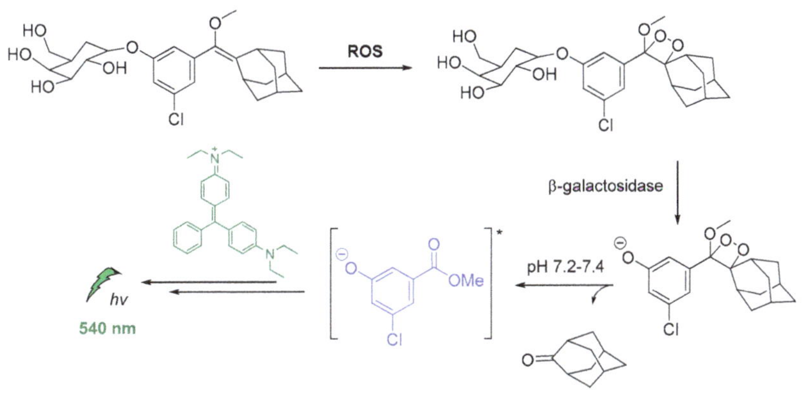

Figure 13. Indirect CL via energy transfer from activated by β-galactosidase 1,2-dioxetane to the Emerald[TM] enhancer emitting at 540 nm.

Other modified 1,2-dioxetane probes were also synthesized and used for *in vivo* imaging of hydrogen sulfide (H_2S). The reduction of the azide on the *para*-azidobenzyl carbonate moiety triggered the self-immolative cleavage, affording the chemiexcited phenolate. Then, the energy transfer via CRET to the Emerald II[TM] CL enhancer [121] allowed detection at 545 nm (Figure 14) [122].

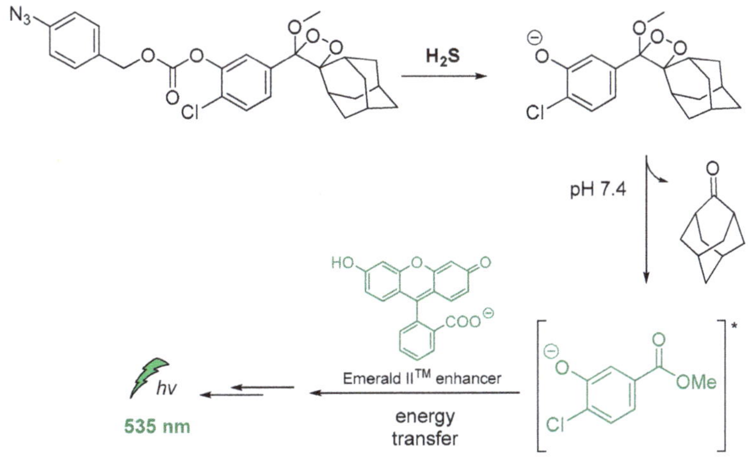

Figure 14. Indirect CL via energy transfer from H_2S-activated 1,2-dioxetane to the Emerald II[TM] enhancer emitting at 535 nm.

Another important study by Lipper et al., described that CL energy transfer from Schaap's dioxetane probe can be efficiently used for ratiometric pH imaging. Their approach was based on energy transfer after the emission of phenolate during radiative relaxation to the pH-sensitive dye carboxy-SNARF-1 in CL-enhancer solutions, providing thee relative intensities at 535 nm, 585 nm and 650 nm due to relaxation of the protonated and deprotonated dye, respectively (Figure 15) [116].

Recently, a supramolecular strategy was reported, achieving a highly efficient CL-initiated cascade FRET by following a three-criteria rule: (i) D emission and the E absorption spectral overlapping; (ii) 10 nm or less as the D–E distance; and (iii) orientation of the D and E dipoles (see also Scheme 14). For this purpose, a nanoassemble β-cyclodextrin (β-CD) was used in an aqueous luminol/H_2O_2 solution in the presence of various fluorophores, which enable an accelerated multicolor CL with flexible emission wavelength in the range of 410–610 nm. Hydroxypropyl methylcellulose was utilized to slow down the diffusion rate, while the reaction was buffered by addition of solid $Ca(OH)_2$. Moreover, it was found that Co^{2+}-chitosan increased the CL emission of the system when merged with the β-CD nanoassembly system [126].

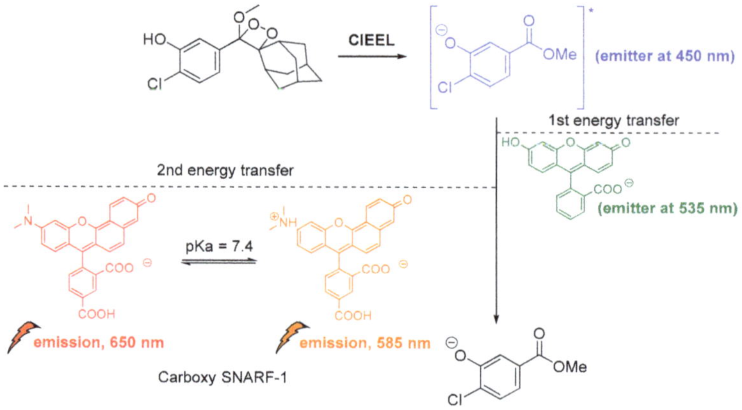

Figure 15. Indirect CL via energy transfer from 1,2-dioxetane to the Emerald II^{TM} enhancer and subsequently to the carboxy-SNARF-1 dye (pKa = 7.4), emitting at 585 nm and 650 nm depending on the pH.

Another interesting CL system of cyclic peroxides was catalytically produced by tetrahydrofuran hydrogen peroxide in the presence of BSA-stabilized Gold Nanoclusters (Au@BSA NCs). Gold nanoclusters were found to accelerate the decomposition of the THF-hydroperoxides and promoted CRET between the chemiexcited dioxetane derivatives and the Au@BSA NCs (Figure 16). The CL emission (λ_{max} = 650 nm) was found to be further enhanced by the presence of copper ions [127].

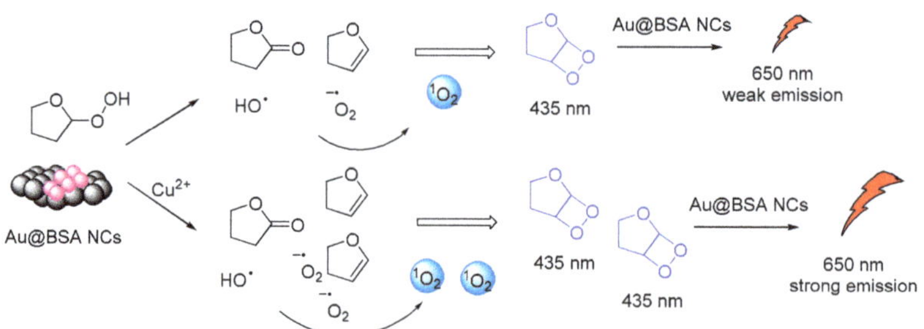

Figure 16. CL mechanism of the Au@BSA NCs-THF-CPO-Cu^{2+} system.

From the investigation into the CL behavior of an amphiphilic conjugate (defined as CLP) NPs, it was revealed that the NPs consisting of chlorine6 (Ce6), conjugated with luminol and polyethylene glycol undergo CRET with energy transfer to Ce6 upon triggering with H_2O_2. The CLP NPs were tested in cancer cells with high expression of H_2O_2 and monitored at 660–740 nm. In addition, excited Ce6 was capable of producing $^1O_2$ in situ PDT against H_2O_2-high tumors, inhibiting lung metastasis. *In vitro* and *in vivo* experiments showed that CLP NPs could be a promising system in nanotheranostics, providing imaging and therapy, simultaneously, particularly in tumors with high levels of H_2O_2 [128].

Black Phosphorene (BP) QDs were initially studied in CL experiments upon reaction with H_2O_2 and HClO by Y. Lv and coworkers [129]. The same group two years later reported on the extension of the study with NH_2-functionalized BP QDs (N-BPQDs) and showed that surface modification dictates the energy gap and electron mobility, thus favoring CL emission at 500 nm after triggering by persulfates [130].

Luminol's analogue, L012, that shows 100 times higher than luminol CL intensity after oxidation by H_2O_2 at pH 7, was also used as a CRET donor. *In vitro* and *in vivo* studies with polyenthylene(glycol)-coated CdSeTe core QDs, used as energy acceptors, demonstrated that they can image H_2O_2 in live cells, pointing out the high potential of the system in the diagnosis of H_2O_2-associated diseases. The chemiexcited QDs were able to emit light at 780 nm, thus allowing monitoring [117].

The same luminol analogue L012 and H_2O_2 were recently studied in the presence of Fe(III) deuteroporphyrin IX chloride-polymer dots (FeDP-Pdots) as CL catalysts towards cancer therapy, providing high concentrations of ROS due to high CRET efficiency (Figure 17) [118].

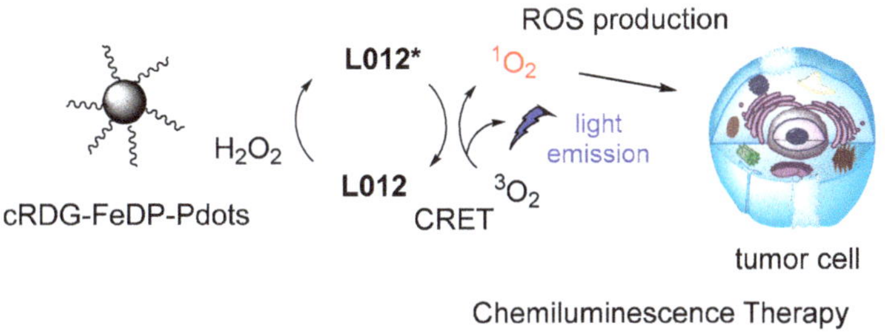

Figure 17. CL dynamic therapy based on L012/H_2O_2 catalyzed by FeDP-Pdots.

CL has been also used in polymerization reactions. Linear polymer chains were synthesized along with cross-linked functional polymeric networks as a result of CL photoinitiated cationic polymerization based on a sulfonium salt (Figure 18). The polymerization reaction initiated CL emission at 430 nm. The photoinitiator decomposition after excitation produced protonic acids that were capable of initiating cationic polymerization of oxirane (epoxides) and vinyl monomers. The presence of iodonium salt in the initiating system was found to increase the polymerization efficiency [119].

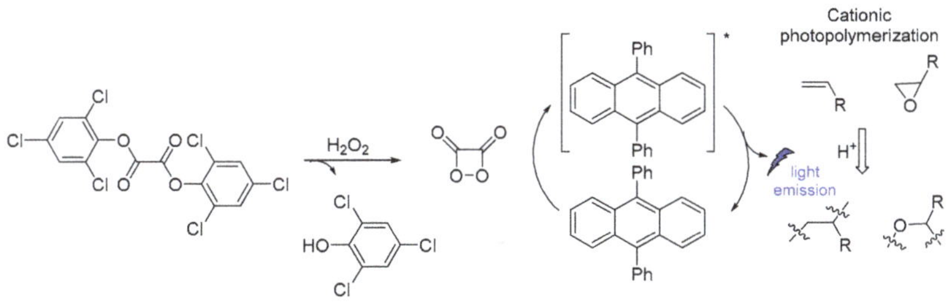

Figure 18. CL induced cationic polymerization by using sulfonium salt as a photoinitiator.

4.2. Intramolecular Chemiluminescence Emission Mode

An intramolecular CRET approach based on chemically initiated electron exchange luminescence (CIEEL) has been recently adapted, while developing the first single molecule ratiometric CL probe for imaging pH in live animals. The proposed direct–indirect emission hybrid methodology, with emissions at 530 nm and 580 nm, relies on an acrylamide 1,2-dioxetane scaffold linked via a piperazine linker to a pH-sensitive carbofluorescein. This methodology withholds many characteristics that make it suitable for applications in quantifications of other important analytes produced in vivo (Figure 19) [123].

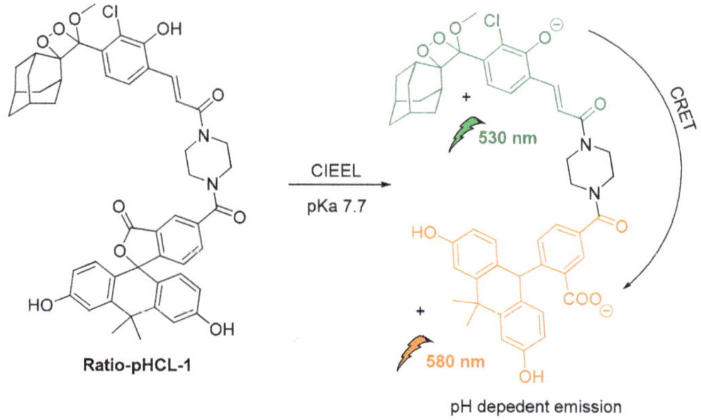

Figure 19. Design and mechanism of ratio-pHCL-1 (CIEEL).

Teranishi proposed a cypridina luciferin analogue for detecting $O_2^{\bullet-}$ via intramolecular CRET to sulforhodamine 101. The energy transfer from the oxidized imidazopyrazinone to the acceptor dye results in light emission at 610 nm (Scheme 15). A similar approach was studied earlier with fluorescein analogue FCLA, emitting at 532 nm [124].

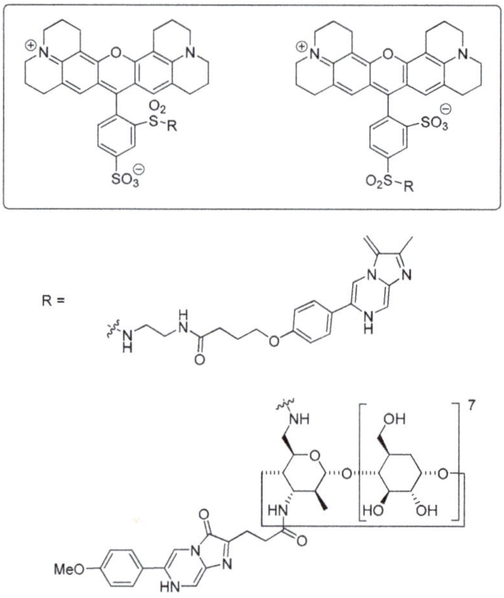

Scheme 15. Chemical structures of chemiluminescent probes emitting light at 610 nm.

A novel on–off CRET system based on phenoxy-dioxetane 7-hydroxycoumarin scaffold for light emission, upon slow chemiexcitation in aqueous solution, was recently reported. The energy donor is covalently linked to a quencher via a peptide linker that is recognizable by metalloproteinases. In the absence of enzymatic activity CRET is on, thus the energy is transferred to the quencher. When the peptide linker is cleaved by enzymes CRET is off, thus light emission is observed (Figure 20). The system was tested in cancer cells, showing good detection levels in matrix metalloproteinase activity [125].

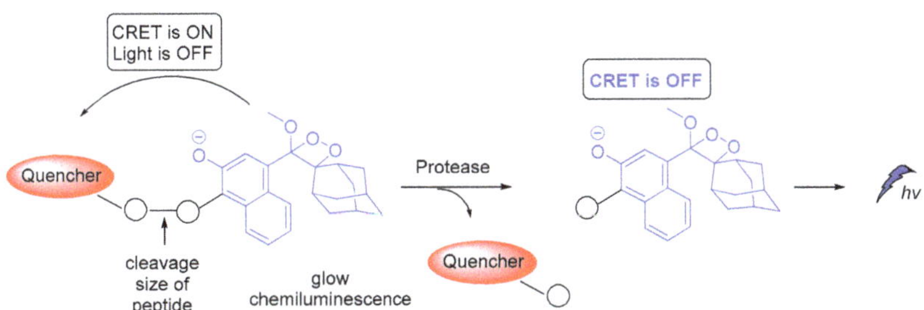

Figure 20. Chemiluminescence activation mechanism of a Glow-CRET system.

5. The Dual Role of Singlet Oxygen in Chemiluminescence

Singlet oxygen was discovered and confirmed spectroscopically in the early 1960s when the H_2O_2/hypochlorite system was at the epicenter of the CL studies [131–138]. Its emission at 1270 nm was attributed to the monomeric singlet oxygen and the emissions at 633 nm and 703 nm, to its dimeric aggregate form (Scheme 16). It is noteworthy that the energy difference between the ground state of O_2 ($^3\Sigma_g$) and its singlet excited state ($^1\Delta_g$, singlet oxygen, $^1O_2$) is 22.5 kcal/mol. Moreover, its lifetime depends significantly on the solvent (e.g., ~59 ms in carbon tetrachloride, 9.8 ± 0.6 μs in methanol-h_4, 77 ± 4 μs acetonitrile and 3.7 ± 0.4 μs in water) [139,140].

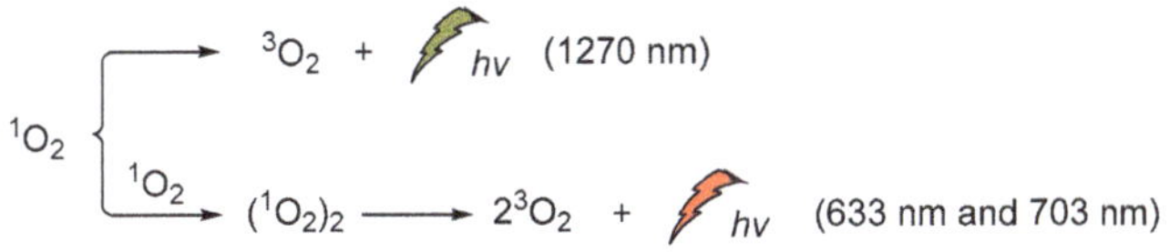

$$^1O_2 \rightarrow {}^3O_2 + h\nu \ (1270\ \text{nm})$$

$$^1O_2 \xrightarrow{^1O_2} (^1O_2)_2 \rightarrow 2\,^3O_2 + h\nu \ (633\ \text{nm and } 703\ \text{nm})$$

Scheme 16. $^1O_2$ chemiluminescence with emissions at 633 nm, 703 nm and 1270 nm.

During recent decades, an increasing number of studies have focused on exploring the role of $^1O_2$ in chemistry, biology, medicine and environmental and material sciences. Lately, the increasing interest of the scientific community in novel therapies has led to further exploration of $^1O_2$ against persisting diseases (i.e., cancer). Thus, a significant amount of the new studies are now focused on the generation modes and the photophysical properties [141], dictating its interaction with other biomolecules in live cells. On the other hand, $^1O_2$ is a very important oxidizing agent in organic chemistry transformations. There are two main approaches for generating $^1O_2$. The first, and more famous one, involves light and a photosensitizer, while the less famous one is hydroperoxide-based and it is often called the "in dark" process, since $^1O_2$ is produced in the absence of light. To avoid confusion between the CL-induced $^1O_2$production and the CL of the $^1O_2$ itself, we should distinguish here the two completely different CL. The energy transfer and radiative processes of the CL energy acceptors or chemiexcited intermediates are illustrated in Figure 21.

The majority of the recent photochemical or CL studies on $^1O_2$ applications employ well-known and established methodologies for generating an efficient amount of the oxidizing agent based on photosensitizers such as TTP, Rose Bengal, Cu^{2+} [142], Ru^{2+} and Ir^{3+} complexes, phthalocyanines [143,144] and other analogues. Kellett and his coworkers have presented a class of di-copper (II) complexes based on the synthetic chemical nuclease $[Cu(Phen)_2]^+$ (where Phen = 1,10-phenathroline), that were selective against solid epithelial cancer cells from line panel NCI-60. Among other ROS, their agents are found to catalyze intracellular $^1O_2$ contributing to oxidative DNA damage [145]. Recently, our team has also explored the role of Cu^+ complexes in the photochemical production of $^1O_2$. We found and presented for the first time that $[Cu(Xantphos)(neoc)]BF_4$ can be used as an alternative and very efficient photosensitizer in $^1O_2$-mediated C–H allylic oxygenations [146].

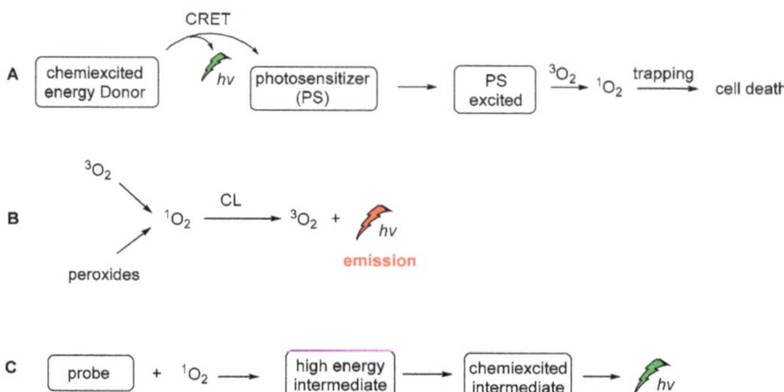

Figure 21. The CL-induced $^1O_2$ production (**A**), the CL emission produced by the radiative decay of singlet oxygen (**B**) and CL emission after the activation of a probe molecule by $^1O_2$ (**C**).

Recently, in our group, the plausible production of $^1O_2$ was studied under different CL reaction conditions and in the presence of various sensitizers. In this study, except for the common solvents used in CL procedures, diethylphthalates, the greener ethyl acetate was found to give unexpectedly good results. In addition, several photosensitizers, i.e., Rose Bengal (RB), Methylene Blue, 9,10-diphenylanthracene (9,10-DPA), Eosin Y and Rhodamine B, were tested for light emission through the CL process, with RB providing the best results based on the light emission duration. In the presence of TCPO, H_2O_2, RB, sodium salicylate and a mixture of ethyl acetate/acetonitrile in a ratio of 2/1, light emission was observed for 40–50 min (Scheme 17). Under these CL conditions and in the presence of 1-phenylcycloalkene, no allylic oxidation product was observed. For comparison, the same solution was bubbled with molecular oxygen and irradiated using a Xenon lamp (300 W, $\lambda > 320$ nm) [147–149], however, no oxidation products were observed. In the absence of TCPO and H_2O_2, the corresponding alkene was oxidized leading to the corresponding allylic hydroperoxides via a $^1O_2$ oxidation pathway, within 20 minutes (Scheme 17). These results indicate that the observed CL phenomenon didn't induce $^1O_2$ production, or the energy transfer and radiative processes of the CL energy acceptors or chemiexcited intermediates are not enough for the oxidation of cycloalkene. Further experiments using different photosensitizers and initial alkenes are in progress to determine this new intermolecular oxidation pathway [150].

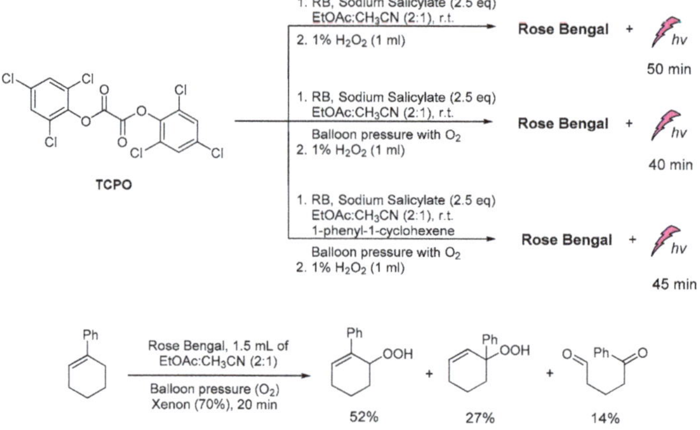

Scheme 17. The CL-induced singlet oxygen production for the oxidation of 1-phenylcyclohexene in "green" solvent ethyl acetate.

6. Future Prospects

Various technologies have emerged during the past few years aiming to fine-tune the molecular scaffolds and the experimental conditions under which CL can be used for the development of more advanced molecular sensors and new materials, in drug delivery, theranostics and polymerization reactions. To that end, several scaffolds such as dioxetanes and luminol-based systems, etc., have already shown great potential and are expected to play a significant role in the years to come. The growing demand for highly efficient PDT platforms that could produce significant amounts of reactive species under CL conditions may also reveal another important research direction that is still unexplored, regarding the replacement of light sources by CL in standard photochemical reactions.

A summary of the major obstacles that cause delay in the expansion of the CL applications include, but are not limited to, (i) the photophysical properties of the CL probe such as the fluorescence quantum yield, emission wavelength, photon flux, CL kinetic profile; (ii) the solubility and stability of the CL probe in the reaction media; (iii) the distance between the energy donor and energy acceptor, plus their emission and absorption spectral overlapping, in the case of direct CL; (iv) the selectivity of the triggering agent and oxidation mechanism of the CL substrates; (v) the efficiency of the emission mechanism (i.e., CIEEL); (vi) the toxicity, when considering bioapplications.

Recently there has been a tremendous increase in the number of new photochemical reactions reported in the literature. Nonetheless, reports of industrial applications utilizing this new knowledge are still scarce and mainly limited to flow-chemistry processes. Maybe nature can inspire the research community to find a solution to this problem through CL?

Author Contributions: M.A.Tz., D.K.G. and M.G.K. collected the literature. M.A.Tz. conducted the experiments with the TCPO and the various photosensitizers, and acquired and analyzed the data. N.V.P. and N.P.K. assisted in the gas chromatographic analysis and data acquisition. M.A.Tz., D.K.G., P.A.A., I.N.L. and M.A.T. wrote the manuscript. All authors have read and agreed to the published version of the manuscript.

Funding: This research was funded by the Hellenic Foundation for Research and Innovation (HFRI) and the General Secretariat for Research and Innovation (GSRI), grant number [776].

Institutional Review Board Statement: Not applicable.

Informed Consent Statement: Not applicable.

Data Availability Statement: Not applicable.

Acknowledgments: The authors kindly acknowledge financial support from the Hellenic Foundation for Research and Innovation (HFRI) and the General Secretariat for Research and Innovation (GSRI) under grant agreement No. [776] "PhotoDaLu" (KA97507).

Conflicts of Interest: The authors declare no conflict of interest.

References

1. Vacher, M.; Fdez Galván, I.; Ding, B.-W.; Schramm, S.; Berraud-Pache, R.; Naumov, P.; Ferré, N.; Liu, Y.-J.; Navizet, I.; Roca-Sanjuán, D.; et al. Chemi- and Bioluminescence of Cyclic Peroxides. *Chem. Rev.* **2018**, *118*, 6927–6974. [CrossRef]
2. McCapra, F. Chemiluminescence and Bioluminescence. *J. Photochem. Photobiol. A Chem.* **1990**, *51*, 21–28. [CrossRef]
3. Viviani, V.R. The Origin, Diversity, and Structure Function Relationships of Insect Luciferases. *CMLS Cell. Mol. Life Sci.* **2002**, *59*, 1833–1850. [CrossRef]
4. White, E.H.; McCapra, F.; Field, G.F.; McElroy, W.D. The structure and synthesis of firefly luciferin. *J. Am. Chem. Soc.* **1961**, *83*, 2402–2403. [CrossRef]
5. Yagur-Kroll, S.; Bilic, B.; Belkin, S. Strategies for Enhancing Bioluminescent Bacterial Sensor Performance by Promoter Region Manipulation: Enhancing Bacterial Sensor Performance. *Microb. Biotechnol.* **2010**, *3*, 300–310. [CrossRef]
6. Goto, T.; Takagi, T. Chemiluminescence of a *Cypridina* Luciferin Analogue, 2-Methyl-6-Phenyl-3,7-Dihydroimidazo[1,2- *a*]Pyrazin-3-One, in the Presence of the Xanthine–Xanthine Oxidase System. *BCSJ* **1980**, *53*, 833–834. [CrossRef]
7. Toya, Y.; Kayano, T.; Sato, K.; Goto, T. Synthesis and Chemiluminescence Properties of 6-(4-Methoxyphenyl)-2-Methylimidazo[1,2-*a*]Pyrazin-3(7*H*)-One and 2-Methyl-6-(2-Naphthyl)Imidazo[1,2-*a*]Pyrazin-3(7*H*)-One. *BCSJ* **1992**, *65*, 2475–2479. [CrossRef]
8. Teranishi, K. Development of Imidazopyrazinone Red-Chemiluminescent Probes for Detecting Superoxide Anions via a Chemiluminescence Resonance Energy Transfer Method. *Luminescence* **2007**, *22*, 147–156. [CrossRef]

9. Sekiya, M.; Umezawa, K.; Sato, A.; Citterio, D.; Suzuki, K. A Novel Luciferin-Based Bright Chemiluminescent Probe for the Detection of Reactive Oxygen Species. *Chem. Commun.* **2009**, 3047–3049. [CrossRef] [PubMed]
10. Marques, S.M.; Esteves da Silva, J.C.G. Firefly Bioluminescence: A Mechanistic Approach of Luciferase Catalyzed Reactions. *IUBMB Life* **2009**, *61*, 6–17. [CrossRef] [PubMed]
11. Kim, J.E.; Kalimuthu, S.; Ahn, B.-C. In Vivo Cell Tracking with Bioluminescence Imaging. *Nucl. Med. Mol. Imaging* **2015**, *49*, 3–10. [CrossRef]
12. Brown, J.E.; De Weer, P.; Salzberg, B.M. Optical Measurement of Changes in Intracellular Calcium. *Biophys. J.* **2020**, *118*, 788–789. [CrossRef] [PubMed]
13. Adam, W.; Kazakov, D.V.; Kazakov, V.P. Singlet-Oxygen Chemiluminescence in Peroxide Reactions. *Chem. Rev.* **2005**, *105*, 3371–3387. [CrossRef]
14. Gnaim, S.; Shabat, D. Activity-Based Optical Sensing Enabled by Self-Immolative Scaffolds: Monitoring of Release Events by Fluorescence or Chemiluminescence Output. *Acc. Chem. Res.* **2019**, *52*, 2806–2817. [CrossRef] [PubMed]
15. Haris, U.; Kagalwala, H.N.; Kim, Y.L.; Lippert, A.R. Seeking Illumination: The Path to Chemiluminescent 1,2-Dioxetanes for Quantitative Measurements and In Vivo Imaging. *Acc. Chem. Res.* **2021**, *54*, 2844–2857. [CrossRef] [PubMed]
16. Chatgilialoglu, C.; Ferreri, C.; Krokidis, M.G.; Masi, A.; Terzidis, M.A. On the Relevance of Hydroxyl Radical to Purine DNA Damage. *Free. Radic. Res.* **2021**, *55*, 384–404. [CrossRef]
17. Chatgilialoglu, C.; Krokidis, M.G.; Masi, A.; Barata-Vallejo, S.; Ferreri, C.; Terzidis, M.A.; Szreder, T.; Bobrowski, K. New Insights into the Reaction Paths of Hydroxyl Radicals with Purine Moieties in DNA and Double-Stranded Oligodeoxynucleotides. *Molecules* **2019**, *24*, 3860. [CrossRef]
18. Chatgilialoglu, C.; Ferreri, C.; Terzidis, M.A. Purine 5′,8-Cyclonucleoside Lesions: Chemistry and Biology. *Chem. Soc. Rev.* **2011**, *40*, 1368. [CrossRef]
19. Terzidis, M.A.; Chatgilialoglu, C. An Ameliorative Protocol for the Quantification of Purine 5′,8-Cyclo-2′-Deoxynucleosides in Oxidized DNA. *Front. Chem.* **2015**, *3*, 47. [CrossRef]
20. McCaughan, J.S. Photodynamic Therapy: A Review. *Drugs Aging* **1999**, *15*, 49–68. [CrossRef]
21. Dolmans, D.E.; Fukumura, D.; Jain, R.K. Photodynamic Therapy for Cancer. *Nat. Rev. Cancer* **2003**, *3*, 380–387. [CrossRef]
22. Zhao, X.; Liu, J.; Fan, J.; Chao, H.; Peng, X. Recent Progress in Photosensitizers for Overcoming the Challenges of Photodynamic Therapy: From Molecular Design to Application. *Chem. Soc. Rev.* **2021**, *50*, 4185–4219. [CrossRef] [PubMed]
23. Gunaydin, G.; Gedik, M.E.; Ayan, S. Photodynamic Therapy for the Treatment and Diagnosis of Cancer–A Review of the Current Clinical Status. *Front. Chem.* **2021**, *9*, 686303. [CrossRef] [PubMed]
24. Yamaguchi, S.; Kishikawa, N.; Ohyama, K.; Ohba, Y.; Kohno, M.; Masuda, T.; Takadate, A.; Nakashima, K.; Kuroda, N. Evaluation of Chemiluminescence Reagents for Selective Detection of Reactive Oxygen Species. *Anal. Chim. Acta* **2010**, *665*, 74–78. [CrossRef] [PubMed]
25. Worsfold, P.; Townshend, A.; Poole, C.F. *Encyclopedia of Analytical Science*, 2nd ed.; Elsevier Academic Press: Amsterdam, The Netherlands; Boston, MA, USA, 2005; ISBN 978-0-12-764100-3.
26. Qi, Y.; Xiu, F.; Weng, Q. A Novel and Convenient Chemiluminescence Sensing of DNA: Nanometer Interface Effect and DNA Action Mechanism. *Sens. Actuators B Chem.* **2018**, *260*, 303–310. [CrossRef]
27. Ling, K.; Jiang, H.; Huang, X.; Li, Y.; Lin, J.; Li, F.-R. Direct Chemiluminescence Detection of Circulating MicroRNAs in Serum Samples Using a Single-Strand Specific Nuclease-Distinguishing Nucleic Acid Hybrid System. *Chem. Commun.* **2018**, *54*, 1909–1912. [CrossRef]
28. Li, W.; Zhang, Q.; Zhou, H.; Chen, J.; Li, Y.; Zhang, C.; Yu, C. Chemiluminescence Detection of a Protein through the Aptamer-Controlled Catalysis of a Porphyrin Probe. *Anal. Chem.* **2015**, *87*, 8336–8341. [CrossRef]
29. Carter, K.P.; Young, A.M.; Palmer, A.E. Fluorescent Sensors for Measuring Metal Ions in Living Systems. *Chem. Rev.* **2014**, *114*, 4564–4601. [CrossRef]
30. Roda, A.; Pasini, P.; Musiani, M.; Girotti, S.; Baraldini, M.; Carrea, G.; Suozzi, A. Chemiluminescent Low-Light Imaging of Biospecific Reactions on Macro- and Microsamples Using a Videocamera-Based Luminograph. *Anal. Chem.* **1996**, *68*, 1073–1080. [CrossRef]
31. Roda, A.; Pasini, P.; Musiani, M.; Baraldini, M. Chemiluminescence Imaging Systems for the Analysis of Macrosamples: Microtiter Format, Blot Membrane, and Whole Organs. In *Methods in Enzymology*; Elsevier: Amsterdam, The Netherlands, 2000; Volume 305, pp. 120–132. ISBN 978-0-12-182206-4.
32. Padoan, A.; Cosma, C.; Sciacovelli, L.; Faggian, D.; Plebani, M. Analytical Performances of a Chemiluminescence Immunoassay for SARS-CoV-2 IgM/IgG and Antibody Kinetics. *Clin. Chem. Lab. Med. (CCLM)* **2020**, *58*, 1081–1088. [CrossRef]
33. Yan, Y.; Wang, X.; Hai, X.; Song, W.; Ding, C.; Cao, J.; Bi, S. Chemiluminescence Resonance Energy Transfer: From Mechanisms to Analytical Applications. *TrAC Trends Anal. Chem.* **2020**, *123*, 115755. [CrossRef]
34. Yan, Y.; Shi, P.; Song, W.; Bi, S. Chemiluminescence and Bioluminescence Imaging for Biosensing and Therapy: In Vitro and In Vivo Perspectives. *Theranostics* **2019**, *9*, 4047–4065. [CrossRef]
35. Delafresnaye, L.; Bloesser, F.R.; Kockler, K.B.; Schmitt, C.W.; Irshadeen, I.M.; Barner-Kowollik, C. All Eyes on Visible-Light Peroxyoxalate Chemiluminescence Read-Out Systems. *Chem. Eur. J.* **2020**, *26*, 114–127. [CrossRef]
36. Hananya, N.; Shabat, D. Recent Advances and Challenges in Luminescent Imaging: Bright Outlook for Chemiluminescence of Dioxetanes in Water. *ACS Cent. Sci.* **2019**, *5*, 949–959. [CrossRef] [PubMed]

37. Albini, A. *Photochemistry: Past, Present and Future*; Springer: Berlin, Germany, 2015; ISBN 3-662-47977-X.
38. Crespi, S.; Protti, S. (Eds.) *Photochemistry: Volume 49*; Photochemistry; Royal Society of Chemistry: Cambridge, UK, 2021; Volume 49, ISBN 978-1-83916-388-3.
39. Powe, A.M.; Fletcher, K.A.; St. Luce, N.N.; Lowry, M.; Neal, S.; McCarroll, M.E.; Oldham, P.B.; McGown, L.B.; Warner, I.M. Molecular Fluorescence, Phosphorescence, and Chemiluminescence Spectrometry. *Anal. Chem.* **2004**, *76*, 4614–4634. [CrossRef]
40. Shimomura, O. *Bioluminescence: Chemical Principles and Methods*; World Scientific: Singapore, 2006; ISBN 978-981-256-801-4.
41. Magalhães, C.M.; Esteves da Silva, J.C.G.; Pinto da Silva, L. Chemiluminescence and Bioluminescence as an Excitation Source in the Photodynamic Therapy of Cancer: A Critical Review. *ChemPhysChem* **2016**, *17*, 2286–2294. [CrossRef] [PubMed]
42. Miao, W. Electrogenerated Chemiluminescence and Its Biorelated Applications. *Chem. Rev.* **2008**, *108*, 2506–2553. [CrossRef]
43. Albrecht, H.O. Über Die Chemiluminescenz Des Aminophthalsäurehydrazids. *Z. Phys. Chem.* **1928**, *136*, 321–330. [CrossRef]
44. Vaughan, W.R. The Chemistry of the Phthalazines. *Chem. Rev.* **1948**, *43*, 447–508. [CrossRef]
45. White, E.H.; Zafiriou, O.; Kagi, H.H.; Hill, J.H. Chemilunimescence of Luminol: The Chemcial Reaction. *J. Am. Chem. Soc.* **1964**, *86*, 940–941. [CrossRef]
46. Yu, Y.; Mallick, S.; Wang, M.; Börjesson, K. Barrier-Free Reverse-Intersystem Crossing in Organic Molecules by Strong Light-Matter Coupling. *Nat Commun.* **2021**, *12*, 3255. [CrossRef]
47. Wada, Y.; Nakagawa, H.; Matsumoto, S.; Wakisaka, Y.; Kaji, H. Organic Light Emitters Exhibiting Very Fast Reverse Intersystem Crossing. *Nat. Photonics* **2020**, *14*, 643–649. [CrossRef]
48. Cui, L.-S.; Gillett, A.J.; Zhang, S.-F.; Ye, H.; Liu, Y.; Chen, X.-K.; Lin, Z.-S.; Evans, E.W.; Myers, W.K.; Ronson, T.K.; et al. Fast Spin-Flip Enables Efficient and Stable Organic Electroluminescence from Charge-Transfer States. *Nat. Photonics* **2020**, *14*, 636–642. [CrossRef]
49. Nakamura, M.; Nakamura, S. One- and Two-Electron Oxidations of Luminol by Peroxidase Systems. *Free. Radic. Biol. Med.* **1998**, *24*, 537–544. [CrossRef]
50. Morin, J.G. Based on a Review of the Data, Use of the Term 'Cypridinid' Solves the Cypridina/Vargula Dilemma for Naming the Constituents of the Luminescent System of Ostracods in the Family Cypridinidae. *Luminescence* **2011**, *26*, 1–4. [CrossRef]
51. Johnson, F.H.; Shimomura, O. Introduction to the Cypridina System. *Methods Enzymol.* **1978**, *57*, 331–364. [CrossRef]
52. Khakhar, A.; Starker, C.G.; Chamness, J.C.; Lee, N.; Stokke, S.; Wang, C.; Swanson, R.; Rizvi, F.; Imaizumi, T.; Voytas, D.F. Building customizable auto-luminescent luciferase-based reporters in plants. *Elife* **2020**, *9*, e52786. [CrossRef]
53. Zheng, Z.; Wang, L.; Tang, W.; Chen, P.; Zhu, H.; Yuan, Y.; Li, G.; Zhang, H.; Liang, G. Hydrazide D-Luciferin for in Vitro Selective Detection and Intratumoral Imaging of Cu^{2+}. *Biosens. Bioelectron.* **2016**, *83*, 200–204. [CrossRef] [PubMed]
54. Yang, M.; Huang, J.; Fan, J.; Du, J.; Pu, K.; Peng, X. Chemiluminescence for Bioimaging and Therapeutics: Recent Advances and Challenges. *Chem. Soc. Rev.* **2020**, *49*, 6800–6815. [CrossRef]
55. Li, Z.; Chen, X.; Teng, X.; Lu, C. Chemiluminescence as a New Indicator for Monitoring Hydroxylated Intermediates in Persulfate-Based Advanced Oxidation Processes. *J. Phys. Chem. C* **2019**, *123*, 21704–21712. [CrossRef]
56. Vakh, C.; Kuzmin, A.; Sadetskaya, A.; Bogdanova, P.; Voznesenskiy, M.; Osmolovskaya, O.; Bulatov, A. Cobalt-Doped Hydroxyapatite Nanoparticles as a New Eco-Friendly Catalyst of Luminol–H2O2 Based Chemiluminescence Reaction: Study of Key Factors, Improvement the Activity and Analytical Application. *Spectrochim. Acta Part A Mol. Biomol. Spectrosc.* **2020**, *237*, 118382. [CrossRef]
57. Zambrano, G.; Nastri, F.; Pavone, V.; Lombardi, A.; Chino, M. Use of an Artificial Miniaturized Enzyme in Hydrogen Peroxide Detection by Chemiluminescence. *Sensors* **2020**, *20*, 3793. [CrossRef] [PubMed]
58. Yang, C.P.; He, L.; Huang, C.Z.; Li, Y.F.; Zhen, S.J. Continuous Singlet Oxygen Generation for Persistent Chemiluminescence in Cu-MOFs-Based Catalytic System. *Talanta* **2021**, *221*, 121498. [CrossRef]
59. Nishinaka, Y.; Aramaki, Y.; Yoshida, H.; Masuya, H.; Sugawara, T.; Ichimori, Y. A New Sensitive Chemiluminescence Probe, L-012, for Measuring the Production of Superoxide Anion by Cells. *Biochem. Biophys. Res. Commun.* **1993**, *193*, 554–559. [CrossRef] [PubMed]
60. Yang, J.; Yin, W.; Van, R.; Yin, K.; Wang, P.; Zheng, C.; Zhu, B.; Ran, K.; Zhang, C.; Kumar, M.; et al. Turn-on Chemiluminescence Probes and Dual-Amplification of Signal for Detection of Amyloid Beta Species in Vivo. *Nat Commun.* **2020**, *11*, 4052. [CrossRef] [PubMed]
61. Green, O.; Gnaim, S.; Blau, R.; Eldar-Boock, A.; Satchi-Fainaro, R.; Shabat, D. Near-Infrared Dioxetane Luminophores with Direct Chemiluminescence Emission Mode. *J. Am. Chem. Soc.* **2017**, *139*, 13243–13248. [CrossRef] [PubMed]
62. Gnaim, S.; Scomparin, A.; Das, S.; Blau, R.; Satchi-Fainaro, R.; Shabat, D. Direct Real-Time Monitoring of Prodrug Activation by Chemiluminescence. *Angew. Chem. Int. Ed.* **2018**, *57*, 9033–9037. [CrossRef]
63. Das, S.; Ihssen, J.; Wick, L.; Spitz, U.; Shabat, D. Chemiluminescent Carbapenem-Based Molecular Probe for Detection of Carbapenemase Activity in Live Bacteria. *Chem. Eur. J.* **2020**, *26*, 3647–3652. [CrossRef]
64. Gutkin, S.; Green, O.; Raviv, G.; Shabat, D.; Portnoy, O. Powerful Chemiluminescence Probe for Rapid Detection of Prostate Specific Antigen Proteolytic Activity: Forensic Identification of Human Semen. *Bioconjug. Chem.* **2020**, *31*, 2488–2493. [CrossRef]
65. Yue, L.; Liu, Y.-T. Mechanistic Insight into PH-Dependent Luminol Chemiluminescence in Aqueous Solution. *J. Phys. Chem. B* **2020**, *124*, 7682–7693. [CrossRef] [PubMed]
66. Kamidate, T.; Katayama, A.; Ichihashi, H.; Watanabe, H. Characterization of Peroxidases in Luminol Chemiluminescence Coupled with Copper-Catalysed Oxidation of Cysteamine. *J. Biolumin. Chemilumin.* **1994**, *9*, 279–286. [CrossRef]

67. Gorsuch, J.D.; Hercules, D.M. Studies on the chemiluminescence of luminol in dimethylsulfoxide and dimethylsulfoxide-water mixtures. *Photochem. Photobiol.* **1972**, *15*, 567–583. [CrossRef]
68. Lee, J.; Seliger, H.H. Quantum yields of the luminol chemiluminescence reaction in aqueous and aprotic solvents. *Photochem. Photobiol.* **1972**, *15*, 227–237. [CrossRef]
69. Cormier, M.J.; Prichard, P.M. An Investigation of the Mechanism of the Luminescent Peroxidation of Luminol by Stopped Flow Techniques. *J. Biol. Chem.* **1968**, *243*, 4706–4714. [CrossRef]
70. Li, L.; Arnold, M.A.; Dordick, J.S. Mathematical Model for the Luminol Chemiluminescence Reaction Catalyzed by Peroxidase: Peroxidase-catalyzed luminol chemiluminescence. *Biotechnol. Bioeng.* **1993**, *41*, 1112–1120. [CrossRef] [PubMed]
71. Cercek, B.; Cercek, B.; Roby, K.; Cercek, L. Effect of Oxygen Abstraction on the Peroxidase-Luminol-Perborate System: Relevance to the HRP Enhanced Chemiluminescence Mechanism. *J. Biolumin. Chemilumin.* **1994**, *9*, 273–277. [CrossRef]
72. Mikroulis, T.; Cuquerella, M.C.; Giussani, A.; Pantelia, A.; Rodríguez-Muñiz, G.M.; Rotas, G.; Roca-Sanjuán, D.; Miranda, M.A.; Vougioukalakis, G.C. Building a Functionalizable, Potent Chemiluminescent Agent: A Rational Design Study on 6,8-Substituted Luminol Derivatives. *J. Org. Chem.* **2021**, *86*, 11388–11398. [CrossRef]
73. Brundrett, R.B.; White, E.H. Synthesis and Chemiluminescence of Derivatives of Luminol and Isoluminol. *J. Am. Chem. Soc.* **1974**, *96*, 7497–7502. [CrossRef]
74. Karatani, H. Microenvironmental Effects of Water-Soluble Polymers on the Chemiluminescence of Luminol and Its Analogs. *BCSJ* **1987**, *60*, 2023–2029. [CrossRef]
75. Baader, W.J.; Stevani, C.V.; Bastos, E.L. Chemiluminescence of Organic Peroxides The Authors Dedicate This Chapter to the Late Prof. Dr. Giuseppe Cilento (Universidade de São Paulo), Whose Untimely Demise in 1994 Saddened Everyone Who Knew Him as an Excellent Scientist and Outstanding Human Being. This Chapter Is Also Dedicated to Prof. Waldemar Adam (Universität Würzburg, Now 'Retired' in Puerto Rico), Who Was Directly or Indirectly Responsible for the Research Interests of the Authors. In *PATAI'S Chemistry of Functional Groups*; Rappoport, Z., Ed.; John Wiley & Sons, Ltd.: Chichester, UK, 2009; ISBN 978-0-470-68253-1.
76. Tonkin, S.A.; Bos, R.; Dyson, G.A.; Lim, K.F.; Russell, R.A.; Watson, S.P.; Hindson, C.M.; Barnett, N.W. Studies on the Mechanism of the Peroxyoxalate Chemiluminescence Reaction. *Anal. Chim. Acta* **2008**, *614*, 173–181. [CrossRef]
77. Farahani, P.; Baader, W.J. Unimolecular Decomposition Mechanism of 1,2-Dioxetanedione: Concerted or Biradical? That Is the Question! *J. Phys. Chem. A* **2017**, *121*, 1189–1194. [CrossRef] [PubMed]
78. Bos, R.; Tonkin, S.A.; Hanson, G.R.; Hindson, C.M.; Lim, K.F.; Barnett, N.W. In Search of a Chemiluminescence 1,4-Dioxy Biradical. *J. Am. Chem. Soc.* **2009**, *131*, 2770–2771. [CrossRef]
79. Orosz, G. The Role of Diaryl Oxalates in Peroxioxalate Chemiluminescence. *Tetrahedron* **1989**, *45*, 3493–3506. [CrossRef]
80. Stevani, C.V.; Lima, D.F.; Toscano, V.G.; Baader, W.J. Kinetic Studies on the Peroxyoxalate Chemiluminescent Reaction: Imidazole as a Nucleophilic Catalyst. *J. Chem. Soc. Perkin Trans. 2* **1996**, 989–995. [CrossRef]
81. Silva, S.M.; Casallanovo, F.; Oyamaguchi, K.H.; Ciscato, L.F.; Stevani, C.V.; Baader, W.J. Kinetic Studies on the Peroxyoxalate Chemiluminescence Reaction: Determination of the Cyclization Rate Constant. *Luminescence* **2002**, *17*, 313–320. [CrossRef] [PubMed]
82. Souza, G.A.; Lang, A.P.; Baader, W.J. Mechanistic Studies on the Peroxyoxalate Chemiluminescence Using Sodium Salicylate as Base Catalyst. *Photochem. Photobiol.* **2017**, *93*, 1423–1429. [CrossRef] [PubMed]
83. Jonsson, T.; Irgum, K. New Nucleophilic Catalysts for Bright and Fast Peroxyoxalate Chemiluminescence. *Anal. Chem.* **2000**, *72*, 1373–1380. [CrossRef] [PubMed]
84. Cabello, M.C.; Souza, G.A.; Bello, L.V.; Baader, W.J. Mechanistic Studies on the Salicylate-Catalyzed Peroxyoxalate Chemiluminescence in Aqueous Medium. *Photochem. Photobiol.* **2020**, *96*, 28–36. [CrossRef] [PubMed]
85. Augusto, F.A.; Bartoloni, F.H.; Cabello, M.C.; dos Santos, A.P.F.; Baader, W.J. Kinetic Studies on 2,6-Lutidine Catalyzed Peroxyoxalate Chemiluminescence in Organic and Aqueous Medium: Evidence for General Base Catalysis. *J. Photochem. Photobiol. A Chem.* **2019**, *382*, 111967. [CrossRef]
86. Shah, S.N.A.; Shah, A.H.; Dou, X.; Khan, M.; Lin, L.; Lin, J.-M. Radical-Triggered Chemiluminescence of Phenanthroline Derivatives: An Insight into Radical–Aromatic Interaction. *ACS Omega* **2019**, *4*, 15004–15011. [CrossRef]
87. Shah, S.N.A.; Khan, M.; Rehman, Z.U. A Prolegomena of Periodate and Peroxide Chemiluminescence. *TrAC Trends Anal. Chem.* **2020**, *122*, 115722. [CrossRef]
88. Watanabe, N.; Takatsuka, H.; Ijuin, H.K.; Matsumoto, M. Highly Effective and Rapid Emission of Light from Bicyclic Dioxetanes Bearing a 3-Hydroxyphenyl Substituted with a 4-p-Oligophenylene Moiety in an Aqueous System: Two Different Ways for the Enhancement of Chemiluminescence Efficiency. *Tetrahedron* **2020**, *76*, 131203. [CrossRef]
89. Schaap, A.P.; Chen, T.-S.; Handley, R.S.; DeSilva, R.; Giri, B.P. Chemical and Enzymatic Triggering of 1,2-Dioxetanes. 2: Fluoride-Induced Chemiluminescence from Tert-Butyldimethylsilyloxy-Substituted Dioxetanes. *Tetrahedron Lett.* **1987**, *28*, 1155–1158. [CrossRef]
90. Schaap, A.P.; Handley, R.S.; Giri, B.P. Chemical and Enzymatic Triggering of 1,2-Dioxetanes. 1: Aryl Esterase-Catalyzed Chemiluminescence from a Naphthyl Acetate-Substituted Dioxetane. *Tetrahedron Lett.* **1987**, *28*, 935–938. [CrossRef]
91. Schaap, A.P.; Sandison, M.D.; Handley, R.S. Chemical and Enzymatic Triggering of 1,2-Dioxetanes. 3: Alkaline Phosphatase-Catalyzed Chemiluminescence from an Aryl Phosphate-Substituted Dioxetane. *Tetrahedron Lett.* **1987**, *28*, 1159–1162. [CrossRef]
92. Augusto, F.A.; de Souza, G.A.; de Souza Júnior, S.P.; Khalid, M.; Baader, W.J. Efficiency of Electron Transfer Initiated Chemiluminescence. *Photochem. Photobiol.* **2013**, *89*, 1299–1317. [CrossRef] [PubMed]

93. Adam, W.; Bronstein, I.; Trofimov, A.V.; Vasil'ev, R.F. Solvent-Cage Effect (Viscosity Dependence) as a Diagnostic Probe for the Mechanism of the Intramolecular Chemically Initiated Electron-Exchange Luminescence (CIEEL) Triggered from a Spiroadamantyl-Substituted Dioxetane. *J. Am. Chem. Soc.* **1999**, *121*, 958–961. [CrossRef]
94. Ryan, L.S.; Nakatsuka, A.; Lippert, A.R. PhotoactivaTable 1,2-Dioxetane Chemiluminophores. *Results Chem.* **2021**, *3*, 100106. [CrossRef]
95. Zadykowicz, B.; Czechowska, J.; Ożóg, A.; Renkevich, A.; Krzymiński, K. Effective Chemiluminogenic Systems Based on Acridinium Esters Bearing Substituents of Various Electronic and Steric Properties. *Org. Biomol. Chem.* **2016**, *14*, 652–668. [CrossRef]
96. Czechowska, J.; Kawecka, A.; Romanowska, A.; Marczak, M.; Wityk, P.; Krzymiński, K.; Zadykowicz, B. Chemiluminogenic Acridinium Salts: A Comparison Study. Detection of Intermediate Entities Appearing upon Light Generation. *J. Lumin.* **2017**, *187*, 102–112. [CrossRef]
97. Nakazono, M.; Nanbu, S.; Akita, T.; Hamase, K. Synthesis, Chemiluminescence, and Application of 2,4-Disubstituted Phenyl 10-Methyl-10λ4-Acridine-9-Carboxylates. *Dyes Pigments* **2019**, *170*, 107628. [CrossRef]
98. Ren, L.; Cui, H. Chemiluminescence Accompanied by the Reaction of Acridinium Ester and Manganese (II): Reaction of Acridinium Ester and Manganese(II). *Luminescence* **2014**, *29*, 929–932. [CrossRef]
99. Natrajan, A.; Wen, D. Effect of Branching in Remote Substituents on Light Emission and Stability of Chemiluminescent Acridinium Esters. *RSC Adv.* **2014**, *4*, 21852–21863. [CrossRef]
100. Krzymiński, K.; Ożóg, A.; Malecha, P.; Roshal, A.D.; Wróblewska, A.; Zadykowicz, B.; Błażejowski, J. Chemiluminogenic Features of 10-Methyl-9-(Phenoxycarbonyl)Acridinium Trifluoromethanesulfonates Alkyl Substituted at the Benzene Ring in Aqueous Media. *J. Org. Chem.* **2011**, *76*, 1072–1085. [CrossRef]
101. McCapra, F. The Chemiluminescence of Organic Compounds. *Q. Rev. Chem. Soc.* **1966**, *20*, 485. [CrossRef]
102. Weeks, I.; Beheshti, I.; McCapra, F.; Campbell, A.K.; Woodhead, J.S. Acridinium Esters as High-Specific-Activity Labels in Immunoassay. *Clin. Chem.* **1983**, *29*, 1474–1479. [CrossRef] [PubMed]
103. Natrajan, A.; Sharpe, D. Synthesis and Properties of Differently Charged Chemiluminescent Acridinium Ester Labels. *Org. Biomol. Chem.* **2013**, *11*, 1026. [CrossRef]
104. Pieńkos, M.; Zadykowicz, B. Computational Insights on the Mechanism of the Chemiluminescence Reaction of New Group of Chemiluminogens—10-Methyl-9-Thiophenoxycarbonylacridinium Cations. *IJMS* **2020**, *21*, 4417. [CrossRef]
105. Nakazono, M.; Nanbu, S.; Akita, T.; Hamase, K. Chemiluminescence of Methoxycarbonylphenyl 10-Methyl-10λ4 -2,7-Disubstituted Acridine-9-Carboxylate Derivatives. *J. Photochem. Photobiol. A Chem.* **2020**, *403*, 112851. [CrossRef]
106. Huang, X.; Li, L.; Qian, H.; Dong, C.; Ren, J. A Resonance Energy Transfer between Chemiluminescent Donors and Luminescent Quantum-Dots as Acceptors (CRET). *Angew. Chem. Int. Ed.* **2006**, *45*, 5140–5143. [CrossRef]
107. Tanielian, C.; Mechin, R.; Seghrouchni, R.; Schweitzer, C. Mechanistic and Kinetic Aspects of Photosensitization in the Presence of Oxygen†§. *Photochem. Photobiol.* **2000**, *71*, 12. [CrossRef]
108. Greer, A. Christopher Foote's Discovery of the Role of Singlet Oxygen [$^1O_2$ ($^1\Delta_g$)] in Photosensitized Oxidation Reactions. *Acc. Chem. Res.* **2006**, *39*, 797–804. [CrossRef]
109. Stratakis, M.; Orfanopoulos, M. Regioselectivity in the Ene Reaction of Singlet Oxygen with Alkenes Bearing an Electron Withdrawing Group at β- Position. *Tetrahedron Lett.* **1997**, *38*, 1067–1070. [CrossRef]
110. You, Y. Chemical Tools for the Generation and Detection of Singlet Oxygen. *Org. Biomol. Chem.* **2018**, *16*, 4044–4060. [CrossRef]
111. Tanielian, C. Decatungstate Photocatalysis. *Coord. Chem. Rev.* **1998**, *178*, 1165–1181. [CrossRef]
112. Tzirakis, M.D.; Lykakis, I.N.; Orfanopoulos, M. Decatungstate as an Efficient Photocatalyst in Organic Chemistry. *Chem. Soc. Rev.* **2009**, *38*, 2609. [CrossRef]
113. Ravelli, D.; Protti, S.; Fagnoni, M. Decatungstate Anion for Photocatalyzed "Window Ledge" Reactions. *Acc. Chem. Res.* **2016**, *49*, 2232–2242. [CrossRef]
114. Meyer, S.; Tietze, D.; Rau, S.; Schäfer, B.; Kreisel, G. Photosensitized Oxidation of Citronellol in Microreactors. *J. Photochem. Photobiol. A Chem.* **2007**, *186*, 248–253. [CrossRef]
115. Samia, A.C.S.; Chen, X.; Burda, C. Semiconductor Quantum Dots for Photodynamic Therapy. *J. Am. Chem. Soc.* **2003**, *125*, 15736–15737. [CrossRef]
116. An, W.; Mason, R.P.; Lippert, A.R. Energy Transfer Chemiluminescence for Ratiometric PH Imaging. *Org. Biomol. Chem.* **2018**, *16*, 4176–4182. [CrossRef]
117. Lee, E.S.; Deepagan, V.G.; You, D.G.; Jeon, J.; Yi, G.-R.; Lee, J.Y.; Lee, D.S.; Suh, Y.D.; Park, J.H. Nanoparticles Based on Quantum Dots and a Luminol Derivative: Implications for in Vivo Imaging of Hydrogen Peroxide by Chemiluminescence Resonance Energy Transfer. *Chem. Commun.* **2016**, *52*, 4132–4135. [CrossRef] [PubMed]
118. Teng, Y.; Li, M.; Huang, X.; Ren, J. Singlet Oxygen Generation in Ferriporphyrin-Polymer Dots Catalyzed Chemiluminescence System for Cancer Therapy. *ACS Appl. Bio Mater.* **2020**, *3*, 5020–5029. [CrossRef]
119. Wang, C.; Meng, X.; Li, Z.; Li, M.; Jin, M.; Liu, R.; Yagci, Y. Chemiluminescence Induced Cationic Photopolymerization Using Sulfonium Salt. *ACS Macro Lett.* **2020**, *9*, 471–475. [CrossRef]
120. Liu, L.; Mason, R.P. Imaging β-Galactosidase Activity in Human Tumor Xenografts and Transgenic Mice Using a Chemiluminescent Substrate. *PLoS ONE* **2010**, *5*, e12024. [CrossRef]

121. Cao, J.; Campbell, J.; Liu, L.; Mason, R.P.; Lippert, A.R. In Vivo Chemiluminescent Imaging Agents for Nitroreductase and Tissue Oxygenation. *Anal. Chem.* **2016**, *88*, 4995–5002. [CrossRef]
122. Cao, J.; Lopez, R.; Thacker, J.M.; Moon, J.Y.; Jiang, C.; Morris, S.N.S.; Bauer, J.H.; Tao, P.; Mason, R.P.; Lippert, A.R. Chemiluminescent Probes for Imaging H_2S in Living Animals. *Chem. Sci.* **2015**, *6*, 1979–1985. [CrossRef]
123. Ryan, L.S.; Gerberich, J.; Haris, U.; Nguyen, D.; Mason, R.P.; Lippert, A.R. Ratiometric PH Imaging Using a 1,2-Dioxetane Chemiluminescence Resonance Energy Transfer Sensor in Live Animals. *ACS Sens.* **2020**, *5*, 2925–2932. [CrossRef]
124. Suzuki, N.; Suetsuna, K.; Mashiko, S.; Yoda, B.; Nomoto, T.; Toya, Y.; Inaba, H.; Goto, T. Reaction Rates for the Chemiluminescence of *Cypridina* Luciferin Analogues with Superoxide: A Quenching Experiment with Superoxide Dismutase. *Agric. Biol. Chem.* **1991**, *55*, 157–160. [CrossRef]
125. Hananya, N.; Press, O.; Das, A.; Scomparin, A.; Satchi-Fainaro, R.; Sagi, I.; Shabat, D. Persistent Chemiluminescent Glow of Phenoxy-dioxetane Luminophore Enables Unique CRET-Based Detection of Proteases. *Chem. Eur. J.* **2019**, *25*, 14679–14687. [CrossRef]
126. Song, Q.; Yan, X.; Cui, H.; Ma, M. Efficient Cascade Resonance Energy Transfer in Dynamic Nanoassembly for Intensive and Long-Lasting Multicolor Chemiluminescence. *ACS Nano* **2020**, *14*, 3696–3702. [CrossRef]
127. Zhang, K.; Sun, M.; Song, H.; Su, Y.; Lv, Y. Synergistic Chemiluminescence Nanoprobe: Au Clusters-Cu^{2+} -Induced Chemiexcitation of Cyclic Peroxides and Resonance Energy Transfer. *Chem. Commun.* **2020**, *56*, 3151–3154. [CrossRef] [PubMed]
128. An, H.; Guo, C.; Li, D.; Liu, R.; Xu, X.; Guo, J.; Ding, J.; Li, J.; Chen, W.; Zhang, J. Hydrogen Peroxide-Activatable Nanoparticles for Luminescence Imaging and In Situ Triggerable Photodynamic Therapy of Cancer. *ACS Appl. Mater. Interfaces* **2020**, *12*, 17230–17243. [CrossRef]
129. Liu, H.; Sun, M.; Su, Y.; Deng, D.; Hu, J.; Lv, Y. Chemiluminescence of Black Phosphorus Quantum Dots Induced by Hypochlorite and Peroxide. *Chem. Commun.* **2018**, *54*, 7987–7990. [CrossRef] [PubMed]
130. Liu, H.; Su, Y.; Sun, T.; Deng, D.; Lv, Y. Engineering the Energy Gap of Black Phosphorene Quantum Dots by Surface Modification for Efficient Chemiluminescence. *Chem. Commun.* **2020**, *56*, 1891–1894. [CrossRef]
131. Zhang, D.; Zheng, Y.; Dou, X.; Lin, H.; Shah, S.N.A.; Lin, J.-M. Heterogeneous Chemiluminescence from Gas–Solid Phase Interactions of Ozone with Alcohols, Phenols, and Saccharides. *Langmuir* **2017**, *33*, 3666–3671. [CrossRef]
132. Gao, H.-Y.; Mao, L.; Li, F.; Xie, L.-N.; Huang, C.-H.; Shao, J.; Shao, B.; Kalyanaraman, B.; Zhu, B.-Z. Mechanism of Intrinsic Chemiluminescence Production from the Degradation of Persistent Chlorinated Phenols by the Fenton System: A Structure–Activity Relationship Study and the Critical Role of Quinoid and Semiquinone Radical Intermediates. *Environ. Sci. Technol.* **2017**, *51*, 2934–2943. [CrossRef]
133. Zou, F.; Zhou, W.; Guan, W.; Lu, C.; Tang, B.Z. Screening of Photosensitizers by Chemiluminescence Monitoring of Formation Dynamics of Singlet Oxygen during Photodynamic Therapy. *Anal. Chem.* **2016**, *88*, 9707–9713. [CrossRef] [PubMed]
134. Ghogare, A.A.; Greer, A. Using Singlet Oxygen to Synthesize Natural Products and Drugs. *Chem. Rev.* **2016**, *116*, 9994–10034. [CrossRef] [PubMed]
135. Yang, Z.; Cao, Y.; Li, J.; Lu, M.; Jiang, Z.; Hu, X. Smart CuS Nanoparticles as Peroxidase Mimetics for the Design of Novel Label-Free Chemiluminescent Immunoassay. *ACS Appl. Mater. Interfaces* **2016**, *8*, 12031–12038. [CrossRef]
136. Shu, J.; Qiu, Z.; Zhou, Q.; Lin, Y.; Lu, M.; Tang, D. Enzymatic Oxydate-Triggered Self-Illuminated Photoelectrochemical Sensing Platform for Portable Immunoassay Using Digital Multimeter. *Anal. Chem.* **2016**, *88*, 2958–2966. [CrossRef]
137. Zhang, H.; Zhang, X.; Dong, S. Enhancement of the Carbon Dots/$K_2S_2O_8$ Chemiluminescence System Induced by Triethylamine. *Anal. Chem.* **2015**, *87*, 11167–11170. [CrossRef]
138. Liang, Y.-F.; Wang, X.; Yuan, Y.; Liang, Y.; Li, X.; Jiao, N. Ligand-Promoted Pd-Catalyzed Oxime Ether Directed C–H Hydroxylation of Arenes. *ACS Catal.* **2015**, *5*, 6148–6152. [CrossRef]
139. Bregnhøj, M.; Westberg, M.; Jensen, F.; Ogilby, P.R. Solvent-Dependent Singlet Oxygen Lifetimes: Temperature Effects Implicate Tunneling and Charge-Transfer Interactions. *Phys. Chem. Chem. Phys.* **2016**, *18*, 22946–22961. [CrossRef] [PubMed]
140. Wilkinson, F.; Helman, W.P.; Ross, A.B. Rate Constants for the Decay and Reactions of the Lowest Electronically Excited Singlet State of Molecular Oxygen in Solution. An Expanded and Revised Compilation. *J. Phys. Chem. Ref. Data* **1995**, *24*, 663–677. [CrossRef]
141. Ouyang, H.; Zhang, L.; Jiang, S.; Wang, W.; Zhu, C.; Fu, Z. Co Single-Atom Catalysts Boost Chemiluminescence. *Chem. Eur. J.* **2020**, *26*, 7583–7588. [CrossRef] [PubMed]
142. Bhattacharyya, A.; Dixit, A.; Banerjee, S.; Roy, B.; Kumar, A.; Karande, A.A.; Chakravarty, A.R. BODIPY Appended Copper(II) Complexes for Cellular Imaging and Singlet Oxygen Mediated Anticancer Activity in Visible Light. *RSC Adv.* **2016**, *6*, 104474–104482. [CrossRef]
143. McRae, E.K.S.; Nevonen, D.E.; McKenna, S.A.; Nemykin, V.N. Binding and Photodynamic Action of the Cationic Zinc Phthalocyanines with Different Types of DNA toward Understanding of Their Cancer Therapy Activity. *J. Inorg. Biochem.* **2019**, *199*, 110793. [CrossRef]
144. Binder, S.; Hosikova, B.; Mala, Z.; Zarska, L.; Kolarova, H. Effect of ClAlPcS2 Photodynamic and Sonodynamic Therapy on Hela Cells. *Physiol. Res.* **2019**, *68*, S375–S384. [CrossRef]
145. Slator, C.; Molphy, Z.; McKee, V.; Long, C.; Brown, T.; Kellett, A. Di-Copper Metallodrugs Promote NCI-60 Chemotherapy via Singlet Oxygen and Superoxide Production with Tandem TA/TA and AT/AT Oligonucleotide Discrimination. *Nucleic Acids Res.* **2018**, *46*, 2733–2750. [CrossRef]

146. Kallitsakis, M.G.; Gioftsidou, D.K.; Tzani, M.A.; Angaridis, P.A.; Terzidis, M.A.; Lykakis, I.N. Selective C–H Allylic Oxygenation of Cycloalkenes and Terpenoids Photosensitized by [Cu(Xantphos)(Neoc)]BF_4. *J. Org. Chem.* **2021**, *86*, 13503–13513. [CrossRef]
147. Kallitsakis, M.G.; Ioannou, D.I.; Terzidis, M.A.; Kostakis, G.E.; Lykakis, I.N. Selective Photoinduced Reduction of Nitroarenes to *N* -Arylhydroxylamines. *Org. Lett.* **2020**, *22*, 4339–4343. [CrossRef] [PubMed]
148. Symeonidis, T.S.; Athanasoulis, A.; Ishii, R.; Uozumi, Y.; Yamada, Y.M.A.; Lykakis, I.N. Photocatalytic Aerobic Oxidation of Alkenes into Epoxides or Chlorohydrins Promoted by a Polymer-Supported Decatungstate Catalyst. *ChemPhotoChem* **2017**, *1*, 479–484. [CrossRef]
149. Symeonidis, T.S.; Tamiolakis, I.; Armatas, G.S.; Lykakis, I.N. Green Photocatalytic Organic Transformations by Polyoxometalates vs. Mesoporous TiO_2 Nanoparticles: Selective Aerobic Oxidation of Alcohols. *Photochem. Photobiol. Sci.* **2015**, *14*, 563–568. [CrossRef] [PubMed]
150. Tzani, M.A.; (Department of Chemistry, Aristotle University of Thessaloniki, University Campus, 54124 Thessaloniki, Greece); Terzidis, M.A.; (Department of Nutritional Sciences and Dietetics, International Hellenic University, Sindos Campus, 57400 Thessaloniki, Greece); Lykakis, I.N.; (Department of Chemistry, Aristotle University of Thessaloniki, University Campus, 54124 Thessaloniki, Greece). Unpublished work. 2021.

Article

Disproportionation of H_2O_2 Mediated by Diiron-Peroxo Complexes as Catalase Mimics

Dóra Lakk-Bogáth, Patrik Török, Flóra Viktória Csendes, Soma Keszei, Beatrix Gantner and József Kaizer *

Research Group of Bioorganic and Biocoordination Chemistry, University of Pannonia, H-8201 Veszprém, Hungary; lakkd@almos.uni-pannon.hu (D.L.-B.); patriktrk6@gmail.com (P.T.); csvflora@gmail.com (F.V.C.); kornflakes09@gmail.com (S.K.); gantnerbea@gmail.com (B.G.)

* Correspondence: kaizer@almos.uni-pannon.hu; Tel.: +36-88-62-4720

Abstract: Heme iron and nonheme dimanganese catalases protect biological systems against oxidative damage caused by hydrogen peroxide. Rubrerythrins are ferritine-like nonheme diiron proteins, which are structurally and mechanistically distinct from the heme-type catalase but similar to a dimanganese KatB enzyme. In order to gain more insight into the mechanism of this curious enzyme reaction, non-heme structural and functional models were carried out by the use of mononuclear $[Fe^{II}(L_{1-4})(solvent)_3](ClO_4)_2$ (**1–4**) (L_1 = 1,3-bis(2-pyridyl-imino)isoindoline, L_2 = 1,3-bis(4′-methyl-2-pyridyl-imino)isoindoline, L_3 = 1,3-bis(4′-Chloro-2-pyridyl-imino)isoindoline, L_4 = 1,3-bis(5′-chloro-2-pyridyl-imino)isoindoline) complexes as catalysts, where the possible reactive intermediates, diiron-perroxo $[Fe^{III}_2(\mu\text{-}O)(\mu\text{-}1,2\text{-}O_2)(L_1\text{-}L_4)_2(Solv)_2]^{2+}$ (**5–8**) complexes are known and well-characterized. All the complexes displayed catalase-like activity, which provided clear evidence for the formation of diiron-peroxo species during the catalytic cycle. We also found that the fine-tuning of iron redox states is a critical issue, both the formation rate and the reactivity of the diiron-peroxo species showed linear correlation with the Fe^{III}/Fe^{II} redox potentials. Their stability and reactivity towards H_2O_2 was also investigated and based on kinetic and mechanistic studies a plausible mechanism, including a rate-determining hydrogen atom transfer between the H_2O_2 and diiron-peroxo species, was proposed. The present results provide one of the first examples of a nonheme diiron-peroxo complex, which shows a catalase-like reaction.

Keywords: catalase mimics; peroxide; diiron-peroxo complexes; structure/activity; kinetic studies

Citation: Lakk-Bogáth, D.; Török, P.; Csendes, F.V.; Keszei, S.; Gantner, B.; Kaizer, J. Disproportionation of H_2O_2 Mediated by Diiron-Peroxo Complexes as Catalase Mimics. *Molecules* **2021**, *26*, 4501. https://doi.org/10.3390/molecules26154501

Academic Editor: Chryssostomos Chatgilialoglu

Received: 7 July 2021
Accepted: 24 July 2021
Published: 26 July 2021

1. Introduction

Nonheme manganese catalases (MnCAT), as alternatives to the heme-containing catalases, contain a binuclear active sites and deplete hydrogen peroxide in cells through a ping-pong mechanism interconventing between Mn_2(II,II) and Mn_2(III,III) states in a two-electron catalytic cycle [1,2]. Manganese catalase enzymes such as *Lactobacillus plantarum* (LPC) [3,4], *Thermus thermophilus* (TTC) [5,6], *Thermoleophilium album* (TAC) [7], and *Pyrobaculum calidifontis VA1* (PCC) [8] have been isolated and characterized from both bacteria and archae. Despite their large numbers and importance, few catalase enzyme structures are known and they can also be sub-divided into two groups, each with a different active site.

Among them both LPC and TTC have a dimanganese catalytic core with Glu_3His_2 ligands coupled by two single-atom solvent bridges (Scheme 1A) [4,9]. In contrast to the above structure, the cyanobacterial manganese catalase (KatB) from *Anabaena* shows a markedly different active site with canonical Glu_4His_2 coordination geometry and two terminal water molecules (Scheme 1B). However, it closely resembles the active site of the ferritin-like ruberythrin (RBR), which is a nonheme diiron protein [10]. The Mn-Mn distances (~3.8 Å) in KatB are similar to that was observed for the iron-containing ribonucleotide reductase R2 (3.6–3.7 Å), but much shorter distances have been found for LPC and TTC (3.0–3.1 Å). The H_2O_2 binds terminally to one of the Mn center in LPC/TTC, while in

KatB and RBR, the symmetrical active sites and the relatively long M-M distances favor its μ-1,2 binding mode [11]. These enzymes, coupled with superoxide dimutates (SOD), are responsible for the full detoxification of $O_2^{\bullet -}$ through the formation of H_2O_2 and its redox disproportionation into water and dioxygen. H_2O_2 can also be classified as a ROS particle, since it readily reacts with reduced transition metals and results in highly reactive $HO^{\bullet}$ radical via Fenton-like reactions. Many rationally designed synthetic biomimetics are being developed against oxidative stress to treat various diseases, including Alzheimer's, cancer, aging, neurodegenerative, inflammatory and heart diseases [12–21]. The identification and study of the structure of the above enzymes can greatly contribute to the design of new catalase mimetics for potential therapeutic use. The structural and functional investigations on low molecular-weight copper, manganese and iron containing model compounds that are focused mainly on the effect of the used metals and their ligand environment (donor sites) on redox potential and through it on reactivity. Based on the literature data, various Schiff base, aminopyridine and azamacrocyclic derivatives have proven to be the most popular and practical ligands [12–21]. Considering the reactivity of manganese catalases (KatB) toward H_2O_2, including a ping-pong mechanism interconventing between Mn_2(II,II) and Mn_2(III,III) states in a two-electron catalytic cycle, a similar interaction with H_2O_2 has not been reported for diiron(III) peroxo intermediates which may be proposed in the catalytic cycles of RBR. Progress in recent years has made a number of diiron(III)-peroxo complexes available that can be generated by stoichiometric amount of H_2O_2 [22–27]. Therefore, the opportunity has arisen to investigate both electrophilic and nucleophilic reactivity towards various substrates, such as phenols, *N,N*-dimethylanilines, and various aldehydes, respectively [25,26]. The differences in reactivity observed in the investigated systems, as well as the stability of the peroxo intermediates can in principle be explained by the disproportionation of H_2O_2 as a competing process. Clarification of the relationship between catalase-like activity and catalytic oxidation is an important factor in the design and development of high efficiency catalytic systems. In addition, the investigation of the metal-based disproportionation of H_2O_2 may help us to understand the nature of the catalytically active species present both, in the enzymatic and biomimetic systems. We have reported that the $[Fe^{II}(L_1)(solvent)_3](ClO_4)_2$ (**1**) is an efficient catalyst for the oxidation of thioanisoles and benzyl alcohols by the use of H_2O_2 as cooxidant, where a metastabile green species (**5**) with a $Fe^{III}(\mu\text{-}O)(\mu\text{-}1,2\text{-}O_2)Fe^{III}$ core was observed and characterized by UV-Vis, EPR, rRaman, and X-ray absorption spectroscopic measurements [24]. The reactivity of **5** was also published both in nucleophilic (deformylation of aldehydes) and electrophilic (oxidation of phenols) reactions [27].

As a possible functional model of RBR, we recently investigated the reactivity of the peroxo adduct $[Fe_2(\mu\text{-}O_2)(MeBzim\text{-}Py)_4(CH_3CN)]^{4+}$ (MeBzim-Py = (2-(2′-pyridyl)-*N*-methylbenzimidazole) with H_2O_2 and found direct kinetic and computational evidence for the formation of low-spin oxoiron(IV) via dissociation process [22,23]. As a continuation of this tudy, the previously reported and fully characterized $Fe^{III}{}_2(\mu\text{-}O)(\mu\text{-}1,2\text{-}O_2)(IndH)_2(Solv)_2]^{2+}$ (IndH = L_1 = 1,3-bis(2-pyridyl-imino)-isoindolines) [24,25] and their substituted isoindolin-containing derivatives (L_2–L_4, **2**–**4**) were chosen as model compounds, in which the dissociation of the diiron(III) core can be ruled out due to the μ-oxo-bridge, and investigated its reactivity towards H_2O_2 to gain insight into the mechanism of RBR enzymes (Scheme 2).

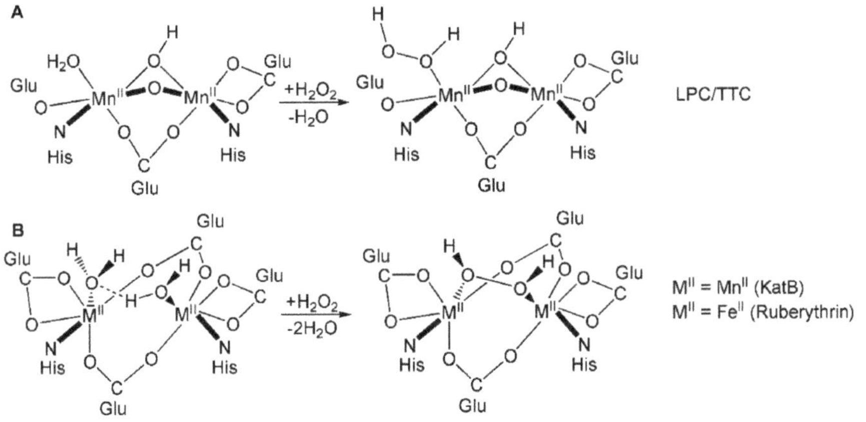

Scheme 1. Binding mode of H_2O_2 in nonheme dimanganese and diiron enzymes.

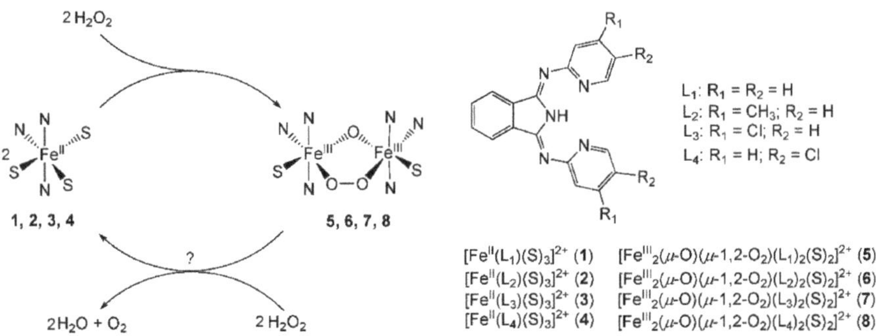

Scheme 2. Structural formulae for the ligands and their complexes used in the disproportionation reaction of H_2O_2.

2. Results and Discussions

In view of the above results, further studies were planned to elucidate the mechanism of both the diiron(III)-peroxo formation and its decay mediated by H_2O_2. Complex **5** from its precursor complex **1** can be generated even at room temperature by the use of excess of H_2O_2 in CH_3CN. Its formation and decay process was followed as an increase and decrease in absorbance at 690 nm (ε = 1500 M^{-1} cm^{-1} per Fe ion and assuming 100% conversion), which can be assigned as a charge transfer between Fe^{III} and the $O_2{}^{2-}$ ligand (Figure 1a) [24]. Similar spectral feature can be observed for the complexes **6**–**8**: 691 nm (ε = 1404 M^{-1} cm^{-1}), 692 nm (ε = 1460 M^{-1} cm^{-1}), and 693 nm (ε = 1308 M^{-1} cm^{-1}), respectively. It is worth noting that the green species (**5**), after decomposition, can be regenerated 2–3 times by adding additional H_2O_2 without significant decrease on the yields (Table 1, Figure 1b and inset in Figure 2a). Similar behavior and yields were observed for **6**, but much lower values were obtained for **7** and **8**. These results are consistent with the stability and/or the reactivity of the intermediates at high concentrations of added water, providing further information for the possible role of the binding water during the formation of the reactive oxidant.

Preliminary kinetic studies were performed at 20 °C to establish the role of the observed diiron(III)-peroxo intermediates in the catalytic disproportionation reaction of H_2O_2. The formation and decomposition of **5** in the presence of H_2O_2 was monitored by following the rise and fall of its visible chromophore (Tables 2 and 3), as well as the appearance of the dioxygen formed by gas volumetric methods (Figure 2b). Dioxygen formation started after a short lag phase during which the peroxo-adduct (**5**) accumulated. The lag phase was followed by a linear dioxygen formation period during which time **5** persisted in a pseudo-steady-state. This profile is similar to that found for the water-assisted decay of $Fe^{III}(H_2O)(OOH)$ intermediate into the reactive $Fe^{V}(O)(OH)$ species [28]. Upon depletion of H_2O_2, **5** decomposed rapidly and the dioxygen formation ceased. The yield of dioxygen, TON (turnover number = mol H_2O_2/mol catalyst) and the TOF (turnover frequency = mol H_2O_2/mol catalyst/h) values were determined to be Yield = 85%, TON = 10.4 and TOF = 288 h^{-1} as a result of volumetric method via the measurements of dioxygen evolution from the reaction mixture ($[H_2O_2]$ = 0.243 M and [**1**] = 0.002 M) at 20 °C in CH_3CN. Similar values were obtained for the complexes **2**, **3** and **4** under the same conditions (Table 4). The results clearly indicate that diiron-peroxo intermediates play a key role in the formation of the reactive species responsible for the H_2O_2 oxidation.

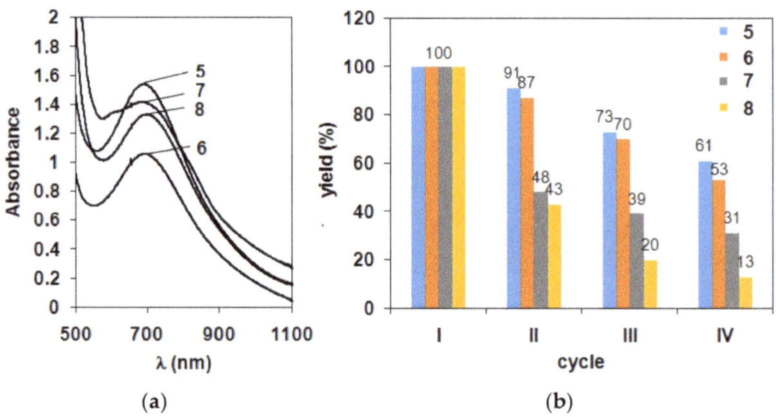

Figure 1. (**a**) Characteristic UV-Vis spectra of $Fe^{III}_2(\mu\text{-}O)(\mu\text{-}1,2\text{-}O_2)(L_{1\text{–}4})_2(Solv)_2]^{2+}$ (**5–8**) intermediates in CH_3CN. (**b**) Dependence of the yield of peroxo complexes (**5–8**) on the number of cycles. $[Fe]_0$ = 2 mM, $[H_2O_2]_0$ = 24 mM (4x) in CH_3CN at 20 °C.

Table 1. Dependence of the yield of peroxo complexes (**5–8**) and the evolved oxygen (parantheses) on the number of cycles. $[Fe]_0$ = 2 mM, $[H_2O_2]_0$ = 24 mM (4×) in CH_3CN at 20 °C.

Cycles	Yield (5) and (O_2)/%	Yield (6)/%	Yield (7)/%	Yield (8)/%
I.	100 (95)	100	100	100
II.	91 (92)	87	48	43
III.	73 (78)	70	39	20
IV.	61 (64)	53	31	13
V.	41 (27)	36	26	12

2.1. *Kinetic Studies on the Formation of Diiron-Peroxo Complexes* (**5–8**)

Detailed kinetic studies on the reaction of **1** (2 mM) with H_2O_2 (24–98 mM) were carried out in CH_3CN at 15 °C, and the formation of the green species **5** was followed by UV-vis spectroscopy as an increase in absorbance at 690 nm under pseudo-first-order conditions (excess of H_2O_2) (Table 2). At constant $[\mathbf{1}]_0$ = 2 mM, the formation of **5** exhibits a first-order dependence, and the first-order rate constants show a good linear dependence with $[H_2O_2]_0$ (Figure 3a), affording the second-order rate constant k_2 = 0.93 M^{-1} s^{-1} at 15 °C and 1.09 M^{-1} s^{-1} at 20 °C. These values are about 5–6 times smaller than those obtained for $[Fe^{II}(MeBzim\text{-}Py)]^{2+}$ (5.5 M^{-1} s^{-1}), $Fe^{II}(MeBzim\text{-}Py)^{2+}$ (6.6 M^{-1} s^{-1}) and $Fe^{II}(TBA)^{2+}$ (6.54 M^{-1} s^{-1}) complexes under the same conditions at 20 °C [23,26]. It is worth noting that adding large volumes of water (~0.5 M) resulted in nearly twice the rate of the diiron-peroxo formation. Furthermore, a kinetic isotope effect (KIE) of 2.99 and 3.07

were observed for the formation of **6** and **7**, when the experiments were carried out in the presence of added H_2O or D_2O. This value may represent the result of much more multiple effects compared to the elemental step, but clearly indicate the key role of the water during the diiron-peroxo complex formation.

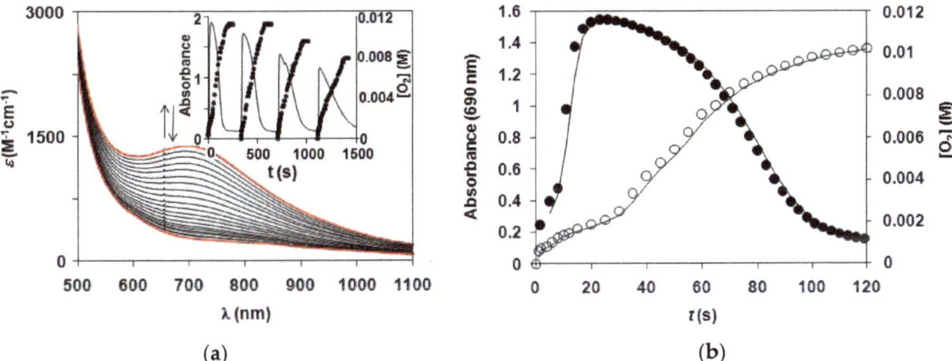

Figure 2. (**a**) UV-Vis spectral changes of **5** during the **1**-catalyzed disproportionation of H_2O_2. The inset shows the the time course of the formation and decay of **5** and the parallel formation of dioxygen for 5 cycles. $[\mathbf{5}]_0 = 2$ mM, $[H_2O_2]_0 = 24$ mM (4×) in CH_3CN at 20 °C (**b**) Time course for the H_2O_2 (24 mM) oxidation by **5** (2 mM), monitored at 690 nm in CH_3CN at 20 °C and by volumetrically determined amounts of evolved dioxygen.

Table 2. Kinetic data for the formation of intermediates **5**, **6**, **7** and **8** in CH_3CN.

Nr.	Complex	T/°C	$[\mathbf{1}\text{–}\mathbf{4}]_0$/mM	$[H_2O_2]_0$/M	$k_{obs}10^{-2}$ s^{-1}	k_2/M^{-1} s^{-1}
1	**1**	15	2	0.0243	2.12	0.87 ± 0.03
2	**1**	15	2	0.0367	3.15	0.86 ± 0.04
3	**1**	15	2	0.0489	4.07	0.83 ± 0.03
4	**1**	15	2	0.0734	6.76	0.92 ± 0.04
5	**1**	15	2	0.0979	9.04	0.92 ± 0.03
6	**2**	15	2	0.0243	4.45	1.83 ± 0.09
7	**2**	15	2	0.0367	5.83	1.58 ± 0.05
8	**2**	15	2	0.0489	8.44	1.72 ± 0.06
9	**2**	15	2	0.0734	10.5	1.43 ± 0.04
10	**2**	15	2	0.0979	14.45	1.48 ± 0.04
11	**3**	15	2	0.0243	9.22	3.79 ± 0.15
12	**3**	15	2	0.0367	12.64	3.44 ± 0.13
13	**3**	15	2	0.0489	16.95	3.46 ± 0.14
14	**3**	15	2	0.0734	24.61	3.35 ± 0.14
15	**3**	15	2	0.0979	31.07	3.17 ± 0.11
16	**4**	15	2	0.0243	9.25	3.81 ± 0.15
17	**4**	15	2	0.0367	13.17	3.59 ± 0.14
18	**4**	15	2	0.0489	18.48	3.78 ± 0.16
19	**4**	15	2	0.0734	27.07	3.69 ± 0.15
20	**4**	15	2	0.0979	35.83	3.66 ± 0.15

Table 3. Kinetic data for the disproportionation of H_2O_2 mediated by **5**, **6**, **7** and **8** in CH_3CN.

Nr.	Complex	T/°C	[1–4]$_0$/mM	$[H_2O_2]_0$/M	$k_{obs}/10^{-3}$ s^{-1}	$k_2/10^{-2}$ M^{-1} s^{-1}
1	**1**	15	0.738	0.0243	0.358	1.47 ± 0.06
2	**1**	15	1.48	0.0243	0.387	1.59 ± 0.07
3	**1**	15	2	0.0243	0.370	1.52 ± 0.06
4	**1**	5	2	0.0243	0.116	0.477 ± 0.02
5	**1**	10	2	0.0243	0.160	0.658 ± 0.03
6	**1**	20	2	0.0243	0.523	2.15 ± 0.11
7	**1**	25	2	0.1	3.33	3.33 ± 0.11
8	**1**	25	2	0.2	5.57	2.79 ± 0.08
9	**1**	25	2	0.3	10.1	3.36 ± 0.09
10	**1**	25	2	0.6	17.3	2.89 ± 0.07
11	**1**	25	2	0.9	27.1	3.01 ± 0.06
12	**1**	25	0.3	0.3	10.3	3.42 ± 0.12
13	**1**	25	0.6	0.3	10.1	3.38 ± 0.11
14	**1**	25	1	0.3	11.1	3.68 ± 0.14
15	**1**	25	1.4	0.3	9.47	3.16 ± 0.13
16	**1**	10	2	0.3	1.99	0.663 ± 0.02
17	**1**	15	2	0.3	3.63	1.21 ± 0.05
18	**1**	20	2	0.3	6.12	2.04 ± 0.08
19	**2**	15	0.738	0.0243	0.823	3.39 ± 0.11
20	**2**	15	1.07	0.0243	0.692	2.85 ± 0.09
21	**2**	15	1.48	0.0243	0.772	3.18 ± 0.10
22	**2**	15	1.77	0.0243	0.744	3.06 ± 0.09
23	**2**	15	2	0.0243	0.730	3.00 ± 0.11
24	**2**	5	2	0.0243	0.332	1.37 ± 0.04
25	**2**	10	2	0.0243	0.413	1.70 ± 0.05
26	**2**	20	2	0.0243	1.09	4.49 ± 0.27
27	**3**	15	0.738	0.0243	3.92	16.1 ± 0.72
28	**3**	15	1.48	0.0243	4.71	19.4 ± 0.62
29	**3**	15	2	0.0243	4.83	19.9 ± 0.68
30	**3**	5	2	0.0243	2.42	9.96 ± 0.35
31	**3**	10	2	0.0243	3.41	14.01 ± 0.63
32	**3**	20	2	0.0243	8.90	36.6 ± 2.01
33	**4**	15	0.738	0.0243	5.49	22.6 ± 0.79
34	**4**	15	1.48	0.0243	5.44	22.4 ± 0.83
35	**4**	15	2	0.0243	5.75	23.8 ± 0.85
36	**4**	5	2	0.0243	2.89	11.9 ± 0.48
37	**4**	10	2	0.0243	4.20	17.3 ± 0.67
38	**4**	20	2	0.0243	10.2	42.01 ± 1.77

In the next step, we investigated the effect of the the ligand modification by introducing electron withdrawing (Cl) or electron releasing (CH_3) substituents on the pyridine arms in the fourth and fifth positions with emphasis on the redox potential and Lewis acidity of the precursor complex. Since the redox potential of the substituted complexes can be measured under the same conditions, it can be used as excellent reactivity descriptor. On the cyclic voltammogram of the Fe(II)-isoindoline complexes (**1**–**4**) irreversible reduction and oxidation peaks can be recognized within the range of −440 to +1160 mV (vs. Fc/Fc$^+$). The huge peak separations ($\Delta E = E_{pa} - E_{pc}$, Table 4) indicates the irreversibility of the redox processes and suggests chemical reactions coupled to the electrode reaction. The results, obtained at the scan rate of 100 mV/s, show that oxidation potential is mainly influenced by the electronic properties of the ligands, since the anodic peak potentials (E_{pa}) of the complexes are increasing in the order of electron releasing ability of the substituent on the pyridine moiety (Figure 4). The E_{pa} spans a 128 mV range from 655 mV (**1**) to 783 mV (**4**), which is not too large but informative, in terms of the effect of the substituents. However, the fact that the anodic peak of complex **1** can be found at lower potentials than it could be expected according to electronic effects, suggest that steric effects are also influencing the oxidation reactions. One can assume that the steric effect of the methylated and chlorinated ligands can cause an anodic shift in the oxidation peak potential of the complexes.

Table 4. Rate constants, electrochemical data and activation parameters for the formation of $Fe^{III}_2(\mu$-O)(μ-1,2-O_2)(L_1-L_4)$_2$(Solv)$_2$]$^{2+}$ (**5–8**) complexes and their catalase-like reactions.

	1 (5)	2 (6)	3 (7)	4 (8)
$E_{pa}(Fe^{II}/Fe^{III})$/mV vs. Fc/Fc$^+$	655	683	765	783
$E_{pc}(Fe^{II}/Fe^{III})$/(mV vs. Fc/Fc$^+$	−90	−223	−155	−143
ΔE/mV	745	905	920	925
yield/%	85	100	92	85
TON ([S]/[Fe])	10.4	12.2	11.3	10.4
TOF ([S]/[Fe]/h)	288	418	354	234
k_2(cat)/10^{-2} M^{-1} s^{-1} (15 °C)	1.52	3.05	19.9	23.8
E_A/kJ mol^{-1}	72.3	55.9	57.5	55.5
$\Delta S^{\neq}$/J mol^{-1} K^{-1}	−37.7	−88.1	−65.9	−71.4
$\Delta H^{\neq}$/kJ mol^{-1}	70	53.5	55	53.1
$\Delta G^{\neq}$/kJ mol^{-1}	81	78.9	74	73.7
k_{decay}/10^{-2}s^{-1} (15 °C)	0.13	0.15	1.22	1.41
k_{form}/M^{-1} s^{-1} (15 °C)	0.93	1.43	3.16	3.66

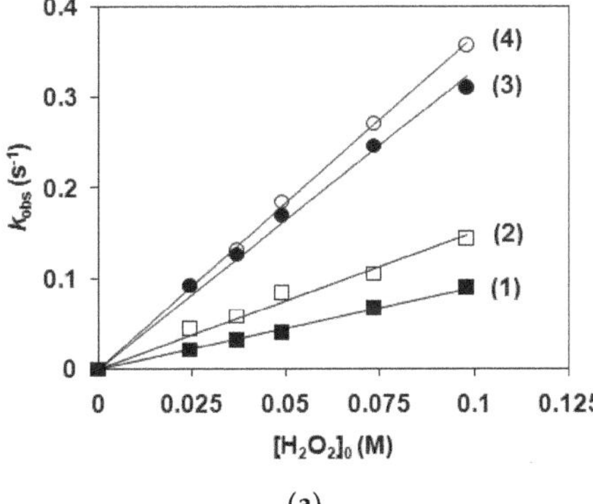

(a)

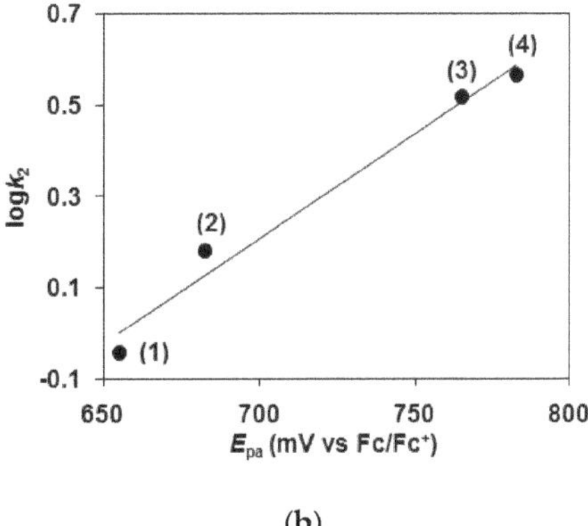

(b)

Figure 3. (**a**) Formation of diiron-peroxo (**5–8**) complexes. Plots of k_{obs} against H_2O_2 concentration to determine a second-order rate constant. $[$**1–4**$]_0$ = 2 mM, in CH_3CN at 15 °C. (**b**) Dependence of the formation rate of peroxo-diiron complexes (**5–8**) on the oxidation potential (E_{pa}) of the $[Fe^{II}(L_{1-4})(solvent)_3](ClO_4)_2$ (**1–4**) complexes.

The four iron complexes, **1–4** have been compared to investigate the effect of the various aryl substituents. Figure 3a shows the results of the activity for the precursor complexes (**1–4**) during the formation of their appropriate diiron-peroxo intermediates (**5–8**). This demonstrates only small differences exist, based on the first-order rate constants obtained under the same conditions. It was found that complex **4** with 5-chloropyridyl side chains is the most reactive complex with k_{obs} of 3.66 M^{-1} s^{-1}, whilst complex **1** with unsubstituted side chains was proved to be the least active with k_{obs} of 0.93 M^{-1} s^{-1} at 15 °C. Based on the CV data, we have got a clear evidence that the formation rate of the diiron-peroxo species (k_2) increases almost linearly with the redox potential (E_{pa} for Fe^{III}/Fe^{II}) of the precursor complex (Figure 3b). In summary, the higher the redox potentials of Fe^{III}/Fe^{II} redox couple the higher is the activity of the precursor complex towards to H_2O_2. Complexes with 4- and 5-chloro-pyridine side chains (**3** and **4**), whose metal sites are electron-deficient are faster in their reaction with H_2O_2 than electron-rich derivatives **1** and **2** with more Lewis basic side chains.

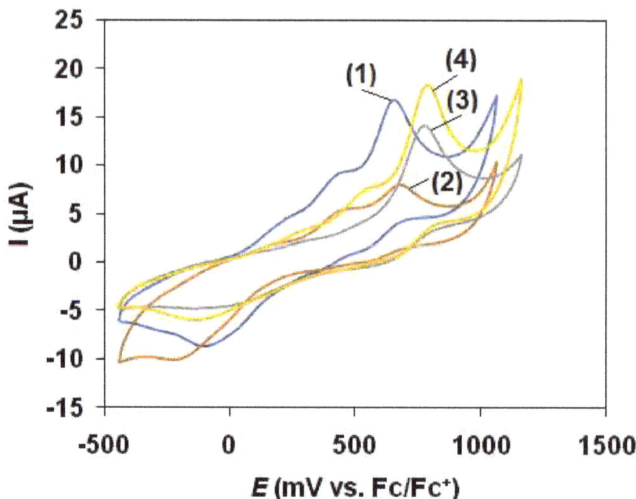

Figure 4. Cyclic voltammograms of $[Fe^{II}(L_{1-4})(solvent)_3](ClO_4)_2$ complexes (**1–4**) (5 mM in acetonitrile, supporting electrolyte: tetrabutylammonium-perchlorate (0.1 M), scan rate: 100 mV/s).

2.2. Catalase-Like Activity of Peroxo-Diiron Complexes

The ability of the in situ generated diiron-peroxo species (**5**) to mediate the disproportionation reaction of H_2O_2 into O_2 and H_2O was tested in CH_3CN by UV-vis spectroscopy as a decrease in absorbance at 690 nm under pseudo-first-order conditions (excess of H_2O_2) (Table 3). The initial rate of decay of diiron-peroxo species was measured as a function of the complex and substrate concentrations in CH_3CN. At constant $[H_2O_2]_0$ = 300 mM, the disproportionation reaction shows first-order kinetic behavior on [5], and the first-order rate constant $k_1 = 4.968 \times 10^{-3}\ s^{-1}$ was obtained at 20 °C from the slope of the plot of V_i versus $[\mathbf{5}]_0$ (Figure 5a). At constant $[\mathbf{5}]_0$ = 2 mM, initial rates (V_i) exhibit a linear dependence with $[H_2O_2]_0$ (Figure 5b) affording first-order rate constant $k_{1'} = 29.63 \times 10^{-3}\ s^{-1}$. The second-order rate constant, $k_2 = 3.4\ M^{-1}\ s^{-1}$ was obtained from either $k_1/[H_2O_2]_0$ or $k_{1'}/[\mathbf{5}]_0$. This value is much smaller than that was found for the analogue $[Mn(L_1)]^{2+}$ complex ($k_2(k_{cat}/K_M) = 79 \pm 4\ M^{-1}\ s^{-1}$) in protic solution, where $Mn^{IV}(O)$ intermediate was proposed as reactive species [29]. A kinetic isotope effect (KIE) of 1.73 and 1.66 were observed for the decay of **6** and **7**, when the experiments were carried out in the presence of added H_2O or D_2O. These results indicate the key role of the water during the diiron-peroxo mediated disproportionation reaction.

In the next step, we investigated the effect of the ligand modification by introducing electron donating (-CH_3) and electron withdrawing (-Cl) phenyl-ring substituents. The substituted ligand-containing diiron-peroxo intermediates **5**, **6**, **7** and **8** show an increasing rate in the listed order (Figure 6a). It was found that complex **8** with 5-chloropyridine side chains is the most efficient oxidant with the fastest rate $k_2 = 23.8 \times 10^{-2}\ M^{-1}\ s^{-1}$, while complexes **5** and **6** with unsubstituted and 4-methylpyridine side chains are the less efficient oxidants with $k_2 = 1.52 \times 10^{-2}\ M^{-1}\ s^{-1}$ and $k_2 = 3.05 \times 10^{-2}\ M^{-1}\ s^{-1}$, respectively. These results give clear evidence that substituents affect the redox potential and the catalase activity of the complexes. Table 4 and Figure 6b show that the redox potentials (E_{pa}) of the complexes and their activity increase with increasing electron-withdrawing ability of the substituent.

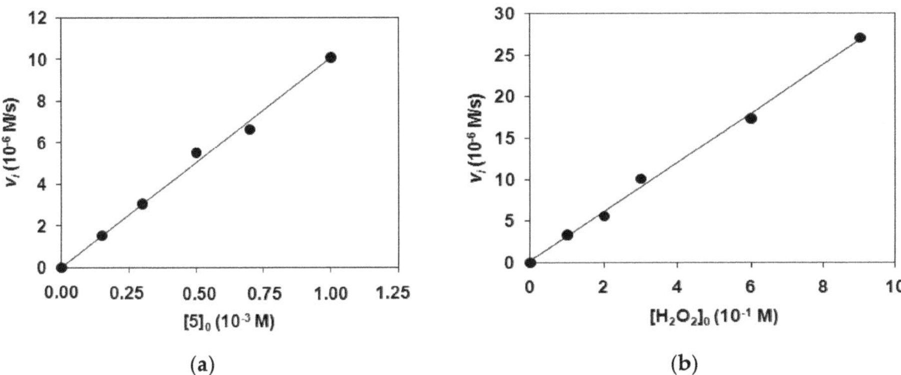

Figure 5. **5**-mediated disproportionation of H_2O_2 in CH_3CN at 25 °C. (**a**) Dependence of the reaction rate (V_i) for H_2O_2 oxidation on the diiron-peroxo complex **5** concentration. $[H_2O_2]_0$ = 300 mM (**b**) Dependence of the reaction rate (V_i) for H_2O_2 oxidation on the H_2O_2 concentration. $[\mathbf{5}]_0$ = 1 mM.

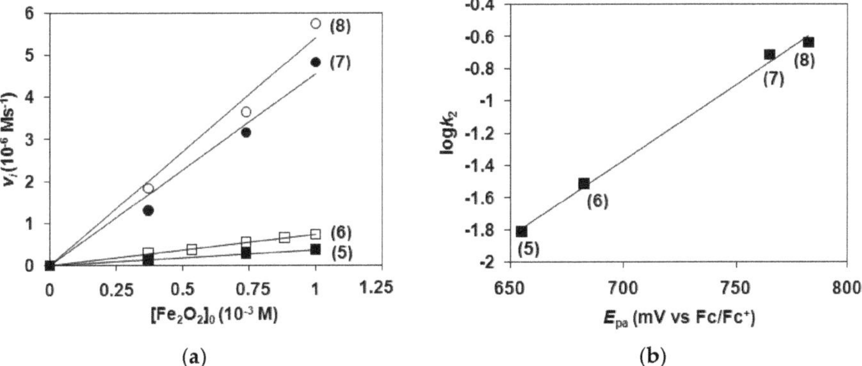

Figure 6. Peroxo-diiron complexes-mediated disproportionation of H_2O_2 in CH_3CN at 15 °C. (**a**) Dependence of the reaction rate (V_i) for H_2O_2 oxidation on the diiron-peroxo complex **5–8** concentration. $[H_2O_2]_0$ = 24 mM (**b**) Dependence of the decay rate of peroxo-diiron complexes (**5–8**) on the oxidation potential (E_{pa}) of the $[Fe^{II}(L_{1-4})(solvent)_3](ClO_4)_2$ (**1–4**) complexes.

Based on the temperature dependence of the reactivity of diiron-peroxo complexes (Figure 7a), the experimentally determined the difference between $\Delta G^{\neq}$ values was 7.3 kJ mol^{-1}. Significantly lower free activation energy value was obtained for the **8**-mediated disproportionation of H_2O_2 in comparison to **5**. The calculated Gibbs energy values correlate very well with the reaction rates (Figure 7b). The values of $-T\Delta S^{\neq}$ were lower than $\Delta H^{\neq}$ in the investigated temperature range, indicating an enthalpy-controlled reactions. As a result of a compensation effect increasing activation enthalpies are offset by increasingly poitive entropies yielding $\Delta H^{\neq}$ = 81 kJ mol^{-1} at the intercept (Figure 8), which value is a little bit higher than that was obtained for the conversion alkylperoxo-iron(III) intermediates to oxoiron(IV) through O-O bond homolysis ($\Delta H^{\neq}$ = 61.3 kJ mol^{-1}) [30].

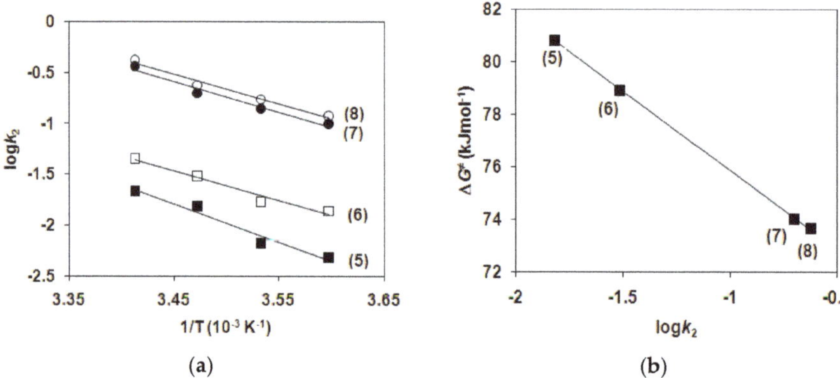

Figure 7. (**a**) Arrhenius plots of logk_2 versus 1/T for the $Fe^{III}_2(\mu\text{-}O)(\mu\text{-}1,2\text{-}O_2)(L_{1\text{-}4})_2(Solv)_2]^{2+}$ (**5–8**)-mediated disproportionation of H_2O_2 in CH_3CN. $[\mathbf{5\text{-}8}]_0$ = 2 mM, $[H_2O_2]_0$ = 24 mM. (**b**) Plot of $\Delta G^{\neq}$ versus lgk_2 for the reaction of $[Fe^{III}_2(\mu\text{-}O)(\mu\text{-}1,2\text{-}O_2)(L_{1\text{-}4})_2(Solv)_2]^{2+}$ (**5–8**) with H_2O_2.

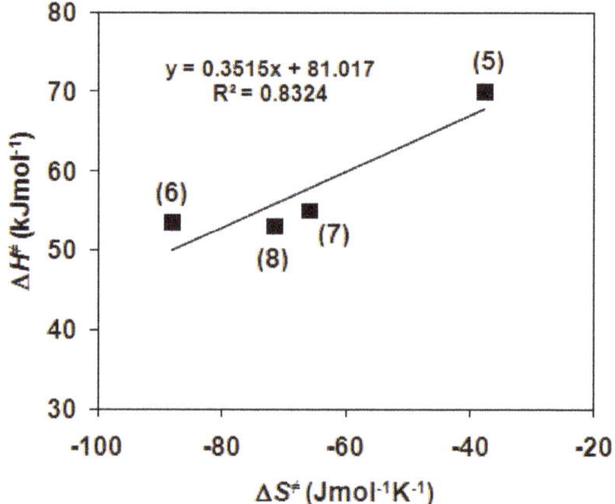

Figure 8. Isokinetic plot for the for the $Fe^{III}_2(\mu\text{-}O)(\mu\text{-}1,2\text{-}O_2)(L_{1\text{-}4})_2(Solv)_2]^{2+}$ (**5–8**)-mediated disproportionation of H_2O_2 in CH_3CN.

3. Experimental

3.1. Materials and Methods

The ligands 1,3-bis(2′-pyridylimino)isoindolines (L_1 = indH, L_2 = 4Me-indH, L_3 = 4Cl-indH, L_4 = 5Cl-indH,) and their complexes (**1–4**) were prepared according to published procedures [31]. All manipulations were performed under a pure argon atmosphere using standard Schlenk-type inert-gas techniques. Solvents used for the reactions were purified by literature methods and stored under argon. The starting materials for the ligand are commercially available and they were purchased from Sigma Aldrich. The UV-visible spectra were recorded on an Agilent 8453 diode-array spectrophotometer using quartz cells. IR spectra were recorded using a Thermo Nicolet Avatar 330 FT-IR instrument (Thermo Nicolet Corporation, Madison, WI, USA), Samples were prepared in the form of KBr pellets. Microanalyses elemental analysis was done by the Microanalytical Service of the University of Pannonia.

3.2. Characterization of Ligands and Their Complexes

indH: FT-IR (ATR) ν = 3199 (w), 3061 (w), 1622 (s), 1606 (m), 1577 (s), 1550 (s), 1454 (s), 1427 (s), 1373 (m), 1306 (m), 1259 (s), 1217 (s), 1139 (m), 1097 (m), 1042 (m), 991 (w), 885 (w),

875 (w), 859 (w), 839 (w), 808 (m), 792 (s), 783 (s), 769 (s), 737 (s), 708 (m), 696 (s), 689 (s), 627 (w). UV/Vis (DMF): λ_{max} (logε) = 317 (3.09), 332 (3.18), 348 (3.21), 368 (3.28), 386 (3.34), 410 (3.11) nm.

4Me-ind: FT-IR (ATR) ν = 3209 (w), 3049 (w), 1645 (m), 1627 (s), 1591 (s), 1541 (m), 1460 (s), 1408 (w), 1364 (m), 1305 (m), 1281 (m), 1242 (s), 1186 (m), 1132 (w), 1114 (w), 1101 (m), 1036 (m), 997 (w), 929 (m), 891 (m), 848 (w), 810 (s), 779 (m), 742 (w), 694 (s), 660 (w). UV/Vis (DMF): λ_{max} (logε) = 316 (2.99), 332 (3.08), 348 (3.12), 366 (3.18), 387 (3.21), 411 (2.98) nm.

4Cl-ind: FT-IR (ATR) ν = 3213 (w), 1743 (w), 1637 (m), 1624 (m), 1565 (s), 1539 (m), 1448 (s), 1359 (m), 1309 (m), 1300 (m), 1281 (w), 1219 (m), 1184 (w), 1089 (m), 1040 (m), 989 (m), 895 (s), 877 (s), 804 (m), 779 (m), 735 (s), 701 (s) 687 (m), 656 (w). UV/Vis (DMF): λ_{max} (logε) = 315 (3.16), 332 (3.27), 348 (3.29), 367 (3.31), 385 (3.35), 409 (3.12) nm.

5Cl-ind: FT-IR (ATR) ν = 3317 (w), 3228 (w), 1745 (m), 1633 (s), 1572 (s), 1470 (m), 1447 (s), 1359 (m), 1311 (m), 1256 (w), 1244 (w), 1228 (m), 1217 (m), 1184 (w), 1105 (s), 1034 (s), 1007 (m), 955 (w), 924 (w), 880 (w), 843 (s), 773 (s), 754 (m), 725 (w), 702 (s), 691 (s), 638 (m), 619 (w), 602 (m). UV/Vis (DMF): λ_{max} (logε) = 338 (3.15), 354 (3.15), 376 (3.19), 395 (3.25), 421 (3.02) nm.

1: FT-IR (ATR) ν = 3329 (w), 1728 (w), 1654 (m), 1626 (s), 1614 (m), 1593 (w), 1554 (w), 1524 (s), 1485 (s), 1467 (m), 1433 (m), 1373 (w), 1304 (w), 1262 (w), 1207 (m), 1055 (s), 929 (w), 860 (w), 779 (s), 714 (m), 621 (m). $C_{24}H_{20}Cl_2FeN_8O_8$ (675.22): calcd. C 42.69, H 2.99, Cl 10.15, N 16.60; found C 42.82, H 2.93, Cl 10.11, N 16.42. UV/Vis (CH_3CN): λ_{max} (logε) = 315 (3.36), 329 (3.45), 345 (3.48), 365 (3.56), 382 (3.61), 406 (3.37) nm.

2: FT-IR (ATR) ν = 3456 (w), 1734 (w), 1665 (m), 1654 (s), 1605 (s), 1551 (m), 1520 (s), 1493 (s), 1477 (m), 1367 (w), 1313 (w), 1279 (m), 1244 (w), 1215 (w), 1049 (s), 941 (w), 827 (w), 777 (w), 752 (w), 710 (m), 621 (m). $C_{26}H_{27}Cl_2FeN_8O_8$ (706.29): calcd. C 44.21, H 3.85, Cl 10.04, N 15.87; found C 43.99, H 3.64, Cl 9.58, N 15.79. UV/Vis (CH_3CN): λ_{max} (logε) = 315 (3.48), 328 (3.58), 345 (3.62), 365 (3.71), 383 (3.76), 406 (3.50) nm.

3: FT-IR (ATR) ν = 3366 (w), 3217 (w), 1773 (w), 1724 (w), 1670(m), 1628 (m), 1564 (m), 1543 (m), 1516 (s), 1483 (m), 1460 (s), 1362 (w), 1306 (w), 1228 (w), 1205 (m), 1090 (s), 1051 (s), 910 (m), 867 (w), 823 (w), 744 (s), 710 (s), 621 (m). $C_{24}H_{21}Cl_4FeN_8O_8$ (747.13): calcd. C 38.58, H 2.83, Cl 18.98, N 15.00; found C 38.25, H 2.95, Cl 18.63, N 15.52. UV/Vis (CH_3CN): λ_{max} (logε) = 328 (3.10), 344 (3.09), 363 (3.04), 383 (2.99), 406 (2.71) nm.

4: FT-IR (ATR) ν = 3323 (w), 3228 (w), 1773 (w), 1717 (w), 1651(w), 1628 (m), 1612 (m), 1546 (m), 1521 (s), 1477 (s), 1458 (s), 1391 (w), 1366 (m), 1335 (w), 1298 (w), 1261 (w), 1236 (m), 1205 (m), 1096 (s), 1076 (s), 922 (m), 851 (w), 835 (m), 781 (w), 752 (m), 711 (s), 650 (w), 621 (m). $C_{24}H_{21}Cl_4FeN_8O_8$ (747.13): calcd. C 38.58, H 2.83, Cl 18.98, N 15.00; found C 38.81, H 3.04, Cl 18.63, N 15.39. UV/Vis (CH_3CN): λ_{max} (logε) = 335 (2.64), 353 (2.59), 373 (2.47), 395 (2.36), 418 (2.04) nm.

3.3. Cyclic Voltammetry

Cyclic voltammetry experiments were performed on a Radiometer Analytical PGZ-301 potentiostat with a conventional three-electrode configuration, consisting of a glassy carbon working electrode (ID = 3 mm), a platinum wire auxiliary electrode, and Ag/AgCl reference electrode. Potentials are referenced to the Fc/Fc^+ redox couple. Ferrocene was added as an internal standard at the end of the experiments. Procedure for the cyclic voltammetry experiment: 0.05 mmol of **1–4** was dissolved in 10 mL dry acetonitrile, containing 349.1 mg (1 mmol) of tetrabutylammonium-perchlorate, which served as supporting electrolyte. The solution was bubbled with argon to remove dissolved gas residuals and to ensure inert atmosphere during measurements. The working electrode was wet polished on 0.5 μm alumina slurry or emery paper grade 500, after each measurement. Cyclic voltammograms were recorded with a scan rate of 100 mV s^{-1}.

3.4. *Generation of Diiron-Peroxo Complexes* (**5–8**)

Precursor complexes (**1–4**) were dissolved in 1.5 mL acetonitrile and added 4 equivalent H_2O_2, in different temperature (0, 5, 10 15 °C) and the reactions were followed with UV-Vis spectroscopy at 680 nm, the cuvette length was 1 cm.

3.5. *Diiron-Peroxo-Mediated Disproportionaton of H_2O_2*

Reactions were carried out by mixing the substrate (H_2O_2) with the in situ generated complex **5–8** (1 mM) in 1.5 mL acetonitrile in different temperature (0, 5, 10 15 °C) and the reactions were followed with UV-Vis spectroscopy at 680 nm, the cuvette length was 1 cm. To ensure the reliability of the measurements, 2–3 parallel measurements were performed.

Catalytic reactions were carried out at 20 °C in a 30 cm^3 reactor containing stirring bar under air. In a typical experiment the complex was dissolved dissolved in 20 cm^3 CH_3CN, and the flask was closed with a rubber septum. H_2O_2 was injected by syringe through the septum. The reactor was connected to a graduated burette filled with oil, and the evolved dioxygen was measured volumetrically at time intervals of 2 s.

4. Conclusions

In summary, the results of the reaction kinetics and CV measurements are shown in Table 4. Based on these data the following reaction mechanism, including the diiron-peroxo complex formation, and its reaction with H_2O_2, was proposed (Scheme 3). We obtained clear evidence for the role of the peroxo-intermediate in diiron catalase mimics. We also found that the higher the redox potentials of the Fe^{III}/Fe^{II} redox couple, the higher the catalase-like activity, and the complexes with electron-deficient metal sites are significantly more reactive. It can be explained by their electrophilic character. The results obtained may contribute to the elucidation of the mechanism of both dimanganase and diiron-containing catalase enzymes.

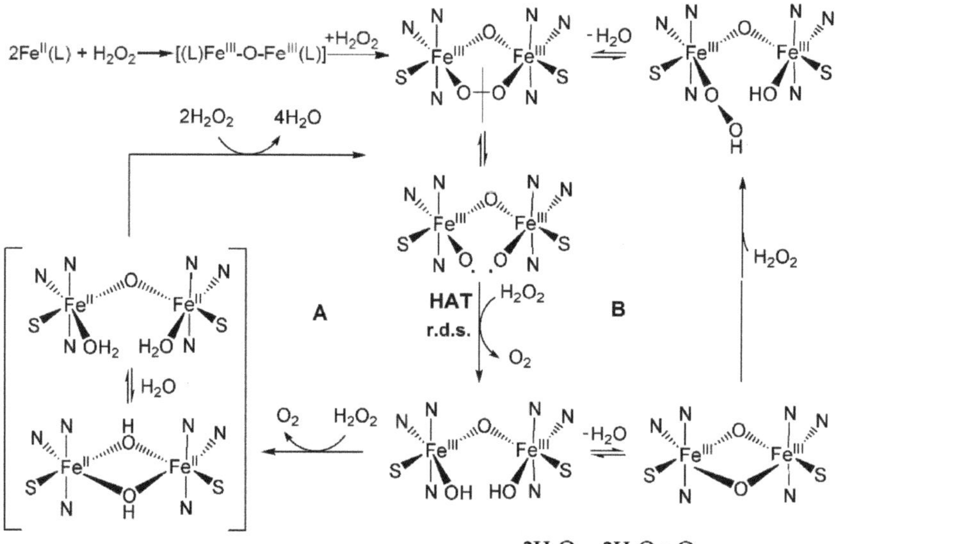

$$2Fe^{II}(L) + H_2O_2 \rightarrow [(L)Fe^{III}\text{-}O\text{-}Fe^{III}(L)] \xrightarrow{+H_2O_2}$$

$2H_2O_2 = 2H_2O + O_2$

Scheme 3. Proposed mechanism for the formation of $Fe^{III}{}_2(\mu\text{-}O)(\mu\text{-}1,2\text{-}O_2)(L_1\text{-}L_4)_2(Solv)_2]^{2+}$ (**5–8**) complexes and their catalase-like reactions.

Author Contributions: Conceptualization, J.K.; resources, D.L.-B., P.T., F.V.C., S.K. and B.G., writing—original draft preparation, J.K., writing—review and editing, J.K. All authors have read and agreed to the published version of the manuscript.

Funding: Financial support of the GINOP-2.3.2-15-2016-00049 (J.K.) and TKP2020-IKA-07 (J.K. and D.L.-B.) are gratefully acknowledged.

Institutional Review Board Statement: Not applicable.

Informed Consent Statement: Not applicable.

Data Availability Statement: Not available.

Conflicts of Interest: The authors declare no conflict of interest.

Sample Availability: Not available.

References

1. Beyer, W.F.; Fridovich, I. Catalases-with and without heme. *Basic Life Sci.* **1988**, *49*, 651–661. [PubMed]
2. Nicholls, P.; Fita, I.; Loewen, P.C. Enzymology and structure of catalases. *Adv. Inorg. Chem.* **2001**, *51*, 51–106. [CrossRef]
3. Kono, Y.; Fridovich, I. Isolation and characterization of the pseudocatalase of *Lactobacillus plantarum*. *J. Biol. Chem.* **1983**, *258*, 6015–6019. [CrossRef]
4. Barynin, V.V.; Whittaker, M.M.; Antonyuk, S.V.; Lamzin, V.S.; Harrison, P.M.; Artymiuk, P.J.; Whittaker, J.W. Crystal Structure of Manganese Catalase from *Lactobacillus plantarum*. *Structure* **2001**, *9*, 725–738. [CrossRef]
5. Antonyuk, S.V.; Melik-Adman, V.R.; Popov, A.N.; Lamzin, V.S.; Hempstead, P.D.; Harrison, P.M.; Artymyuk, P.J.; Barynin, V.V. Three-dimensional structure of the enzyme dimanganese catalase from *Thermus thermophilus* at 1 Å resolution. *Crystallogr. Rep.* **2000**, *45*, 105–113. [CrossRef]
6. Barynin, V.V.; Grebenko, A.I. T-catalase is nonheme catalase of the extremely thermophilic bacterium *Thermus thermophilus* HB8. *Dokl. Akad. Nauk SSSR* **1986**, *286*, 461–464.
7. Allgood, G.S.; Perry, J.J. Characterization of a manganese-containing catalase from the obligate thermophile *Thermoleophilum album*. *J. Bacteriol.* **1986**, *168*, 563–567. [CrossRef] [PubMed]
8. Amo, T.; Atomi, H.; Imanaka, T. Unique Presence of a Manganese Catalase in a Hyperthermophilic Archaeon, *Pyrobaculum calidifontis* VA1. *J. Bacteriol.* **2002**, *184*, 3305–3312. [CrossRef] [PubMed]
9. Whittaker, J.W. Non-heme manganese catalase-the "other" catalase. *Arch. Biochem. Biophys.* **2012**, *525*, 111–120. [CrossRef]
10. Cardenas, J.P.; Quatrini, R.; Holmes, D.S. Aerobic Lineage of the Oxidative Stress Response Protein Rubrerythrin Emerged in an Ancient Microaerobic, (Hyper)Thermophilic Environment. *Front. Microbiol.* **2016**, *7*, 1822–1831. [CrossRef] [PubMed]
11. Bihani, S.C.; Chakravarty, D.; Ballal, A. KatB, a cyanobacterial Mn-catalase with unique active site configuration: Implications for enzyme function. *Free Radic. Biol. Med.* **2016**, *93*, 118–129. [CrossRef]
12. Signorella, S.; Palopoli, C.; Ledesma, G. Rationally designed mimics of antioxidant manganoenzymes: Role of structural features in the quest for catalysts with catalaseand superoxide dismutase activity. *Coord. Chem. Rev.* **2018**, *365*, 75–102. [CrossRef]
13. Wu, A.J.; Penner-Hahn, J.E.; Pecoraro, V.L. Structural, Spectroscopic, and Reactivity Models for the Manganese Catalases. *Chem. Rev.* **2004**, *104*, 903–938. [CrossRef] [PubMed]
14. Tovmasyan, A.; Maia, C.G.C.; Weitner, T.; Carballal, S.; Sampaio, R.S.; Lieb, D.; Ghazaryan, R.; Ivanovic-Burmazovic, I.; Radi, R.; Reboucas, J.S.; et al. A comprehensive evaluation of catalase-like activity of different classes of redox-active therapeutics. *Free Radic. Biol. Med.* **2015**, *86*, 308–321. [CrossRef] [PubMed]
15. Batinic-Haberle, I.; Tovmasyan, A.; Spasojevic, I. An educational overview of the chemistry, biochemistry and therapeutic aspects of Mn porphyrins—From superoxide dismutation to HO-driven pathways. *Redox Biol.* **2015**, *5*, 43–65. [CrossRef]
16. Kaizer, J.; Baráth, G.; Speier, G.; Réglier, M.; Giorgi, M. Synthesis, structure and catalase mimics of novel homoleptic manganese(II) complexes of 1,3-bis(2′-pyridylimino)isoindoline, Mn(4R-ind)2 (R = H, Me). *Inorg. Chem. Commun.* **2007**, *10*, 292–294. [CrossRef]
17. Kaizer, J.; Kripli, B.; Speier, G.; Párkányi, L. Synthesis, structure, and catalase-like activity of a novel manganese(II) complex: Dichloro[1,3-bis(2′-benzimidazolylimino)isoindoline]manganese(II). *Polyhedron* **2009**, *28*, 933–936. [CrossRef]
18. Kaizer, J.; Csay, T.; Kővári, P.; Speier, G.; Párkányi, L. Catalase mimics of a manganese(II) complex: The effect of axial ligands and pH. *J. Mol. Catal. A Chem.* **2008**, *280*, 203–209. [CrossRef]
19. Kripli, B.; Garda, Z.; Sólyom, B.; Tircsó, G.; Kaizer, J. Formation, stability and catalase-like activity of mononuclear manganese(II) and oxomanganese(IV) complexes in protic and aprotic solvents. *NJC* **2020**. [CrossRef]
20. Stadtman, E.R.; Berlett, B.S.; Chock, P.B. Manganese-dependent disproportionation of hydrogen peroxide in bicarbonate buffer. *Proc. Natl. Acad. Sci. USA* **1990**, *87*, 384–388. [CrossRef]
21. Kripli, B.; Sólyom, B.; Speier, G.; Kaizer, J. Stability and Catalase-Like Activity of a Mononuclear Non-Heme Oxoiron(IV) Complex in Aqueous Solution. *Molecules* **2019**, *24*, 3236. [CrossRef]
22. Szávuly, M.I.; Surducan, M.; Nagy, E.; Surányi, M.; Speier, G.; Silaghi-Dumitrescu, R.; Kaizer, J. Functional models of nonheme enzymes: Kinetic and computational evidence for the formation of oxoiron(IV) species from peroxo-diiron(III) complexes, and their reactivity towards phenols and H_2O_2. *Dalton Trans.* **2016**, *45*, 14709–14718. [CrossRef] [PubMed]

23. Pap, J.S.; Draksharapu, A.; Giorgi, M.; Browne, W.R.; Kaizer, J.; Speier, G. Stabilisation of mu-peroxido-bridged Fe(III) intermediates with non-symmetric bidentate N-donor ligands. *Chem. Commun.* **2014**, *50*, 1326–1329. [CrossRef] [PubMed]
24. Pap, J.S.; Cranswick, M.A.; Balogh-Hergovich, É.; Baráth, G.; Giorgi, M.; Rohde, G.T.; Kaizer, J.; Speier, G.; Que, L., Jr. An Iron(II)[1,3-bis(2′-pyridylimino)isoindoline] Complex as a Catalyst for Substrate Oxidation with H_2O_2—Evidence for a Transient Peroxidodiiron(III) Species. *Eur. J. Inorg. Chem.* **2013**, 3858–3866. [CrossRef]
25. Kripli, B.; Csendes, F.V.; Török, P.; Speier, G.; Kaizer, J. Stoichiometric Aldehyde Deformylation Mediated by Nucleophilic Peroxo-diiron(III) Complex as a Functional Model of Aldehyde Deformylating Oxygenase. *Chem. Eur. J.* **2019**, *25*, 14290–14294. [CrossRef]
26. Török, P.; Unjaroen, D.; Csendes, F.V.; Giorgi, M.; Browne, W.R.; Kaizer, J. A nonheme peroxo-diiron(III) complex exhibiting both nucleophilic and electrophilic oxidation of organic substrates. *Dalton Trans.* **2021**, *50*, 7181–7185. [CrossRef] [PubMed]
27. Kripli, B.; Szávuly, M.; Csendes, F.V.; Kaizer, J. Functional models of nonheme diiron enzymes: Reactivity of the mu-oxo-mu-1,2-peroxo-diiron(III) intermediate in electrophilic and nucleophilic reactions. *Dalton Trans.* **2020**, *49*, 1742–1746. [CrossRef] [PubMed]
28. Oloo, W.N.; Fielding, A.J.; Que, L., Jr. Rate determining water assisted O-O bond cleavage of a FeIII-OOH intermediate in a bioinspired nonheme iron-catalyzed oxidation. *J. Am. Chem. Soc.* **2013**, *135*, 6438–6441. [CrossRef]
29. Meena, B.I.; Kaizer, J. Design and Fine-Tuning Redox Potentials of Manganese(II) Complexes with Ioindoline-Based Ligands: H_2O_2 Oxidation and Oxidative Bleaching Performance in Aqueous Solution. *Catalysts* **2020**, *10*, 404. [CrossRef]
30. Jensen, M.P.; Payeras, A.M.I.; Fiedler, A.T.; Costas, M.; Kaizer, J.; Stubna, A.; Münck, E.; Que, L., Jr. Kinetic Analysis of the Conversion of Nonheme (Alkylperoxo)iron(III) Species to Iron(IV) Complexes. *Inorg. Chem.* **2007**, *46*, 2398–2408. [CrossRef]
31. Csonka, R.; Speier, G.; Kaizer, J. Isoindoline-derived ligands and applications. *RSC Adv.* **2015**, *5*, 18401–18419. [CrossRef]

Article

Biomimetic Ketone Reduction by Disulfide Radical Anion

Sebastian Barata-Vallejo [1,2], Konrad Skotnicki [3], Carla Ferreri [1], Bronislaw Marciniak [4], Krzysztof Bobrowski [3] and Chryssostomos Chatgilialoglu [1,4,*]

1 Istituto per la Sintesi Organica e la Fotoreattività (ISOF), Consiglio Nazionale delle Ricerche (CNR), Via P. Gobetti 101, 40129 Bologna, Italy; sebastian.barata@isof.cnr.it (S.B.-V.); carla.ferreri@isof.cnr.it (C.F.)
2 Departamento de Ciencias Químicas, Facultad de Farmacia y Bioquimíca, Universidad de Buenos Aires, Junin 954, Buenos Aires CP 1113, Argentina
3 Institute of Nuclear Chemistry and Technology, Dorodna 16, 03-195 Warsaw, Poland; k.skotnicki@ichtj.waw.pl (K.S.); kris@ichtj.pl (K.B.)
4 Center for Advanced Technology, Adam Mickiewicz University, Uniwersytetu Poznanskiego 10, 61-614 Poznan, Poland; marcinia@amu.edu.pl
* Correspondence: chrys@isof.cnr.it; Tel.: +390-5-1639-8309

Abstract: The conversion of ribonucleosides to 2′-deoxyribonucleosides is catalyzed by ribonucleoside reductase enzymes in nature. One of the key steps in this complex radical mechanism is the reduction of the 3′-ketodeoxynucleotide by a pair of cysteine residues, providing the electrons via a disulfide radical anion ($RSSR^{\bullet-}$) in the active site of the enzyme. In the present study, the bioinspired conversion of ketones to corresponding alcohols was achieved by the intermediacy of disulfide radical anion of cysteine $(CysSSCys)^{\bullet-}$ in water. High concentration of cysteine and pH 10.6 are necessary for high-yielding reactions. The photoinitiated radical chain reaction includes the one-electron reduction of carbonyl moiety by disulfide radical anion, protonation of the resulting ketyl radical anion by water, and H-atom abstraction from CysSH. The $(CysSSCys)^{\bullet-}$ transient species generated by ionizing radiation in aqueous solutions allowed the measurement of kinetic data with ketones by pulse radiolysis. By measuring the rate of the decay of $(CysSSCys)^{\bullet-}$ at λ_{max} = 420 nm at various concentrations of ketones, we found the rate constants of three cyclic ketones to be in the range of 10^4–10^5 $M^{-1}s^{-1}$ at ~22 °C.

Keywords: biomimetic chemistry; cysteine; ketone reduction; free radicals; pulse radiolysis; kinetics

Citation: Barata-Vallejo, S.; Skotnicki, K.; Ferreri, C.; Marciniak, B.; Bobrowski, K.; Chatgilialoglu, C. Biomimetic Ketone Reduction by Disulfide Radical Anion. *Molecules* **2021**, *26*, 5429. https://doi.org/10.3390/molecules26185429

Academic Editors: Igor Alabugin and Elena G. Bagryanskaya

Received: 6 August 2021
Accepted: 1 September 2021
Published: 7 September 2021

1. Introduction

The comprehension of free radical reactivity taking place in naturally occurring processes can be very important for chemistry in two main ways: (1) inspiring new synthetic methods based on the same mechanisms that nature uses to prepare biomolecules, and (2) designing biomimetic models to simulate the free radical damages and to provide molecular libraries for mechanistic and biomarker discovery. Two successful examples of these approaches from our group are the syntheses of 5′,8-cyclopurine lesions [1–3] and mono-trans PUFA isomers [4–6]. Indeed, there are strict relationships between reactivity involving free radicals and processes taking place in nature [7]. Since biological reactivity occurs in an aqueous environment, free radical chemistry can become attractive for realizing eco-friendly and high-yielding methodologies.

The radical chemistry associated with thiols (RSH) and disulfides (RSSR) plays important roles in nature, particularly in enzymes and proteins. One example is the conversion of ribonucleosides to 2′-deoxyribonucleosides, the monomers required for the construction of DNA, catalyzed by radical enzymes called ribonucleoside reductases (RNRs). There are three forms of the enzyme (classes I, II, and III) that are active in different species, with the former being active in eukaryotes and microorganisms, using ribonucleotide diphosphates as substrates [8–10]. An intriguing step in class I is the transfer from the tyrosyl radical to the thiol moiety in the active site where the ribonucleotide substrate waits for the conversion. Figure 1 summarizes the complex mechanism of this transformation, which initially involves a reversible 3′-hydrogen atom abstraction by the thiyl radical generated from the cysteine residues in the active site (**1**). There follows the elimination of water with translocation of the radical center to C2′ and quenching of the newly formed C2′ radical by another cysteine residue (**2**). This generates the disulfide radical anion ($RSSR^{\bullet -}$) that reduces the 3′-ketodeoxynucleotide (**2**→**3**) [11,12]. To complete the cycle, the C3′ radical abstracts the hydrogen from the initial cysteine residue (**3**→**4**). The restoration of the two cysteine residues occurs by the intervention of thioredoxin reductase (TRR) and NADPH [8].

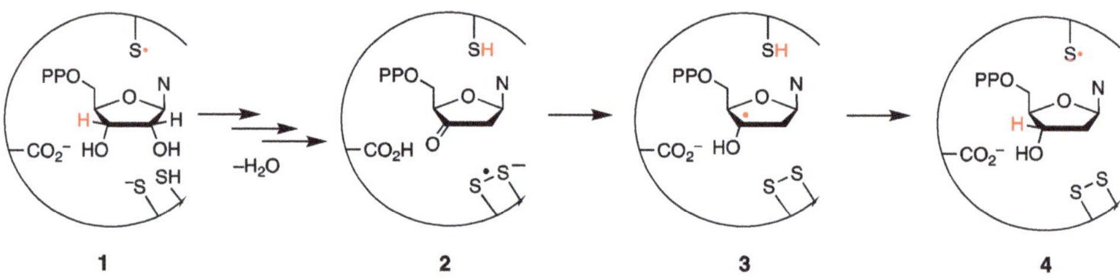

Figure 1. The transformation of ribonucleotides into 2′-ribonucleotides by RNR.

We recently discovered that the sulfur radical species, derived at different pHs from hydrogen sulfide (H_2S), are able to transform 1,2-diols to alcohols via carbonyl reduction [13]. We hypothesized that the intermediate $HSS^{\bullet 2-}$ is able to reduce the carbonyl moiety of ketones by a radical chain reaction. The $HSS^{\bullet 2-}$ species is analogous to $(CysSSCys)^{\bullet -}$, which may be able to reduce ketones to corresponding alcohols under certain conditions. It has been shown that disulfide radical anions $(RSSR)^{\bullet -}$ are strong reductants, with the standard reduction potential (E^0) of the RSSR/$RSSR^{\bullet -}$ redox couple in the range 1.40–1.60 V vs. NHE [14–17]. The reduction potential E^0 (RSSR/$RSSR^{\bullet -}$) = −1.41 V found for the tripeptide glutathione disulfide is an indication of the reduction potential for the cystine residue in proteins [18]. Disulfide radical anions in proteins are stabilized by the tertiary protein structures and do not readily dissociate to thiyl radical and thiolate anion (cf. Structure **2** in Figure 1).

Regarding cysteine (CysSH), which shows three pK_a values with the second one related to side chain thiol group (Figure 2A) [19], the equilibrium between $(CysSSCys)^{\bullet -}$ and thiolate and thiol, namely:

$$CysS^{\bullet} + CysS^{-} \leftrightarrows (CysSSCys)^{\bullet -} \tag{1}$$

Is a complex reaction studied in some detail by different experimental methods [18,20,21]. The forward rate constant (k_1) was measured in the range pH 10–11 and found to be $k_1 = 1.2 \times 10^9$ $M^{-1}s^{-1}$, whereas the reverse rate constant (k_{-1}) strongly depends on whether the amino group is protonated or not, affecting the disulfide radical anion equilibrium (Equation (1)) and the reduction potential of CysSSCys/$CysSSCys^{\bullet -}$ redox couple (Figure 2B). The stability of $(CysSSCys)^{\bullet -}$ increases when protonated amino groups are present and is reflected in equilibrium constants K_1: 438 M^{-1} (with zero protonated amino groups) (**5**), 317 M^{-1} (with one protonated amino group) (**6**), and 8900 M^{-1} (with two protonated amino groups) (**7**). Increases in stability of these species are indicated by progressively less negative reduction potentials: −1.50 V, −1.38 V, and −1.30 V, for 0, 1, and 2 protonated amino groups, respectively [18].

Here, we report the photochemical conditions found to transform ketones to corresponding alcohols by CysSH via a radical chain reaction involving $(CysSSCys)^{\bullet -}$ as reducing species. We also discovered that the reduction of two selected substrates proceeds via a dual radical chain mechanism. Furthermore, we applied time-resolved spectroscopy (pulse radiolysis) to generate $(CysSSCys)^{\bullet-}$ by reaction of $CysS^-$ with $HO^\bullet$ radicals and provided the rate constants of both $(CysSSCys)^{\bullet-}$ and $HO^\bullet$ species with three cyclic ketones.

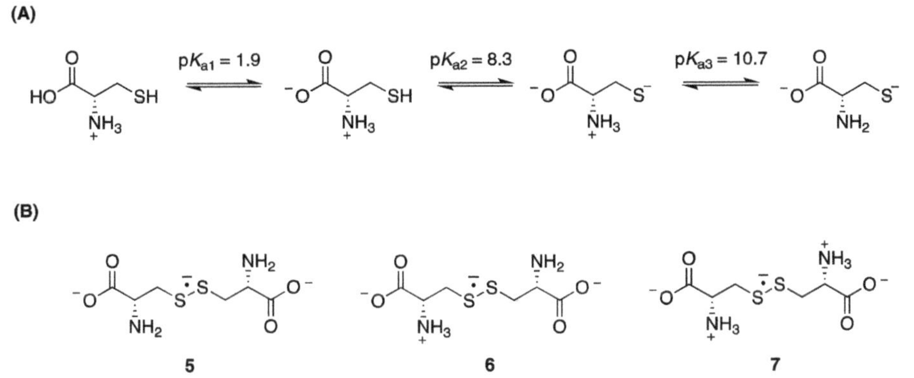

Figure 2. (**A**) The structure of cysteine, CysSH, and associated pK_a values; (**B**) Disulfide radical anion of cysteine $(CysSSCys)^{\bullet-}$ with 0 (**5**), 1 (**6**), and 2 (**7**) protonated amino groups.

2. Results and Discussion

2.1. Ketone Reduction by the Photolysis of Cysteine-Containing Aqueous Solutions

2.1.1. Cyclohexanone Reduction and Optimization Studies at Different pH Values

The photolysis (low-pressure Hg lamp, 5.5 W) of N_2-saturated aqueous solutions of cyclohexanone (**8**, 8.3 mM) containing CysSH (18.3 mM) was monitored for up to 60 min and at various pH values (7.0, 7.5, 8.5, 9.6, 10.6, 11.5, 12.0, and 13.0) adjusted with 5% NaOH. In all experiments, cyclohexanol (**9**) was the only product. Table 1 reports the conversion of **8** to **9** at 30 and 60 min at each pH value. As shown in Table 1, cyclohexanol formation is achieved successfully at all pH values studied, reaching a maximum product yield of 96% at pH 10.6.

Table 1. Photolysis (low-pressure Hg lamp, 5.5 W) of N_2-saturated aqueous solutions of cyclohexanone (**8**, 8.3 mM) containing cysteine (18.3 mM) with formation of cyclohexanol (**9**) at different irradiation times and pH. [1].

pH (Initial)	Time (min)	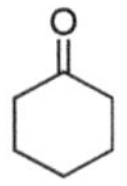 8 (mM)	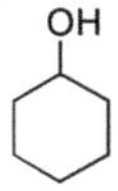 9 (mM)	pH (Final)
7.0	30	6.6	1.7	-
	60	5.1	3.2	6.0
7.5	30	5.5	2.8	-
	60	4.4	3.9	6.5
8.5	30	1.8	6.5	-
	60	1.0	7.3	7.5
9.6	30	1.9	6.4	-
	60	0.9	7.4	8.3
10.6	30	1.2	7.1	-
	60	0.3	8.0	9.5
11.5	30	1.7	6.6	-
	60	0.5	7.8	10.8
12.0	30	1.6	6.7	-
	60	0.6	7.7	11.5

[1] Reaction temperature was constant at 42 ± 1 °C.

It is worth mentioning that other thiol derivatives such as 2-mercaptoethanol or *N*-acetylcysteine afford similar reductions but in lower yields (e.g., 51% and 56% after 60 min at pH 12, respectively) under identical experimental conditions.

2.1.2. Reaction Mechanism: Initiation Steps for a Radical Chain Process

Based on the concentration of cyclohexanone and cysteine as well as their molar absorption coefficients, the light of 253.7 nm is absorbed by approximately 50% by each substrate. Light absorbed by the thiol/thiolate affords the thiyl radical together with $H^{\bullet}/e^{-}$ (Figure 3a) [22]. Light absorbed by the cyclohexanone affords the S_1 (n,π^*) or T_1 (n,π^*) excited states. The T_1 state can react with thiol [23], whereas both S_1 and T_1 states are reported to lead to a classical type I cleavage (ring-opening) [24,25], and we suggest that this biradical intermediate can react with thiol (Figure 3a).

At pH 10.6, where the yield of ketone reduction is higher (cf. Table 1), the $CysS^-$ form is above 99%, being the pK_a value of SH moiety 8.3, although the concentration of CysSH is still ~150 μM. It is known that $CysS^{\bullet}$ adds reversibly to the parent anion to form the dimeric radical anion species, the forward reaction being close to the diffusion control rate (cf. Introduction). Therefore, due to the relatively high concentration of cysteine (18.3 mM), at higher pHs the main reactive species is $(CysSSCys)^{\bullet-}$. We propose that **8** is reduced to ketyl radical anion **10** by the $RSSR^{\bullet-}$, via single-electron transfer (SET). Subsequent protonation **10**→**11** from the aqueous medium and H-atom abstraction from CysSH affords the product **9**, completing the radical chain mechanism (Figure 3b). The low concentration of CysSH (~150 μM) is compensated by fast H-atom abstraction (k_H >10^8 M^{-1} s^{-1} for analogous reactions [26,27]). It is worth mentioning that the pK_a values associated with the hydroxyl protons of ketyl radical **11** and alcohol **9** are ~12.5 and 16, respectively [28,29]. The overall process is perhaps facilitated by a concerted proton-coupled electron transfer (PCET) affording the ketyl radical **11** from **8** [30].

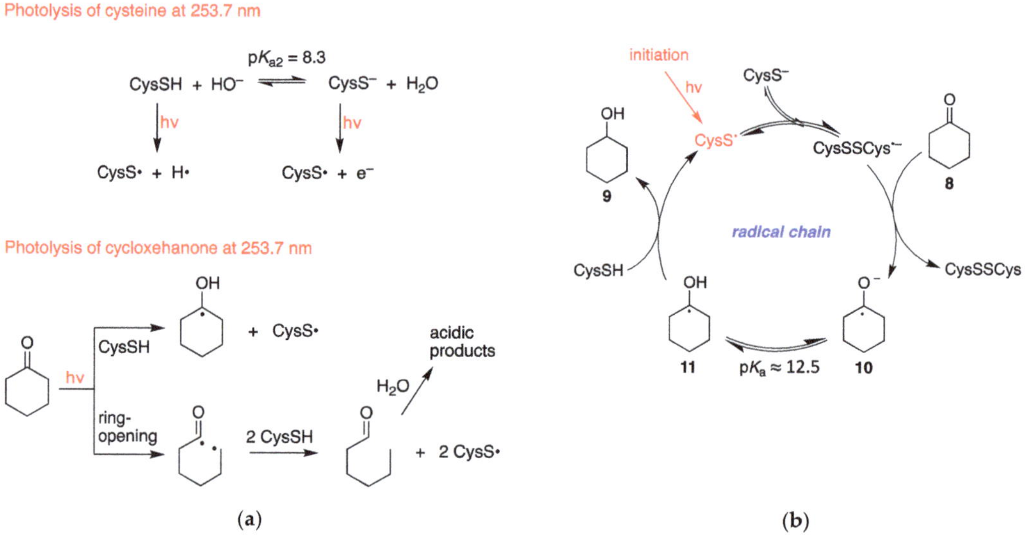

(**a**) (**b**)

Figure 3. Proposed reaction mechanism for the reduction of cyclohexanone to cyclohexanol: (**a**) Possible initiation steps; (**b**) The radical chain reaction.

Next, we considered the reduction of **8** at pH 10.6 by changing the degassing conditions., i.e., N_2O- instead of N_2-saturated solutions. It is well-known that N_2O-saturated solutions (~0.02 M of N_2O) and e^-_{aq} are efficiently transformed into $HO^\bullet$ radicals via Reaction 2 ($k = 9.1 \times 10^9$ $M^{-1}s^{-1}$) [31].

$$e^-_{aq} + N_2O + H_2O \rightarrow HO^\bullet + N_2 + HO^- \quad (2)$$

In N_2O-saturated conditions, the yields of cyclohexanol are 22% and 26% after 30 and 60 min of irradiation, respectively, in comparison to 86% and 96% of cyclohexanol after 30 and 60 min under N_2-saturated conditions. This suggests that the presence of N_2O strongly inhibits the radical chain reaction. Since $(CysSSCys)^{\bullet -}$ does not react with N_2O, we suggest that the e^- generated by photolysis of $CysS^-$ (cf. Figure 3a) undergo very fast hydration into hydrated electrons (e^-_{aq}) and then are trapped by N_2O. This finding indicates that the addition of e^-_{aq} to the carbonyl moiety of ketones is also an initiation step in the radical chain reaction described above.

Another important experimental observation is the decrease in pH during the reaction time (compare Columns 1 and 5 in Table 1). Tentatively, we suggest that after the ring-opening of the ketone triplet state and the reaction of biradical with thiol, the resulting aldehyde is the precursor of acidic products such as hydrated aldehyde. We could not obtain analytical evidences of by-products, probably due to their presence being below the detection limits.

2.1.3. Other Substrates for Ketone Reduction at pH 10.6

Considering the efficiency of the cyclohexanone reduction, the reduction of a few other ketones to the corresponding alcohols mediated by cysteine was further explored. Thus, under the optimized reaction conditions reported in Table 1 and at pH 10.6, ketones **12–15** all produced exclusively the corresponding alcohols, with yield ≥90% (Figure 4).

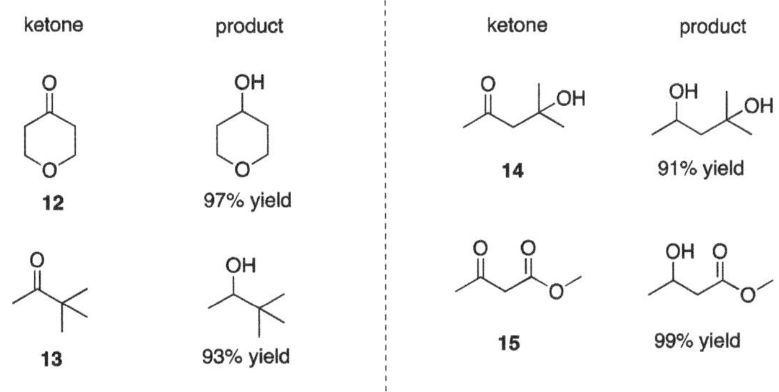

Figure 4. Reduction of ketones **12–15** to the corresponding alcohols. N_2-saturated aqueous solutions of carbonyl compound (8.3 mM) cysteine (18.3 mM), pH adjusted to 10.6 at 42 ± 1 °C, were irradiated for 60 min. Yields by GC analysis based on products formation.

We demonstrated for the first time that the generation of ketyl radical anion via disulfide radical anion can be achieved. It is worth mentioning that: (i) SET reduction of ketones is a classical approach for the formation of ketyl radical anions, and such an approach has been used extensively in organic synthesis, as documented by the recently published tutorial review [32]; and (ii) ketyl radical anions formation have been typically carried out by means of dissolving metals such as Li, Na, and K, usually in liquid ammonia solution and in the presence of a proton source [33]. It can be envisaged that our bioinspired work will foster applications for SET methodology in aqueous environment.

2.1.4. The Reduction of 2-Hydroxycyclohexanone and 2-Cyclopenten-1-one at pH 10.6

For a better understanding of the reaction mechanism, we also considered the carbonyl reduction in 2-hydroxycyclohexanone (**16**) and 2-cyclopenten-1-one (**19**), having α-HO and internal α,β-unsaturated moiety, respectively.

Photolysis (low-pressure Hg lamp, 5.5 W) of N_2-saturated aqueous solutions of 2-hydroxycyclohexanone **16** (8.3 mM) containing CysSH (18.3 mM) was carried out for different times (up to 60 min) at pH 10.6. Cyclohexanone (**8**) and cyclohexanol (**9**) were formed as the only products. Figure 5a displays a graph with the disappearance of **16** (♦) and the formation of **8** (•) and **9** (▲) as a function of the reaction time, showing clearly that the reaction **16**→**9** occurs stepwise. The loss of **16** quantitatively matched the formation of **8** and **9**, with 17% and 82% yields, respectively, after 60 min. The mechanism is depicted in Figure 5b based on a dual radical chain process, starting from the reaction of $(CysSSCys)^{\bullet-}$ with **16** to produce the ketyl radical anion **17**. This species undergoes α,β-C—O scission and HO^- elimination, with the shift of the radical center to produce **18** [27], which completes the catalytic cycle by reacting with CysSH regenerating $CysS^\bullet$. The so-formed cyclohexanone (**8**) undergoes a second radical chain reaction, as described in Figure 3b.

The reduction of 2-cyclopenten-1-one (**19**) was also explored under identical experimental conditions. The time profile of the reduction of (**19**) at pH 10.6 is reported in Figure 6a. In all experiments, cyclopentanone (**22**) and cyclopentanol (**23**) were formed as the only products, and again, the disappearance of **19** (♦) and the formation of **22** (•) and **23** (▲) as a function of the reaction time showed clearly that the reaction **19**→**23** occurs stepwise. The loss of **19** quantitatively matched with the formation of **22** and **23**, with 23% and 77% yields, respectively, after 60 min. The mechanism depicted in Figure 6b is based again on a dual radical chain process, starting from the single-electron transfer from $(CysSSCys)^{\bullet-}$ to **19** give the allyl-type radical **20**. This species undergoes protonation from the aqueous medium with double-bond shift to produce **21**, which completes the first catalytic cycle by reacting with CysSH regenerating $CysS^\bullet$. The so-formed cyclopentanone (**22**) undergoes a second radical chain reaction analogous to the one reported for cyclohexanone in Figure 3b.

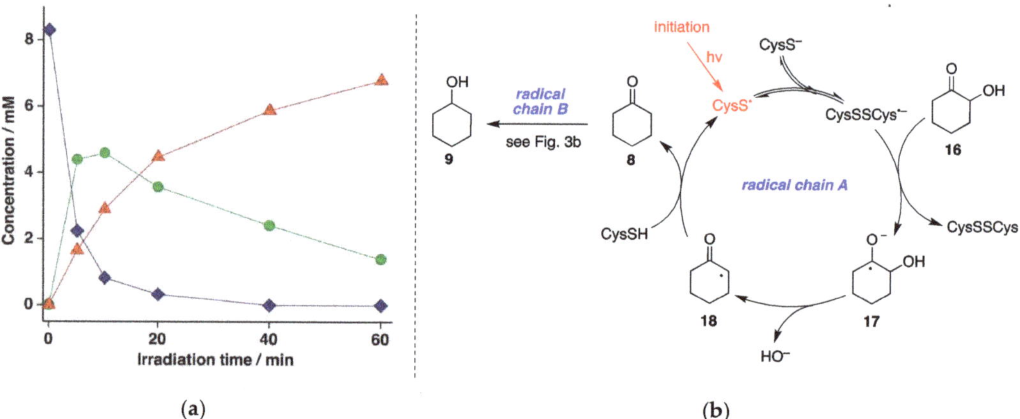

Figure 5. (**a**) Concentration of **16** (♦), **8** (•), and **9** (▲) vs. irradiation time for the photolysis of N_2-saturated 2-hydroxycyclohexanone (**16**) aqueous solutions (8.3 mM) containing CysSH (18.3 mM), pH adjusted to 10.6, at 42 ± 1 °C; (**b**) Proposed reaction mechanism involving dual radical chain reactions A and B.

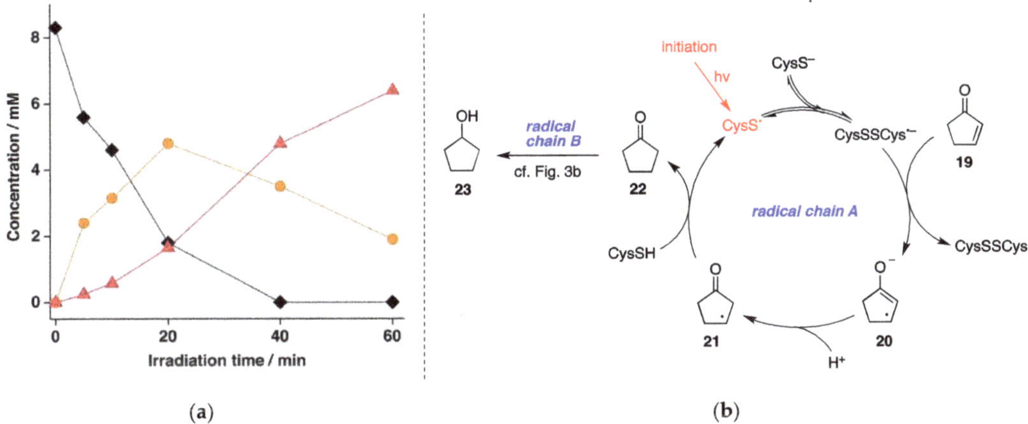

Figure 6. (**a**) Concentration of **19** (♦), **22** (•), and **23** (▲) vs. irradiation time for the photolysis of N_2-saturated 2-cyclopenten-1-one (**19**) aqueous solutions (8.3 mM) containing CysSH (18.3 mM), pH adjusted to 10.6, at 42 ± 1 °C; (**b**) Proposed reaction mechanism involsving dual radical chain reactions A and B.

2.2. Pulse Radiolysis Studies and Rate Constants of SET

Pulse irradiation of water leads to the primary reactive species e^-_{aq}, $HO^\bullet$, and $H^\bullet$, as shown in Reaction 3. The values in brackets represent the radiation chemical yield (*G*) in μmol J^{-1}. In N_2O-saturated solution (~0.02 M of N_2O), e^-_{aq} are efficiently transformed into $HO^\bullet$ radicals via Reaction 1 ($k = 9.1 \times 10^9$ $M^{-1}s^{-1}$), affording $G(HO^\bullet) = 0.56$ μmol J^{-1} [34].

$$H_2O \rightsquigarrow e^-_{aq}\ (0.28),\ HO^\bullet\ (0.28),\ H^\bullet\ (0.06) \quad (3)$$

The reaction of $HO^\bullet$ radicals with cysteine (100 mM) in the absence or in the presence of a ketone (up to 30 mM) was investigated in N_2O-saturated solutions at pH 10.6. These experimental conditions were chosen in order (i) to maximize the formation of thiyl radicals (Reaction 4, $k_4 = (5.35 \pm 0.82) \times 10^9$ $M^{-1}s^{-1}$) [35], (ii) to minimize the quenching of $HO^\bullet$ by ketones (Reaction 5, unknown rate constants), (iii) to convert the thiol moiety of cysteine into thiolate ions, (iv) to maximize conversion of thiyl radicals into $(CysSSCys)^{\bullet-}$ (Reaction 1), and (v) to facilitate the electron transfer between $(CysSSCys)^{\bullet-}$ and ketones (Reaction 6). Furthermore, at pH 10.6, two forms of disulfide radical anion **5** and **6** are present at similar concentrations (Figure 2B).

$$HO^\bullet + CysS^- \rightarrow HO^- + CysS^\bullet \quad (4)$$

$$HO^\bullet + R^1R^2C{=}O \rightarrow \text{H-atom abstraction} \quad (5)$$

$$(CysSSCys)^{\bullet-} + R^1R^2C{=}O \rightarrow CysSSCys + (R^1R^2C{=}O)^{\bullet-} \quad (6)$$

Transient absorption spectrum of $(CysSSCys)^{\bullet-}$ in the range 280–620 nm recorded 10 μs after the electron pulses is shown in Figure 7A, with the absorption maximum at 420 nm and in accord with previous reported ones [35]. The black traces in Figure 7B represent the decay of transient absorption at λ_{max} = 420 nm recorded after the electron pulse. The rate of decay of $(CysSSCys)^{\bullet-}$ was investigated by varying the concentrations of a ketone from 0 to 30 mM. We selected three ketones: cyclohexanone (**8**), 2-hydroxycyclohexanone (**16**), and cyclopentenone (**19**) for the time-resolved studies. The decay kinetics at various concentrations of **8**, **16**, and **19** were recorded at λ_{max}, as shown in Figure 7B, for the presence of 30 mM of **8** (blue), **16** (green) and **19** (red). The pseudo-first-order rate constants of the decay of 420 nm absorption band were plotted as a function of ketone concentration (Figure 7C).

It is clearly seen that the pseudo-first-order rate constants measured at λ = 420 nm show a linear dependence on the concentration of **8**, **16**, and **19** in the full range of concentration studied (Figure 7C). The slopes in Figure 7C represent the second-order rate constants for the decay of $(CysSSCys)^{\bullet -}$ resulting from the reaction of $(CysSSCys)^{\bullet -}$ with **8**, **16**, and **19** (Reaction 6). The obtained values of k_6 for three ketones are collected in Table 2. Both **16** and **19**, having α-HO and internal α,β-unsaturated moieties, increased the rate constant by 3.7- and 5.9-folds, respectively, due to extra stabilization of ketone radical anion.

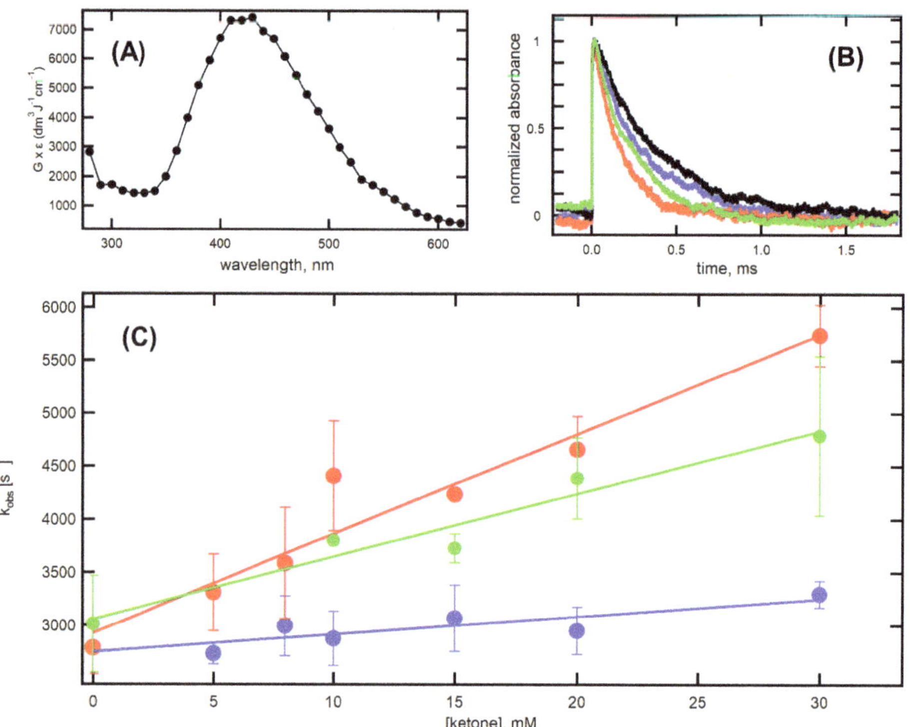

Figure 7. (**A**) Absorption spectrum of $(CysSSCys)^{\bullet -}$ recorded 10 μs after the electron pulse in N_2O-saturated aqueous solutions at pH = 10.6 containing 100 mM of cysteine. (**B**) Normalized time profiles representing decay of transient absorption at λ = 420 nm, recorded after the electron pulse in N_2O-saturated aqueous solutions at pH = 10.6, containing 100 mM of cysteine in the absence of ketones (black) and in the presence of 30 mM of **8** (blue), **16** (green) and **19** (red). (**C**) Plots of the observed pseudo-first-order rate constants of the decay of the 420 nm absorption as a function of **8** (•), **16** (•) and **19** (•) concentration in N_2O-saturated aqueous solutions at pH = 10.6, containing 100 mM of cysteine.

Table 2. Rate constants ($M^{-1}s^{-1}$) for the reactions of $(CysSSCys)^{\bullet -}$ and $HO^{\bullet}$ with the three cyclic ketones. $(CysSSCys)^{\bullet -}$ reacts by one-electron transfer and $HO^{\bullet}$ reacts by H-atom abstraction. [1].

k ($M^{-1}s^{-1}$)/Radical	8	16	19
$k_6/(CysSSCys)^{\bullet -}$	$(1.6 \pm 0.3) \times 10^4$	$(5.9 \pm 0.5) \times 10^4$	$(9.4 \pm 0.2) \times 10^4$
$k_5/HO^{\bullet}$	7.2×10^7	8.6×10^7	1.3×10^8

[1] At room temperature (~22 °C).

By applying the common competition kinetics method [36], the rate constants for the reaction of $HO^{\bullet}$ radicals with ketones (Reaction 5) can also be estimated. To our best knowledge, these rate constants were not measured earlier. Indeed, from the plot of the reciprocal of [$CysSSCys^{\bullet-}$] versus the [CysSH]/[Ketone] ratio and considering $k_4 = (5.35 \pm 0.82) \times 10^9\ M^{-1}s^{-1}$, we obtained the values of k_5 reported in Table 2.

3. Materials and Methods

3.1. Materials and Instrumentation

All commercial chemicals were used as received unless otherwise noted. L-cysteine $\geq$ 98.5%, cyclohexanone $\geq$ 99.5%, cyclohexanol 99%, tetrahydro-4H-pyran-4-one 99%, tetrahydro-4-pyranol 98%, 3,3-dimethyl-2-butanone 97%, 3,3-dimethyl-2-butanol 98%, 4-hydroxy-4-methyl-2-pentanone 99%, hexylene glycol 99%, methyl acetoacetate 99%, methyl 3-hydroxybutyrate $\geq$ 95%, 2-hydroxycyclohexanone 99%, 2-cyclopenten-1-one 98%, cyclopentanone 99%, cyclopentanol 99%, 2-mercaptoethanol $\geq$ 99%, *N*-acetyl-L-cysteine European Pharmacopoeia Reference Standard, and phosphoric acid 85 wt% in water were obtained from Sigma-Aldrich (San Louis, MO, USA). Diethyl ether (HPLC grade) was obtained from Carlo Erba (Milan, Italy) and distilled before use. Water was purified with a Millipore system. Absorption spectra were recorded with a PerkinElmer Lambda 950 spectrophotometer (PerkinElmer, Shelton, USA). UV irradiations were performed in a micro-photochemical reaction assembly with quartz well (Ace Glass) using a 5.5 W cold cathode, low-pressure, mercury arc, and gaseous discharge lamp (corresponds to λ = 250–260 nm) made of double-bore quartz (Ace Glass). Ethyl ether extracts of the irradiated samples were analyzed by GC-MS (Thermo Scientific Trace 1300, Waltham, MA, USA) equipped with a 15m $\times$ 0.25mm $\times$ 0.25μm TG-SQC 5% phenyl methyl polysiloxane column, with helium as carrier gas, coupled to a mass selective detector (Thermo Scientific ISQ, Waltham, MA, USA).

3.2. Ketone Reduction

General procedure for the irradiation experiments. L-cysteine (13.3 mg, 0.11 mmol) was dissolved in 5 mL of N_2-saturated water while continuously bubbling N_2 in the photolysis apparatus. Cyclohexanone (4.9 mg, 0.05 mmol), tetrahydro-4*H*-pyran-4-one (5.0 mg, 0.05 mmol), 3,3-dimethyl-2-butanone (5.0 mg, 0.05 mmol), 4-hydroxy-4-methyl-2-pentanone (5.8 mg, 0.05 mmol), methyl acetoacetate (5.8 mg, 0.05 mmol), 2-hydroxycyclohex anone (5.7 mg, 0.05 mmol), or 2-cyclopenten-1-one (4.1 mg, 0.05 mmol) were added to the photolysis apparatus. The pH was adjusted to the specific value using a NaOH 5% *m/v* or a H_3PO_4 5% *v/v* solution in deoxygenated water, and the final volume of the reaction was adjusted with N_2-saturated water in order to reach 6 mL. The UV irradiation proceeded by inserting a 5.5 W low-pressure mercury arc lamp into a micro photochemical reaction assembly with quartz well, and N_2 was smoothly flushing throughout the irradiation time at 42 $\pm$ 1 °C. After the irradiation was completed, 1 mL of irradiated sample was diluted with 200 mL of a NaCl-saturated aqueous solution and extracted with 6 $\times$ 0.5 mL diethyl ether. The organic layers (3 mL) were gathered, dried over Na_2SO_4 anhydrous, and analyzed by GC-MS.

GC-MS analyses of the diethyl ether extract of the irradiated samples were performed. The diethyl ether extracts were analysed by GC-MS equipped with a 15 m $\times$ 0.25 mm $\times$ 0.25 μm TG-SQC 5% phenyl methyl polysiloxane column, with helium as carrier gas, coupled to a mass-selective detector following a specific oven program (see below). Injection volume was 0.5 μL. Identification of the reaction products was performed by comparison with the commercially available compounds by GC-MS analysis and spike experiments. Quantification of compounds was performed by multiple-point external standard calibration curves for each analyte of interest. For the reaction of cyclohexanone, 3,3-dimethyl-2-butanone, 2-hydroxycyclohexanone, and 2-cyclopenten-1-one, oven program was as follows: temperature started at 30 °C, maintained for 10 min, increased at a rate of 25.0 °C/min up to 250 °C, and held for 8 min. For the reaction of tetrahydro-4*H*-pyran-4-one, 4-hydroxy-4-methyl-2-pentanone, and methyl acetoacetate, oven program was as follows: temperature started at 40 °C, maintained for 1 min, increased at a rate of 5.0 °C/min up to 110 °C, increased at a rate of 20.0 °C/min up to 250 °C, and held for 8 min.

3.3. Pulse Radiolysis

The pulse radiolysis experiments were performed with the LAE-10 linear accelerator at the Institute of Nuclear Chemistry and Technology in Warsaw, Poland, with a typical electron pulse length of 10 ns and 10 MeV of energy. A detailed description of the experimental setup has been given elsewhere along with basic details on the equipment and its data collection system [37–39].

Absorbed dose per pulse was in the order of 3.5 Gy (1 Gy = 1 J kg^{-1}). Experiments were performed with a continuous flow of sample solutions using a standard quartz cell with optical length 1 cm at room temperature (~22 °C). Solutions were purged for at least 20 min per 250 mL sample with N_2O before pulse irradiation.

The dosimetry was based on N_2O-saturated solutions of 10^{-2} M KSCN, which, following radiolysis, produces $(SCN)_2^{\bullet-}$ radicals that have a molar absorption coefficient of 7580 $M^{-1}cm^{-1}$ at λ = 472 nm and are produced with a yield of G = 0.635 μmol J^{-1} [31].

4. Conclusions

The reduction of ketones via disulfide radical anion has been obtained for the first time in an aqueous environment, following a bioinspired process connected to the known mechanism of transformation of ribonucleotides to deoxyribonucleotides. Disulfide radical anion from cysteine, $(CysSSCys)^{\bullet-}$, is a very good $1e^-$ reductant, and experimental conditions were first set to obtain high-yield conversion of representative substrates. Our results can be considered as an example of bimolecular/associative reductant upconversion that has been recently reviewed [40]. Using time-resolved studies, we measured the rate constants of three cyclic aliphatic ketones to be in the range of 10^4–10^5 $M^{-1}s^{-1}$ at ~22 °C. Expansion of this methodology to other substrates and applications of disulfide radical anion in photoredox catalysis can be envisaged.

Author Contributions: Conceptualization, S.B.-V. and C.C.; methodology, S.B.-V. and K.B.; investigation, S.B.-V. and K.S.; resources, C.F. and K.B.; writing—original draft preparation, S.B.-V., K.B., and C.C.; writing—review and editing, C.F., B.M., K.B., and C.C.; supervision, K.B. and C.C. All authors have read and agreed to the published version of the manuscript.

Funding: This research received no external funding.

Institutional Review Board Statement: Not applicable.

Informed Consent Statement: Not applicable.

Data Availability Statement: The data presented in this study are available on request from the corresponding author.

Acknowledgments: Authors would like to thank Tomasz Szreder (INCT, Poland) for his continuous efforts at implementing improvements to the pulse radiolysis setup in the INCT over the last few years.

Conflicts of Interest: The authors declare no conflict of interest.

Sample Availability: Not available.

References

1. Chatgilialoglu, C.; Ferreri, C.; Geacintov, N.E.; Krokidis, M.G.; Liu, Y.; Masi, A.; Shafirovich, V.; Terzidis, M.A.; Tsegay, P.S. 5′,8-Cyclopurine lesions in DNA damage: Chemical, analytical, biological and diagnostic significance. *Cells* **2019**, *8*, 513. [CrossRef] [PubMed]
2. Chatgilialoglu, C.; Eriksson, L.A.; Krokidis, M.G.; Masi, A.; Wang, S.-D.; Zhang, R. Oxygen dependent purine lesions in double-stranded oligodeoxynucleotides: Kinetic and computational studies highlight the mechanism for 5′,8-cyplopurine formation. *J. Am. Chem. Soc.* **2020**, *142*, 5825–5833. [CrossRef]
3. Chatgilialoglu, C.; Ferreri, C.; Krokidis, M.G.; Masi, A.; Terzidis, M.A. On the Relevance of Hydroxyl Radical to Purine DNA Damage. *Free Radic. Res.* **2021**. [CrossRef]

4. Chatgilialoglu, C.; Ferreri, C.; Melchiorre, M.; Sansone, A.; Torreggiani, A. Lipid Geometrical Isomerism: From Chemistry to Biology and Diagnostics. *Chem. Rev.* **2014**, *114*, 255–284. [CrossRef] [PubMed]
5. Menounou, G.; Giacometti, G.; Scanferlato, R.; Dambruoso, P.; Sansone, A.; Tueros, I.; Amézaga, J.; Chatgilialoglu, C.; Ferreri, C. Trans Lipid Library: Synthesis of Docosahexaenoic Acid (DHA) Monotrans Isomers and Regioisomer Identification in DHA-Containing Supplements. *Chem. Res. Toxicol.* **2018**, *31*, 191–200. [CrossRef] [PubMed]
6. Vetica, F.; Sansone, A.; Meliota, C.; Batani, G.; Roberti, M.; Chatgilialoglu, C.; Ferreri, C. Free Radical-Mediated Formation of Trans-Cardiolipin isomers, Analytical Approaches for Lipidomics and Consequences for the Structural Organization of Membranes. *Biomolecules* **2020**, *10*, 1189. [CrossRef]
7. Chatgilialoglu, C.; Studer, A. (Eds.) *Encyclopedia of Radical in Chemistry, Biology and Materials*; Wiley: Chichester, UK, 2012.
8. Minnihan, E.C.; Nocera, D.G.; Stubbe, J. Reversible, Long-Range Radical Transfer in E. coli Class Ia Ribonucleotide Reductase. *Acc. Chem. Res.* **2013**, *46*, 2524–2535. [CrossRef] [PubMed]
9. Greene, B.L.; Taguchi, A.T.; Stubbe, J.; Nocera, D.G. Conformationally Dynamic Radical Transfer within Ribonucleotide Reductase. *J. Am. Chem. Soc.* **2017**, *139*, 16657–16665. [CrossRef]
10. Buckel, W.; Golding, B.T. Radicals Enzymes. In *Encyclopedia of Radical in Chemistry, Biology and Materials*; Chatgilialoglu, C., Studer, A., Eds.; Wiley: Chichester, UK, 2012; Volume 3, Chapter 52; pp. 1501–1546.
11. Lawrence, C.C.; Bennati, M.; Obvias, H.V.; Bar, G.; Griffin, R.G.; Stubbe, J. High-field EPR detection of a disulfide radical anion in the reduction of cytidine 5′-diphosphate by the E441Q R1 mutant of Escherichia coli ribonucleotide reductase. *Proc. Natl. Acad. Sci. USA* **1999**, *96*, 8979–8984. [CrossRef]
12. Zipse, H.; Artin, E.; Wnuk, S.; Lohman, G.J.S.; Martino, D.; Griffin, R.G.; Kacprzak, S.; Kaupp, M.; Hoffman, B.; Bennati, M.; et al. Structure of the Nucleotide Radical Formed during Reaction of CDP/TTP with the E441Q-$\alpha2\beta2$ of E. coli Ribonucleotide Reductase. *J. Am. Chem. Soc.* **2008**, *131*, 200–211. [CrossRef]
13. Barata-Vallejo, S.; Ferreri, C.; Golding, B.T.; Chatgilialoglu, C. Hydrogen Sulfide: A Reagent for pH-Driven Bioinspired 1,2-Diol Mono-deoxygenation and Carbonyl Reduction in Water. *Org. Lett.* **2018**, *20*, 4290–4294. [CrossRef]
14. Ahmad, R.; Armstrong, D.A. The effect of pH and complexation on redox reactions between RS radicals and flavins. *Can. J. Chem.* **1984**, *62*, 171–177. [CrossRef]
15. Surdhar, P.S.; Armstrong, D.A. Redox-Potentials of Some Sulfur-Containing Radicals. *J. Phys. Chem.* **1986**, *90*, 5915–5917. [CrossRef]
16. Surdhar, P.S.; Armstrong, D.A. Reduction Potentials and Exchange Reaction of Thiyl Radicals and Disulfide Anion Radicals. *J. Phys. Chem.* **1987**, *91*, 6532–6537. [CrossRef]
17. Armstrong, D.A. Applications of Pulse Radiolysis for the Study of Short-lived Sulphur Species. In *Sulfur-Centered Reactive Intermediates in Chemistry and Biology*; Chatgilialoglu, C., Asmus, K.-D., Eds.; Plenum Press: New York, NY, USA, 1990.
18. Mezyk, S.P.; Armstrong, D.A. Disulfide anion radical equilibria: Effects of NH_3^+, $-CO_2^-$, -NHC(O)- and $-CH_3$ groups. *J. Chem. Soc. Perkin Trans.* **1999**, *2*, 1411–1419. [CrossRef]
19. Sober, H.A. (Ed.) *Handbook of Biochemistry: Selected Data for Molecular Biology*, 2nd ed.; Chemical Rubber Publishing Co.: Cleveland, OH, USA, 1970.
20. Mezyk, S.P. Determination of the Rate Constant for the Reaction iof Hydroxyl and Oxide Radicals with Cysteine in Aqueous Solution. *Radiat. Res.* **1996**, *145*, 102–106. [CrossRef]
21. Mezyk, S.P. Direct rate constant measurement of radical disulphide anion formation for cysteine and cysteamine in aqueous solution. *Chem. Phys. Lett.* **1995**, *235*, 89–93. [CrossRef]
22. Knight, A.R. Photochemistry of thiols. In *The Chemistry of Thiol Group*; Patai, S., Ed.; Wiley: London, UK, 1974; Part 1; pp. 455–479.
23. Turro, N.J. *Modern Molecular Photochemistry*; University Science Books: Sausalito, CA, USA, 1991.
24. Xia, S.-H.; Liu, X.-Y.; Fang, Q.; Cui, G. Excited-State Ring-Opening Mechanism of Cyclic Ketones: A MS-CASPT2//CASSCF Study. *J. Phys. Chem. A* **2015**, *119*, 3569–3576. [CrossRef] [PubMed]
25. Shemesh, D.; Nizkorodov, S.A.; Gerber, R.B. Photochemical Reactions of Cyclohexanone: Mechanisms and Dynamics. *J. Phys. Chem. A* **2016**, *120*, 7112–7120. [CrossRef]
26. Reid, D.L.; Shustov, G.V.; Armstrong, A.A.; Rauk, A.; Schuchmann, M.N.; Akhlaq, M.S.; von Sonntag, C. H-atom abstraction from thiols by C-centered radicals. A theoretical and experimental study of reaction rates. *Phys. Chem. Chem. Phys.* **2002**, *4*, 2965–2974. [CrossRef]
27. Von Sonntag, C. *Free-Radical-Induced DNA Damage and Its Repair, A Chemical Perspective*; Springer: Berlin, Germany, 2006; p. 144.
28. Laroff, G.P.; Fessenden, R.W. Equilibrium and Kinetics of the Acid Dissociation of Several hydroxyalkyl Radicals. *J. Phys. Chem.* **1973**, *77*, 1283–1288. [CrossRef]
29. Xu, L.; Jin, J.; Lai, M.; Daublain, P.; Newcomb, M. Compatible Injection and Detection Systems for Studing the Kinetics of Excess Electron Transfer. *Org. Lett.* **2007**, *9*, 1837–1840. [CrossRef] [PubMed]
30. Yayla, H.G.; Knowles, R.R. Proton-Coupled Electron Transfer in Organic Synthesis: Novel Homolytic Bond Activation and Catalytic Asymmetyric Reactions with Free Radicals. *Synlett* **2014**, *25*, 2819–2826. [CrossRef]
31. Janata, E.; Schuler, R.H. Rate constant for scavenging e^-_{aq} in N_2O-saturated solutions. *J. Phys. Chem.* **1982**, *86*, 2078–2084. [CrossRef]
32. Péter, A.; Agasti, S.; Knowles, O.; Pye, E.; Procter, D.J. Recent advances in the chemistry of ketyl radicals. *Chem. Soc. Rev.* **2021**, *50*, 5349–5365. [CrossRef] [PubMed]
33. Huffman, J.W. Metal-Ammonia Reductions of Cyclic Aliphatic Ketones. *Acc. Chem. Res.* **1983**, *16*, 399–405. [CrossRef]

34. Buxton, G.V.; Greenstock, C.L.; Helman, W.P.; Ross, A.B. Critical review of rate constants for reactions of hydrated electrons, hydrogen atoms and hydroxyl radicals (OH/O^-) in aqueous solution. *J. Phys. Chem. Ref. Data* **1988**, *17*, 513–886. [CrossRef]
35. Hoffman, M.Z.; Hayon, E. One-Electron Reduction of the Disulfide Linkage in Aqueous Solution. Formation, Protonation, and Decay Kinetics of the RSSR- Radical. *J. Am. Chem. Soc.* **1972**, *94*, 795–7957. [CrossRef]
36. Schaefer, T.; Herrmann, H. Competition kinetics of OH radical reactions with oxygenated organic compounds in aqueous solution: Rate constants and internal optical absorption effects. *Phys. Chem. Chem. Phys.* **2018**, *20*, 10939–10948. [CrossRef]
37. Bobrowski, K. Free radicals in chemistry, biology, and medicine: Contribution of radiation chemistry. *Nukleonika* **2005**, *50* (Suppl. 3), S67–S76.
38. Mirkowski, J.; Wisniowski, P.; Bobrowski, K. *INCT Annual Report 2000*; INCT: Warsaw, Poland, 2001.
39. Pedzinski, T.; Grzyb, K.; Skotnicki, K.; Filipiak, P.; Bobrowski, K.; Chatgilialoglu, C.; Marciniak, B. Radiation- and Photo-Induced Oxidation Pathways of Methionine in Model Peptide Backbone under Anoxic Conditions. *Int. J. Mol. Sci.* **2021**, *22*, 4773. [CrossRef] [PubMed]
40. Syroeshkin, M.A.; Kuriakose, F.; Saverina, E.A.; Timofeeva, V.A.; Egorov, M.P.; Alabugin, I.V. Upconversion of Reductants. *Angew. Chem. Int. Ed.* **2019**, *58*, 5532–5550. [CrossRef] [PubMed]

MDPI
St. Alban-Anlage 66
4052 Basel
Switzerland
Tel. +41 61 683 77 34
Fax +41 61 302 89 18
www.mdpi.com

Molecules Editorial Office
E-mail: molecules@mdpi.com
www.mdpi.com/journal/molecules

www.ingramcontent.com/pod-product-compliance
Lightning Source LLC
LaVergne TN
LVHW070106170726
843515LV00007B/1621
9783036543505